Medicinal Plants and Natural Product Research

Medicinal Plants and Natural Product Research

Special Issue Editor
Milan S. Stankovic

MDPI • Basel • Beijing • Wuhan • Barcelona • Belgrade

Special Issue Editor
Milan S. Stankovic
University of Kragujevac
Serbia

Editorial Office
MDPI
St. Alban-Anlage 66
4052 Basel, Switzerland

This is a reprint of articles from the Special Issue published online in the open access journal *Plants* (ISSN 2223-7747) from 2017 to 2018 (available at: https://www.mdpi.com/journal/plants/special_issues/medicinal_plants).

For citation purposes, cite each article independently as indicated on the article page online and as indicated below:

LastName, A.A.; LastName, B.B.; LastName, C.C. Article Title. *Journal Name* **Year**, *Article Number*, Page Range.

ISBN 978-3-03928-118-3 (Pbk)
ISBN 978-3-03928-119-0 (PDF)

Cover image courtesy of Trinidad Ruiz Téllez.

© 2020 by the authors. Articles in this book are Open Access and distributed under the Creative Commons Attribution (CC BY) license, which allows users to download, copy and build upon published articles, as long as the author and publisher are properly credited, which ensures maximum dissemination and a wider impact of our publications.

The book as a whole is distributed by MDPI under the terms and conditions of the Creative Commons license CC BY-NC-ND.

Contents

About the Special Issue Editor vii

Preface to "Medicinal Plants and Natural Product Research" ix

Carmen X. Luzuriaga-Quichimbo, Míriam Hernández del Barco, José Blanco-Salas, Carlos E. Cerón-Martínez and Trinidad Ruiz-Téllez
Chiricaspi (*Brunfelsia grandiflora*, Solanaceae), a Pharmacologically Promising Plant
Reprinted from: *Plants* **2018**, *7*, 67, doi:10.3390/plants7030067 1

Maryam Malmir, Rita Serrano, Manuela Caniça, Beatriz Silva-Lima and Olga Silva
A Comprehensive Review on the Medicinal Plants from the Genus *Asphodelus*
Reprinted from: *Plants* **2018**, *7*, 20, doi:10.3390/plants7010020 12

Kurt Appel, Eduardo Munoz, Carmen Navarrete, Cristina Cruz-Teno, Andreas Biller and Eva Thiemann
Immunomodulatory and Inhibitory Effect of Immulina®, and Immunloges® in the Ig-E Mediated Activation of RBL-2H3 Cells. A New Role in Allergic Inflammatory Responses
Reprinted from: *Plants* **2018**, *7*, 13, doi:10.3390/plants7010013 29

Karl Egil Malterud
Ethnopharmacology, Chemistry and Biological Properties of Four Malian Medicinal Plants
Reprinted from: *Plants* **2017**, *6*, 11, doi:10.3390/plants6010011 43

Ammar Altemimi, Naoufal Lakhssassi, Azam Baharlouei, Dennis G. Watson and David A. Lightfoot
Phytochemicals: Extraction, Isolation, and Identification of Bioactive Compounds from Plant Extracts
Reprinted from: *Plants* **2017**, *6*, 42, doi:10.3390/plants6040042 56

Harish Chandra, Parul Bishnoi, Archana Yadav, Babita Patni, Abhay Prakash Mishra and Anant Ram Nautiyal
Antimicrobial Resistance and the Alternative Resources with Special Emphasis on Plant-Based Antimicrobials—A Review
Reprinted from: *Plants* **2017**, *6*, 16, doi:10.3390/plants6020016 79

K. D. P. P. Gunathilake, K. K. D. S. Ranaweera and H. P. V. Rupasinghe
Influence of Boiling, Steaming and Frying of Selected Leafy Vegetables on the In Vitro Anti-inflammation Associated Biological Activities
Reprinted from: *Plants* **2018**, *7*, 22, doi:10.3390/plants7010022 90

Motahareh Nobakht, Stephen J. Trueman, Helen M. Wallace, Peter R. Brooks, Klrissa J. Streeter and Mohammad Katouli
Antibacterial Properties of Flavonoids from Kino of the Eucalypt Tree, *Corymbia torelliana*
Reprinted from: *Plants* **2017**, *6*, 39, doi:10.3390/plants6030039 100

Chunpeng Wan, Chuying Chen, Mingxi Li, Youxin Yang, Ming Chen and Jinyin Chen
Chemical Constituents and Antifungal Activity of *Ficus hirta* Vahl. Fruits
Reprinted from: *Plants* **2017**, *6*, 44, doi:10.3390/plants6040044 115

Hanae Naceiri Mrabti, Nidal Jaradat, Ismail Fichtali, Wessal Ouedrhiri, Shehdeh Jodeh, Samar Ayesh, Yahia Cherrah and My El Abbes Faouzi
Separation, Identification, and Antidiabetic Activity of Catechin Isolated from *Arbutus unedo* L. Plant Roots
Reprinted from: *Plants* 2018, 7, 31, doi:10.3390/plants7020031 124

Tamalika Chakraborty, Somidh Saha and Narendra S. Bisht
First Report on the Ethnopharmacological Uses of Medicinal Plants by Monpa Tribe from the Zemithang Region of Arunachal Pradesh, Eastern Himalayas, India
Reprinted from: *Plants* 2017, 6, 13, doi:10.3390/plants6010013 133

Racquel J. Wright, Ken S. Lee, Hyacinth I. Hyacinth, Jacqueline M. Hibbert, Marvin E. Reid, Andrew O. Wheatley and Helen N. Asemota
An Investigation of the Antioxidant Capacity in Extracts from *Moringa oleifera* Plants Grown in Jamaica
Reprinted from: *Plants* 2017, 6, 48, doi:10.3390/plants6040048 145

Mirtha Navarro, Ileana Moreira, Elizabeth Arnaez, Silvia Quesada, Gabriela Azofeifa, Diego Alvarado and Maria J. Monagas
Proanthocyanidin Characterization, Antioxidant and Cytotoxic Activities of Three Plants Commonly Used in Traditional Medicine in Costa Rica: *Petiveria alliaceae* L., *Phyllanthus niruri* L. and *Senna reticulata* Willd.
Reprinted from: *Plants* 2017, 6, 50, doi:10.3390/plants6040050 153

Sarla Saklani, Abhay Prakash Mishra, Harish Chandra, Maria Stefanova Atanassova, Milan Stankovic, Bhawana Sati, Mohammad Ali Shariati, Manisha Nigam, Mohammad Usman Khan, Sergey Plygun, Hicham Elmsellem and Hafiz Ansar Rasul Suleria
Comparative Evaluation of Polyphenol Contents and Antioxidant Activities between Ethanol Extracts of *Vitex negundo* and *Vitex trifolia* L. Leaves by Different Methods
Reprinted from: *Plants* 2017, 6, 45, doi:10.3390/plants6040045 166

Nenad M. Zlatić and Milan S. Stanković
Variability of Secondary Metabolites of the Species *Cichorium intybus* L. from Different Habitats
Reprinted from: *Plants* 2017, 6, 38, doi:10.3390/plants6030038 177

Gerasimia Tsasi, Theofilos Mailis, Artemis Daskalaki, Eleni Sakadani, Panagis Razis, Yiannis Samaras and Helen Skaltsa
The Effect of Harvesting on the Composition of Essential Oils from Five Varieties of *Ocimum basilicum* L. Cultivated in the Island of Kefalonia, Greece
Reprinted from: *Plants* 2017, 6, 41, doi:10.3390/plants6030041 186

Chunpeng Wan, Shanshan Li, Lin Liu, Chuying Chen and Shuying Fan
Caffeoylquinic Acids from the Aerial Parts of *Chrysanthemum coronarium* L.
Reprinted from: *Plants* 2017, 6, 10, doi:10.3390/plants6010010 202

Eva Masiero, Dipanwita Banik, John Abson, Paul Greene, Adrian Slater and Tiziana Sgamma
Genus-Specific Real-Time PCR and HRM Assays to Distinguish *Liriope* from *Ophiopogon* Samples
Reprinted from: *Plants* 2017, 6, 53, doi:10.3390/plants6040053 209

About the Special Issue Editor

Milan S. Stankovic is an Associate Professor of Plant Science at Department of Biology and Ecology, Faculty of Sciences, University of Kragujevac, Republic of Serbia. He completed his Ph.D. on Botany at the same University and postdoctoral research at the Université François-Rabelais de Tours, France. He is the Head of Department of Biology and Ecology. Dr. Stankovic has published over 200 references, including articles in peer-reviewed journals, edited books, book chapters, conference papers, meeting abstracts, etc. He is an editor, editorial board member, and reviewer in several scientific journals. He currently works as an Associate Editor of the *Plants* journal.

Preface to "Medicinal Plants and Natural Product Research"

For a very long time, a large number of plants have been used in medicinal therapy, as well as for food and beverage preparation. It is estimated that over 50,000 plant species are used in pharmaceutical products, as well as that over 50% of available drugs are derived from medicinal plants. Due to their natural origins, the treatment products obtained from medicinal plants are of greater benefit in comparison to synthetic ones. The main carriers of the biological activity of medicinal plants are plant secondary metabolites, as products of a specifically conceived secondary metabolism, which is a continuation of the essential primary metabolism. Plant secondary metabolites are referred to as active substances, which have beneficial physiological effects on living organisms. On the basis of their main roles in plant life, quantitative and qualitative composition of secondary metabolites is in accordance with a variety of environmental influences. Certain active compounds are synthesized in different plant organs in different concentrations. In addition to their role in the process of environmental interaction, secondary metabolites from plants express their biological activity in both in vitro and in vivo conditions. Medicinal plants show promising effects for various health disorders, such as gastrointestinal diseases, throat irritations diseases, colds, coughs, etc. Further, they possess positive protecting activities such as antioxidant, anti-inflammatory, antihyperglycemic, antiseptic, antiviral, anticancer, immunostimulating, sedative, and spasmolytic. There are about half a million plants around the world. Among them, medicinal herbs have a hopeful future because most of them have not yet been studied in medical practice. Therefore, current and future studies on medical activities can be effective in treating diseases. Based on these facts, complex studies of medicinal plants, from habitats to the validation of natural products, are interesting in numerous scientific and practical disciplines, including morphology and anatomy, diversity and phytogeography, physiology and ecology, methodology of cultivation and collection, and traditional and modern folk medicine. In addition, the diversity and quantitative and qualitative analysis as well as isolation and chemical modification of secondary metabolites are of great importance in testing both in vitro and in vivo biological activities. The book entitled "Medicinal Plants and Natural Product Research" fits perfectly into this approach dealing with ethnopharmacological uses of medicinal plants; extraction, isolation, and identification of bioactive compounds from plant extracts; various aspects of biological activity, such as antioxidant, antimicrobial, anticancer, immunomodulatory activity, etc., as well as characterization of plant secondary metabolites as active substances of medicinal plants. I am grateful to all the authors for their contributions as well as the reviewers for their professional suggestions and decisions. I highly thankful to Plants MDPI team for many years of collaboration. Especially, I would like to give special thanks to Shuang Zhao and Sylvia Guo.

Milan S. Stankovic
Special Issue Editor

Article

Chiricaspi (*Brunfelsia grandiflora*, Solanaceae), a Pharmacologically Promising Plant

Carmen X. Luzuriaga-Quichimbo [1], Míriam Hernández del Barco [2], José Blanco-Salas [2,*], Carlos E. Cerón-Martínez [3] and Trinidad Ruiz-Téllez [2]

1. Centro de Investigación Biomédica, Facultad de Ciencias de la Salud Eugenio Espejo, Universidad Tecnológica Equinoccial, Av. Mariscal Sucre y Mariana de Jesús, Quito 170527, Ecuador; luzuriaga.cx@gmail.com
2. Department of Vegetal Biology, Ecology and Earth Science, Faculty of Sciences, University of Extremadura, 06071 Badajoz, Spain; mhernandye@alumnos.unex.es (M.H.d.B.); truiz@unex.es (T.R.-T.)
3. Herbario Alfredo Paredes, QAP, Universidad Central de Ecuador, Quito 170129, Ecuador; cecm57@yahoo.com
* Correspondence: blanco_salas@unex.es; Tel.: +34-924-289-300

Received: 28 June 2018; Accepted: 13 August 2018; Published: 18 August 2018

Abstract: This study's objective was to evaluate the rescued traditional knowledge about the chiricaspi (*Brunfelsia grandiflora* s.l.), obtained in an isolated Canelo-Kichwa Amazonian community in the Pastaza province (Ecuador). This approach demonstrates well the value of biodiversity conservation in an endangered ecoregion. The authors describe the ancestral practices that remain in force today. They validated them through bibliographic revisions in data megabases, which presented activity and chemical components. The authors also propose possible routes for the development of new bioproducts based on the plant. In silico research about new drug design based on traditional knowledge about this species can produce significant progress in specific areas of childbirth, anesthesiology, and neurology.

Keywords: activity; bioproduct; Brunfelsia; Amazonian; Ecuador; ethnobotanic; ayahuasca; validation; drug discovery; scopoletin

---

1. Introduction

Three species of the genus *Brunfelsia* (Solanaceae) used as an additive in the hallucinogenic "ayahuasca" drink have traditionally been very important plants for Ecuadorian indigenous Amazonian cultures such as the Kichwa of the East, Tsa'chi, Cofán, Secoya, Siona, Wao, and Shuar [1]. They are *Brunfelsia chiricaspi* Plowman, *Brunfelsia macrocarpa* Plowman, and *Brunfelsia grandiflora* D.Don, including ssp. *grandiflora* and ssp. *schultesii* Plowman in the variability range. Ancestral knowledge recognizes different applications or degrees of activity for each taxon. *B. grandiflora* is the one with the widest distribution area. It is well known as ornamental in Tropical America and is differentiated from its congeners by morphological characters related to floral and foliar size. This paper presents an ethnobotanical review of *B. grandiflora* ssp. *grandiflora* and *B. grandiflora* ssp. *schultesii*, in order to document its use, to offer arguments for its conservation, and to provide scientific evidence of its activity. The traditional knowledge rescued in a barely contacted Kichwa community in the province of Pastaza (Ecuador) is also presented. The authors describe the ancestral practices that remain in force today. They validated them through bibliographic revisions in data megabases, which presented activity and chemical components. The authors also propose possible routes for the development of new bioproducts based on the plant. In essence, this study's fundamental objective was to conserve, document, and validate the traditional use of *Brunfelsia grandiflora*, as an element for innovation.

2. Results

2.1. Brunfelsia grandiflora: Botanical Description, Chorolorogy, Variability, and Names

Brunfelsia grandiflora (see Figure 1) is a small tree that was described by David Don in 1829 from material collected during the Ruiz and Pavon expeditions undertaken in Peru [2]. It has spread from Central America (Nicaragua, Costa Rica, USA) to the north of South America (Colombia, Brazil, Ecuador, Peru and Bolivia). There, it is well-known in cultivation as ornamental, exhibiting some exceptional forms. In the Amazon region, it is also cultivated for its narcotic and medicinal properties. It grows primarily at elevations of 650–2000 m, mainly on the eastern slopes of the Andes, in the region known as the montana (i.e., a humid, montane rainforest) [3].

Figure 1. *Brunfelsia grandiflora* ssp. *grandiflora* in the Pakayaku rainforest, Ecuador. Photo credit: CX. Luzuriaga-Quichimbo (8 February, 2016).

The author of its monograph [3] gave the following detailed description: "Shrubs or small tree 1–6 (10) m tall. Trunk to 7 cm in diameter near base, much branched. Bark thin, roughish, light to dark brown. Branches slender, ascending or spreading, often subvirgate and arching, leafy, glabrous. Branchlets glabrous, rarely pubescent, green. Leaves 10–23 cm long, 3–8 cm wide, glabrous or sparingly pubescent at midrib; petiole 3–12 mm long. Inflorescence terminal and subterminal, simple or branched, dense or lax, the axis 5–45 mm long. Flowers 5-many, showy, scentless, violet fading to white with age, with rounded, white ring at mouth. Bracts 1–3 per flower, 1–5 (10) mm long, lanceolate to ovate, ciliolate, pubescent or glabrate, caducous. Pedicel 2–10 mm long, glabrous or with few sparse glandular hairs, becoming thicker and corky-verrucose in fruit. Calyx 9–13 mm long, tubular-campanulate, somewhat narrowed toward base, globose to obovate in bud, somewhat inflated or not so, glabrous, rarely punctate or sparsely glandular within, smooth or striate-nerved, light yellow-green to gray-green, firmly membranaceous to subcoriaceous, teeth 2–5 mm long, ovate-lanceolate, blunt to short acuminate, erect or incumbent, recurved slightly with age; calyx in fruit persistent, coriaceous, becoming corky-verrucose especially near base, often splitting on one or more sides. Corolla tube 30–40 mm long, 2.5–3.5 mm in diameter, 2.5–3.5 mm in diameter, the mouth 6–9 mm long; limb 35–52 mm in diameter, spreading. Stamens completely included in upper part of corolla tube; filaments thin, upper pair 4 mm long, lower pair 3 mm long, white; anthers 1–1.5 mm long, orbicular-reniform, light brown. Ovary 1.5–2 mm long, sessile, conical to ovoid, pale yellow; style slender, slightly dilated at apex; stigma about 1 mm long, briefly bifid, unequal, the up-per lobe somewhat larger, obtuse, green. Capsule 8–20 mm long, 8–20 mm in diameter, ovoid to subglobose, obtuse or apiculate at apex, smooth, nitid, dark green turning brownish, with corkypunctate or—verrucose outgrowths, pericarp thin, 0.3 mm thick, crustaceous, drying brittle, tardily dehiscent. Seeds 10–20, 5–7 mm long, 2–3 mm in diameter, variable in shape, ellipsoid to oblong, angular, dark reddish brown, reticulate-pitted. Embryo about 4 mm long; slightly curved, cotyledons 1.5 mm long, ovate-elliptic".

Brunfelsia grandiflora ssp. *schultesii* Plowman, within the species' variability and which has been collected [2] in Venezuela, Colombia, Brazil, Bolivia, Peru, and Ecuador (see Figure 2), was described by Plowman in *Bot. Mus. Leafl.* 23 (6): 259 (1973). It is characterized by its lower altitudinal preferences and smaller dimensions. The aforementioned author [3] described it in detail as follows: "Inflorescence variable, compact or lax. Pedicel 2–6 mm long. Calyx 5–10 mm long, teeth 1–3 mm long, triangular to triangular-ovate. Corolla tube 15–30 mm long, 1–2 mm in diameter, curved toward apex; limb 20–40 mm in diameter, spreading, mouth 3–5 mm long. Capsule 11–16 mm long, 10–16 mm in diameter".

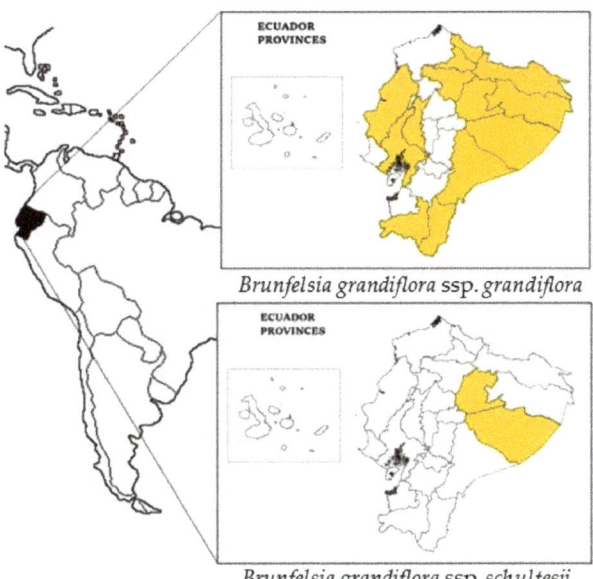

Figure 2. Distribution of *Brunfelsia grandiflora* s.l. in Ecuador.

The species is referenced in Ecuador (see Figure 2) where it is called chiri kaspi, chiri wayusa, chiri wayusa pahu, chiri wayusa panka atu, urku chiri wayusa, wayra panka (Kichwa), uva silvestre (Spanish), i'shan ta'pe, luli ta'pe (Tsafi'ki), tsontimba'cco (A'ingae), jaija'o ujajai (Pai Coca), winemeawe (Wao Tededo), apaj, chirikiasip (Shuar Chicham), paiapia, and simora (unspecified language) [4].

2.2. Compilation of Brunfelsia grandiflora Ethnobotanical Uses

The Kichwa name chiricaspi (which means cold tree) refers to the chills and tingling sensations ("like rain in the ears") that are felt after ingesting the bark. It is widely employed as a hallucinogen, often added to intensify the effect of narcotic drinks. Indigenous groups throughout the northwest Amazon use this plant to treat fever, rheumatism, and arthritis. It is said to act as a tonic over time, giving one strength and resistance to colds [3]. Ethnobotanical uses have been reported from different countries, for example, Bolivia [5], Brazil [6], Venezuela [7,8], Colombia [5,8,9], and Peru [8,10–16]. In Peru, the use of remedies based on this plant is traditionally associated with diet (refrain from eating pepper, i.e., ají, *Capsicum frutescens*), meat, and sexual relations [10,11]. Old thick roots are considered toxic, so only 2–3 roots approximately 1–1.5 cm diameter should be used [12]. Macerated roots are employed to combat rheumatism and syphilis. The leaves are also used to treat colds, arthritis, and snake bites. The bark is utilized against leishmaniasis. It is boiled to obtain a thick liquid that is applied to the affected areas [10,11,17]. In Colombia, both subspecies are commonly known as borrachero [3,9,12]. They are used in that country as an antirheumatic as well. Roots and, less frequently, the leaves are an admixture of ayahuasca. Prepared alone, this plant is used only when the shaman is faced with a particularly difficult or persistent problem, because the toxicity is known by local population. The most common preparation is making a tea from the roots and the bark. Cold water extraction can also be carried out, by shaving the bark from the roots and stems and then allowing them to soak. Alcohol mixture extractions have also been described [7,8]. In the Yabarana tribe (Venezuela), the leaves are routinely dried, crushed, mixed with tobacco and smoked [7,8].

Presented below (see Table 1) is a synthesis of the ethnobotanical knowledge about *Brunfelsia grandiflora* ssp. *grandiflora* and *B. grandiflora* ssp. *schultesii*, obtained from the indigenous communities of Ecuador based in bibliographic revisions [4,18] and our field prospections (see Table 2). Both subspecies seem to be used interchangeably in folk medicine, but ssp. *schultesii* is more widespread in the lowlands and is the form more likely to be employed.

Table 1. Synthesis of the ethnobotanical knowledge about *Brunfelsia grandiflora* ssp. *grandiflora* D. Don and *B. grandiflora* ssp. *schultesii* Plowman * from the indigenous communities of Ecuador.

	Part	Formulation	Traditional Knowledge	Ethnic Group Ecuador Province
Human Consumption				
Edible Fruits/Sweet fruits	F	-	edible	Pichincha
Animal Feeding				
Edible Fruits/Sweet fruits	F *		avian food *	Wao * Orellana *
Building				
Houses, buildings, and agricultural facilities	S		building	Wao Napo
Industry and Crafts				
Cosmetics, perfumery, and cleaning	L	decoction or crushed and mixed with cold water	refreshing baths	Tsa'chi Pichincha
Tools and utensils (working, domestic, hunting, fishing, defense, etc.)	L, S		hunting tools	Kichwa del Oriente Orellana
Personal clothing and ornaments	F	clothing in festivals	personal adornment	Shuar Napo
Medicinal				
Conception, pregnancy, childbirth, and puerperium	L, S, R, B		contraconceptive	Kichwa del Oriente Napo

Table 1. Cont.

	Part	Formulation	Traditional Knowledge	Ethnic Group Ecuador Province
Respiratory system	B	a drop of juice resulting from crushing the bark is applied in the nose	flu	Cofán Sucumbíos
	R, L	coction		Kichwa del Oriente Napo
Musculature and skeleton	S	bark is removed	bloated and aching body	Kichwa del Oriente Orellana
	R	fumes *	rheumatism arthritis	Kichwa del Oriente Napo * Orellana
Skin and subcutaneous cellular tissue	L		burns	Kichwa del Oriente Orellana
	B *, L *	powder and infusion *	wounds and blows *	Kichwa del Oriente * Napo * Orellana *
Nervous system and mental disorders			headache *	Shuar Morona-Santiago
		fumes *		Kichwa del Oriente * Napo *
Symptoms and states of indefinite origin	L	infusion	body weakness	Kichwa del Oriente Pastaza
	L	powder and infusion *	chills and fever *	Kichwa del Oriente * Napo *, Orellana *
	B		chills and fever *	Secoya * Sucumbíos *
		infusion	healthy	Kichwa del Oriente Napo, Orellana
Social, Symbolic, and Ritualistic Uses				
Life Cycle Rituals	R, S		paternity test	Siona Sucumbíos
Rituals of uncertainty, protection, and affliction	R, L	infusion	induce vomiting for body purification	Shuar Napo
	L	infusion for bathing or drinking	improve luck during hunting, attract and tame animals	Kichwa del Oriente Napo, Orellana
	B	infusion for drinking	improve the aim of a blowpipe during hunts	Shuar Napo
		infusion for bathing, mixed with orange, onio, caimito, and achiote	protect against the "evil eye"	Kichwa del Oriente Orellana
	L	infusion	cause chills	Pastaza
		only for shamans (*)	obtain knowledge about new medicines (*)	Cofán * Sucumbíos
Hallucinogen, narcotic, and smoking	B	infusion, only for curacas	"become a tiger"	Cofán Sucumbíos
				Shuar Napo
	S	crushed with cold water, the shaman swallows it	disease diagnosis and to remove evil spirits from the body	Cofán Sucumbíos Shuar Napo
	B, R, S, L	drinks, sometimes mixed with Banisteriosis caapi to intensify the effect	hallucinogen used in rituals	Secoya *, Shuar Kichwa del Oriente Orellana, Sucumbíos Napo, Zamora Chinchipe

Part used: B, bark; R, root; S, stem; L, leaf.

Table 2. Specific medical and cultural uses of *B. grandiflora* ssp. *grandiflora* and *B. grandiflora* ssp. *schultesii* * in Pakayaku (Pastaza, Ecuador).

	Part	Formulation	Traditional Knowledge
Medicinal			
Digestive system	L	a handful of leaves is cooked, from thirty minutes to an hour, in two liters of water; it is taken in small glasses on an empty stomach	stomach pain and diarrhea ###
Conception, pregnancy, childbirth, and puerperium	B	scrape, mix it with a small glass of warm water, let it hover, and drink it	childbirth ###
Insect or other animal bites	B	scrape a piece and put it on the injury (tupe); repeat it when the product becomes dry, until the worm gets out	against tupe ###
Musculature and skeleton	R *	small roots scraped and tied with a rag or bandage and placed twice a day on the affected part *	aching body *
Skin and subcutaneous cellular tissue	B *	the bark is grated, deposited on a rag or bandage and tied to the area that has been hit*	body blows *
	L *	the leaf is crushed, placed on the affected area, tied, and then the patient falls asleep; the process must be repeated as many times as possible*	skin tumors *
Social, Symbolic, and Ritualistic Uses			
Rituals of uncertainty, protection, and affliction	B	it is prepared in a pot and a glassful is swallowed; during the treatment, a diet must be followed (refrain from taking salt or chili, *Capsicum frutescens*), nor should you stay near the candle	improve men's energy when going into the forest
	B	the bark is grated, mixed in a medium recipient called pilchi, with water and taken at midnight; they also inform the authors that consuming this brew produced a lot of cold and chills	ensure strength blowing the blow pump during hunts

Part used: B, bark; R, root; S, stem; L, leaf. ### Recovered.

2.3. Towards a Validation of the Pharmacological Action of B. grandiflora

The most important compounds found [5,7,12,19,20] in the leaves, stems, roots, and root bark are presented in Figure 3: coumarins, as aesculetin and scopoletin; a metilendiamine, as brunfelsamidine; unidentified alcaloids, as manacine and manaceine; tropanic alcaloids, as scopolamine; pirrolidinic alcaloids, as cuscohygrine; and steroidic saponins belonging to the furostan saponins type.

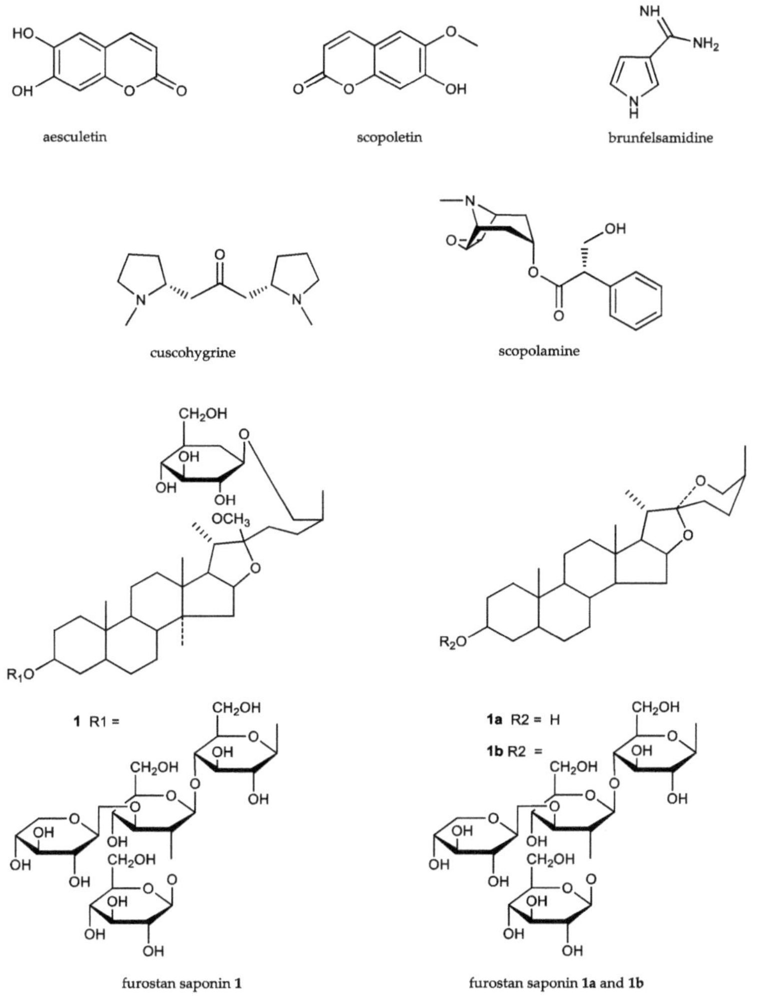

Figure 3. Chemical structures of the principal components of *Brunfelsia grandiflora* s.l.

2.4. Experimental Studies on Activity

The principal evidence elements related to the physiological and pharmacological activities demonstrated by the experimental work carried out in the laboratory were selected and are summarized in Table 3. This table refers only to the activity of molecules that are contained in the species *B. grandiflora*. These activities have been tested and published in literature. The references that appear in the last column belong to the team author of the corresponding research for each case.

Table 3. Biological activity of some chemical compounds present in *Brunfelsia grandiflora*.

Molecules	Activity	References
aesculetin	inhibition of cancer cell migration	[21]
	antileukemia	[22]
brunfelsamidine	convulsant, affects serotonin levels	[19]
cuscohygrine	short-term ganglion-blocking	[23]
scopolamine	anticholinergic	[24]
scopoletin	anti-inflammatory by cytokine suppression	[25,26]
	spasmolytic by inhibition of calcium moving	[27]
	cholinergic in vivo rat brain	[28]
	blood pressure regulator	[29]
	hepatic steatosis protector by enzymatic inhibition	[30]
	antifungal	[29]
	antibacterial	[31]
saponins	antileishmania	[32]

3. Discussion

Brunfelsia grandiflora is an interesting neotropical plant with worthy arguments for its conservation. It is locally cultivated as an ornamental decoration, as a personal decoration during traditional festivals, as an element of construction and even as a singular edible plant. However, its most unique use is cultural, linked to its administration by the local leaders of different ethnic groups. For example, it is used in the Ecuadorian Amazon, among the Cofán, to diagnose diseases or remove evil spirits; among the Shuar, to improve the aim or for inducing vomits; among the Siona to check paternity; among the Kichwa to improve luck during hunting. It has been used as a shamanic beverage in order to obtain knowledge about new medicines to treat diseases. It is ingested in rituals, is a relaxing, hypnotic, or hallucinogenic drug, and it is also prepared in the form of baths, along with orange leaves, caimito, achiote, grapefruit, and onion. As a medicinal plant it is used as an anti-flu medicine, to reduce fever and headache, to treat arthritis and rheumatism, wounds, burns, and even as a contraceptive. The subspecies *grandiflora* has been collected by botanists in the Bobonaza River basin [33], but the subspecies *schultesii* has not even been observed in the biggest province of Ecuador, Pastaza. Our survey revealed that the plant grows in the remote forest, where it was perfectly recognized by local inhabitants; it was currently being used and had a good reputation to combat rheumatism, arthritis, body pain, colds, and fever. Our fieldwork also provided three novel uses that had not been reported before: against stomach pain, as a larvicide against tupe, and as an accelerator of childbirth.

The validation of the pharmacological action of *B. grandiflora* can be supported by interrelating different information published in scientific literature. The anti-inflammatory effect of scopoletin [25,26], fully justifies the abovementioned uses against rheumatism, arthritis, body pain, colds, flu, fever, headache, joint and muscle pain, body blows, and discomfort. It can also explain its use by folk medicine as an anti-snake venom, and in cases of wounds and burns. The hallucinogenic and narcotic properties associated with symbolic cultural rituals clearly depend on the brain and nervous system activities that are mediated by brunfelsamidine [19], cuscohygrine [23], scopolamine [24], and even scopoletin [27–29], the last two sometimes exerting opposite effects. With respect to the traditional custom of associating this plant with increasing energy, wisdom, or marksmanship, the authors had to take into consideration the activity currently being studied in the furostan saponines (i.e., about their capacity to induce pro-sexual and androgenic enhancing effects [34]).

There are promising fields for innovation and scientific research in the future about this taxon. The antiproliferative effects of aesculetin [21,22] open the way in the field of oncology; brunfelsamidine and cuscohygrine [19,23], in the fields of anesthesiology and reanimation; and most relevantly, furostan saponins for the treatment of leishmania and other intracellular parasites that produce malaria [32]. The small molecular size of the main components presented in Table 2 make them very good candidates to be tested in the future "in silico" discovery of new drugs. Neurology seems to be one of the most current specialties, and their use in the childbirth process should also be addressed.

4. Materials and Methods

4.1. Study Area and Voucher Collection

The Kichwa community of Pakayaku (Bobonaza River, Pastaza, Ecuador) lies in an isolated region where bio- and ethnodiversity studies are still lacking. One of the authors (C.X.L.-Q.) was based in the Biological Station Pindo Mirador in the northern Bobonaza River basin (1° 27' 09'' S, 78° 04' 51'' W), and, since 2008, was charge of environmental monitoring and education programs involving the local population.

Plant collection permits were granted by the Ministry of the Environment, following the Convention of Biological Diversity rules [35]. Plant vouchers were deposited at the Herbarium José Alfredo Paredes, Universidad Central de Ecuador, Quito QAP Herbarium as "Ecuador. Pastaza: Sarayaku, Pakayaku, banks of the Bobonaza river, path to the lake by the house belonging to Mr. O. Aranda, 383 m, 01° 39' 0.4'' S, 077° 35' 53'' W, lowland evergreen forest, 9 February 2016, C. X. Luzuriaga-Q & E. Gayas sub subsp. *Grandiflora*: (QAP 93819); and subsp. *Schultesii* Plowman (QAP 93817)." The identification was revised by C. Cerón.

4.2. Ethnobotanical Survey

Collective written research consent was granted by Mrs. Luzmila Gayas, community president of the Assembly of Pakayaku. Prior oral individual consent was obtained from the persons taking part in the survey. Nagoya Protocol Rules were followed [35]. The investigation consisted of a series of planned house visits and walking routes accompanied by Kichwa interpreters and local inhabitants of Pakayaku. Interviews were semi-structured and included a series of open questions aimed to encourage discussion. All interviews were recorded. Ten knowledgeable elders of the Pakayaku community acted as informants and agreed to reveal their wisdom of the plant. The informants answered freely about several topics, namely the Kichwa common name, the part of the plant used, a description of the use, the harvest season, the storage (if any), the concoction, and the treatment target. After the field wok, the data were inserted into an MS Excel spreadsheet. All recorded uses were referred to standard classifications [4,18]. The data provided by the community (see Table 2) were compared with the existing ethnobotanical literature from Ecuador (see Table 1).

4.3. Scientific Validation

A bibliographic study was performed to provide scientific evidence for the medicinal uses of the plant. PRISMA statement reporting items were considered [36]. This included searching databases, removing duplicates, screening records, defining the eligibility criteria for articles, deciding about accessed and excluded articles, including selected articles, and studying the articles. The databases accessed were: Academic Search Complete, Agricola, Agris, Biosis, CAB Abstracts, Cochrane, Cybertesis, Dialnet, Directory of Open Access Journals, Embase, Espacenet, Google Academics, Google Patents, Medline, PubMed, Science Direct, Scopus, Teseo, and Web of Science by the Institute for Scientific Information (ISI). The selected citations are summarized in Table 3. A critical examination of Table 3 was the basis for the discussion of the results and the presentation of a specific and concrete conclusion.

5. Conclusions

This study presents the case of an Amazonian plant for which the approach adopted demonstrates well the value of biodiversity conservation in an endangered ecoregion. In silico new drug design, based on the traditional knowledge about chiricaspi, can produce significant progress in medical fields, such as neurology and anesthesiology.

Author Contributions: Conceptualization, T.R.-T.; methodology, C.E.C.-M., and C.X.L.-Q.; validation, J.B.-S.; formal analysis, M.H.d.B.; investigation, C.X.L.-Q.; data curation, C.E.C.-M. and C.X.L.-Q.; writing and original draft preparation, T.R.-T.; writing, review and editing, J.B.-S.; visualization, M.H.d.B.; supervision, C.X.L.-Q.; project administration, T.R.-T.; funding acquisition, C.X.L.-Q. and T.R.-T.

Funding: This research was partially funded by the Government of Extremadura (Spain) and the European Union through the action "Apoyos a los Planes de Actuación de los Grupos de Investigación Catalogados de la Junta de Extremadura: FEDER GR15080".

Acknowledgments: We are very grateful to the members of the Kichwa community of Pakayaku, Luzmila Gayas, the People's Assembly of Pakayaku, and the collaborating *ayllus* (families), for their cooperation during the field work, and to the anonymous Referee for the evaluation of our paper and for the constructive critics. M.V. Gil Alvarez (Organic Chemistry Department, University of Extremadura) assisted us with the chemical drawing software.

Conflicts of Interest: The authors declare no conflict of interest. The founding sponsors had no role in the design of the study; in the collection, analyses, or interpretation of data; in the writing of the manuscript, and in the decision to publish the results.

References

1. Bennett, B.C. Hallucinogenic Plants of the Shuar and Related Indigenous Groups in Amazonian Ecuador and Peru. *Brittonia* **1992**, *44*, 483–493. [CrossRef]
2. Missouri Botanical Garden. Available online: www.tropicos.org (accessed on 25 June 2018).
3. Plowman, T. Brunfelsia in Ethnomedicine. *Bot. Mus. Lealf. Harv. Univ.* **1977**, *25*, 289–320. [PubMed]
4. De la Torre, L.; Navarrete, H.; Muriel, P.; Marcia, M.; Balslev, H. *Enciclopedia De Plantas Utiles Del Ecuador*; Herbario QCA & Herbario AAU: Quito, Ecuador, 2008; Volume 1.
5. Nordegren, T. *The A-Z Encyclopedia of Alcohol and Drug Abuse*; Brown Walker Press: Parkland, FL, USA, 2002.
6. León, B.; Wiersema, J.H. *World Economic Plants: A Standard Reference*, 2nd ed.; CRC Press: Boca Raton, FL, USA, 2013.
7. Gilman, E.F. Brunfelsia Grandiflora Yesterday, Today, and Tomorrow. FPS77. Available online: http://edis.ifas.ufl.edu/pdffiles/FP/FP07700.pdf (accessed on 23 May 2018).
8. Schultes, R.E.; Hofmann, A.; Rätsch, C. *Plants of the Gods. Their Sacred, Healing and Hallucinogenic Powers*; Healing Arts Press: Rochester, VT, USA, 1998.
9. Pratt, C. *An Encyclopedia of Shamanism, Volume 1*; Rosen Publishing Group: New York, NY, USA, 2007.
10. Santiváñez Acosta, R.; Cabrera Meléndez, J. *Catálogo Florístico de Las Plantas Medicinales Peruanas*; Ministerio de Salud, Instituto Nacional de Salud: Lima, Perú, 2013.
11. Mejía, K.; Reginfo, E. *Plantas Medicinales de Uso Popular En La Amazonía Peruana*; Agencia Española de Cooperación Internacional & Instituto de Investigaciones de la Amazonía Peruana: Lima, Perú, 2000.
12. Voogelbreinder, S. Garden of Eden: The Shamanic Use of Psychoactive Flora and Fauna, and the Study of Consciousness; 2009. Available online: http://docshare01.docshare.tips/files/27747/277478962.pdf (accessed on 23 May 2018).
13. Jauregui, X.; Clavo, Z.M.; Jovel, E.M.; Pardo-De-Santayana, M. "Plantas Con Madre": Plants That Teach and Guide in the Shamanic Initiation Process in the East-Central Peruvian Amazon. *J. Ethnopharmacol.* **2011**, *134*, 739–752. [CrossRef] [PubMed]
14. Kvist, L.P.; Christensen, S.B.; Rasmussen, H.B.; Mejia, K.; Gonzalez, A. Identification and Evaluation of Peruvian Plants Used to Treat Malaria and Leishmaniasis. *J. Ethnopharmacol.* **2006**, *106*, 390–402. [CrossRef] [PubMed]
15. Kloucek, P.; Polesny, Z.; Svobodova, B.; Vlkova, E.; Kokoska, L. Antibacterial Screening of Some Peruvian Medicinal Plants Used in Callería District. *J. Ethnopharmacol.* **2005**, *99*, 309–312. [CrossRef] [PubMed]

16. Polesna, L.; Polesny, Z.; Clavo, M.Z.; Hansson, A.; Kokoska, L. Ethnopharmacological Inventory of Plants Used in Coronel Portillo Province of Ucayali Department, Peru. *Pharm. Biol.* **2011**, *49*, 125–136. [CrossRef] [PubMed]
17. Reginfo, E.; Cerruti, S.; Pinedo, P.M. *Plantas Medicinales de La Amazonía Peruana: Estudio de su uso y cultivo*; Instituto de Investigaciones de la Amazonía Peruana: Iquitos, Perú, 1997.
18. Luzuriaga-Quichimbo, C.X. Estudio Etnobotánico En Comunidades Kichwas Amazónicas de Pastaza (Ecuador). Ph.D. Thesis, Universidad de Extremadura, Badajoz, Spain, 2017.
19. Lloyd, H.A.; Fales, H.M.; Goldman, M.E.; Jerina, D.M.; Plowman, T.; Schultes, R.E. Brunfelsamidine: A Novel Convulsant from the Medicinal Plant Brunfelsia Grandiflora. *Tetrahedron Lett.* **1985**, *26*, 2623–2624. [CrossRef]
20. Kress, H. Brunfelsia Grandiflora. Available online: https://www.henriettes-herb.com/eclectic/kings/franciscea.html (accessed on 23 May 2018).
21. Rubio, V.; García-Pérez, A.I.; Herráez, A.; Tejedor, M.C.; Diez, J.C. Esculetin Modulates Cytotoxicity Induced by Oxidants in NB4 Human Leukemia Cells. *Exp. Toxicol. Pathol.* **2017**, *69*, 700–712. [CrossRef] [PubMed]
22. Gong, J.; Zhang, W.; Feng, X.; Shao, M.; Xing, C. Aesculetin (6,7-Dihydroxycoumarin) Exhibits Potent and Selective Antitumor Activity in Human Acute Myeloid Leukemia Cells (THP-1) via Induction of Mitochondrial Mediated Apoptosis and Cancer Cell Migration Inhibition. *J. BUON* **2017**, *22*, 1563–1569. [PubMed]
23. Minina, S.A.; Astakhova, T.A.; Gromova, E.G.; Vaichageva, Y.V. Preparation and Pharmacological Study of Cuscohygrine bis(methyl benzenesulfonate). *Pharm. Chem. J.* **1977**, *11*, 478–481. [CrossRef]
24. DrugBank. Scopolamine. Available online: http://www.drugbank.ca/drugs/DB00747 (accessed on 23 May 2018).
25. Kim, H.J.; Jang, S.I.; Kim, Y.J.; Chung, H.T.; Yun, Y.G.; Kang, T.H.; Jeong, O.S.; Kim, Y.C. Scopoletin Suppresses Pro-Inflammatory Cytokines and PGE2 from LPS-Stimulated Cell Line, RAW 264.7 Cells. *Fitoterapia* **2004**, *75*, 261–266. [CrossRef] [PubMed]
26. Moon, P.D.; Lee, B.H.; Jeong, H.J.; An, H.J.; Park, S.J.; Kim, H.R.; Ko, S.G.; Um, J.Y.; Hong, S.H.; Kim, H.M. Use of Scopoletin to Inhibit the Production of Inflammatory Cytokines through Inhibition of the IκB/NF-KB Signal Cascade in the Human Mast Cell Line HMC-1. *Eur. J. Pharmacol.* **2007**, *555*, 218–225. [CrossRef] [PubMed]
27. Oliveira, E.J.; Romero, M.A.; Silva, M.S.; Silva, B.A.; Medeiros, I.A. Intracellular Calcium Mobilization as a Target for the Spasmolytic Action of Scopoletin. *Planta Med.* **2001**, *67*, 605–608. [CrossRef] [PubMed]
28. Rollinger, J.M.; Hornick, A.; Langer, T.; Stuppner, H.; Prast, H. Acetylcholinesterase Inhibitory Activity of Scopolin and Scopoletin Discovered by Virtual Screening of Natural Products. *J. Med. Chem.* **2004**, *47*, 6248–6254. [CrossRef] [PubMed]
29. Carpinella, M.C.; Ferrayoli, C.G.; Palacios, S.M. Antifungal Synergistic Effect of Scopoletin, a Hydroxycoumarin Isolated from Melia Azedarach L. Fruits. *J. Agric. Food Chem.* **2005**, *53*, 2922–2927. [CrossRef] [PubMed]
30. Choi, R.Y.; Ham, J.R.; Lee, H.I.; Cho, H.W.; Choi, M.S.; Park, S.K.; Lee, J.; Kim, M.J.; Seo, K.I.; Lee, M.K. Scopoletin Supplementation Ameliorates Steatosis and Inflammation in Diabetic Mice. *Phyther. Res.* **2017**, *31*, 1795–1804. [CrossRef] [PubMed]
31. Kayser, O.; Kolodziej, H. Antibacterial Activity of Extracts and Constituents of Pelargonium Sidoides and Pelargonium Reniforme. *Planta Med.* **1997**, *63*, 508–510. [CrossRef] [PubMed]
32. Fuchino, H.; Sekita, S.; Mori, K.; Kawahara, N.; Satake, M.; Kiuchi, F. A New Leishmanicidal Saponin from Brunfelsia Grandiflora. *Chem. Pharm. Bull.* **2008**, *56*, 93–96. [CrossRef] [PubMed]
33. Borgtoft, H.; Fjeldså, J.; Øllgaard, B. People and Biodiversity: Two Case Studies from the Andean Foothills of Ecuador. Centre for Research on Cultural and Biological Diversity of Andean Rainforests. *DIVA Tech. Rep.* **1998**, *3*, 1–190.
34. Neychev, V.; Mitev, V. Pro-Sexual and Androgen Enhancing Effects of Tribulus Terrestris L.: Fact or Fiction. *J. Ethnopharmacol.* **2016**, *179*, 345–355. [CrossRef] [PubMed]
35. United Nations. Convention on Biological Diversity. Available online: https://www.cbd.int/convention/. (accessed on 23 May 2018).
36. Moher, D.; Liberati, A.; Tetzlaff, J.; Altman, D.G.; PRISMA Group. Preferred Reporting Items for Systematic Reviews and Meta-Analyses: The PRISMA Statement. *BMJ* **2009**, *339*, b2535. [CrossRef] [PubMed]

© 2018 by the authors. Licensee MDPI, Basel, Switzerland. This article is an open access article distributed under the terms and conditions of the Creative Commons Attribution (CC BY) license (http://creativecommons.org/licenses/by/4.0/).

Review

A Comprehensive Review on the Medicinal Plants from the Genus *Asphodelus*

Maryam Malmir [1], Rita Serrano [1], Manuela Caniça [2], Beatriz Silva-Lima [1] and Olga Silva [1,*]

[1] Research Institute for Medicines (iMed.ULisboa), Faculty of Pharmacy, Universidade de Lisboa, Av. Professor Gama Pinto, 1649-003 Lisbon, Portugal; mmalmir@ff.ulisboa.pt (M.M.); rserrano@ff.ulisboa.pt (R.S.); beatrizlima@netcabo.pt (B.S.-L.)

[2] Department of Infectious Diseases, National Reference Laboratory of Antibiotic Resistances and Healthcare Associated Infections, National Institute of Health Doutor Ricardo Jorge, 1649-016 Lisbon, Portugal; Manuela.Canica@insa.min-saude.pt

* Correspondence: odsilva@ff.ulisboa.pt or osilva@campus.ul.pt; Tel.: +351-217-946400; Fax: +351-217-946470

Received: 14 February 2018; Accepted: 10 March 2018; Published: 13 March 2018

Abstract: Plant-based systems continue to play an essential role in healthcare, and their use by different cultures has been extensively documented. *Asphodelus* L. (Asphodelaceae) is a genus of 18 species and of a total of 27 species, sub-species and varieties, distributed along the Mediterranean basin, and has been traditionally used for treating several diseases particularly associated with inflammatory and infectious skin disorders. The present study aimed to provide a general review of the available literature on ethnomedical, phytochemical, and biological data related to the genus *Asphodelus* as a potential source of new compounds with biological activity. Considering phytochemical studies, 1,8-dihydroxyanthracene derivatives, flavonoids, phenolic acids and triterpenoids were the main classes of compounds identified in roots, leaf and seeds which were correlated with their biological activities as anti-microbial, anti-fungal, anti-parasitic, cytotoxic, anti-inflammatory or antioxidant agents.

Keywords: anthracene derivatives; antimicrobial; *Asphodelus*; ethnomedicine; skin diseases

1. Introduction

The genus *Asphodelus* Linnaeus belongs to family Asphodelaceae Jussieu and is native to temperate Europe, the Mediterranean, Africa, the Middle East, and the Indian Subcontinent, and now naturalized in other places (New Zealand, Australia, Mexico, southwestern United States, etc.) [1]. It reaches its maximum diversity in the West of the Mediterranean, particularly in the Iberian Peninsula and in North-West Africa [2].

The family consists of three subfamilies: Asphodeloideae Burnett (including 13 genera), Hemerocallidoideae Lindley (including 19 genera) and Xanthorrhoeoideae M.W. Chase (with only one genus). This botanical family, now called Asphodelaceae, has had a complex history; its circumscription and placement in an order have varied widely. In the Cronquist system of 1981, members of the Asphodelaceae were placed in the order Liliales Perleb [3,4]. Cronquist had difficulty classifying the less obviously delineated lilioid monocots; consequently, he placed taxa from both the modern orders Asparagales Link and Liliales into a single family, Liliaceae Jussieu [5]. The decision to group three formerly separate families, Asphodelaceae, Hemerocallidaceae and Xanthorrhoeaceae, into a single family first occurred in 2003 as an option in the II Angiosperm Phylogeny Group (APG) classification for the orders and families of flowering plants. The name used for the broader family was then Xanthorrhoeaceae Dumortier [6], and the earlier references to this family were related only to subfamily Xanthorrhoeoideae. These changes were a consequence of improvements in molecular and morphological analysis and also a reflection of the increased emphasis on placing families within an appropriate order [5,7,8]. Later in 2009, the APG III classification dropped the option of keeping the three families separate, using only the expanded family, still under the name Xanthorrhoeaceae [7]. Anticipating a decision to conserve the name Asphodelaceae over Xanthorrhoeaceae, the APG IV classification of 2016 used Asphodelaceae as the name for the expanded family [9].

According to the World Checklist of Selected Plant Families (WCSP), there are 32 accepted names with more than 150 homo- and heterotypic synonyms for all species, subspecies and varieties of the genus *Asphodelus* L. namely, *Asphodelus acaulis* Desfontaines, *Asphodelus aestivus* Brotero, *Asphodelus albus* Miller (subsp. *albus*; subsp. *carpetanus* Z. Díaz & Valdés; subsp. *delphinensis* (Grenier & Godron) Z. Díaz & Valdés; subsp. *occidentalis* (Jordan) Z. Díaz & Valdés), *Asphodelus ayardii* Jahandiez & Maire, *Asphodelus bakeri* Breistroffer, *Asphodelus bento-rainhae* P. Silva (subsp. *bento-rainhae*; subsp. *salmanticus* Z. Díaz & Valdés), *Asphodelus cerasiferus* J. Gay, *Asphodelus fistulosus* Linnaeus (subsp. *fistulosus*; subsp. *madeirensis* Simon), *Asphodelus gracilis* Braun-Blanquet & Maire, *Asphodelus lusitanicus* Coutinho (var. *lusitanicus*; var. *ovoideus* (Merino) Z. Díaz & Valdés), *Asphodelus macrocarpus* Parlatore (subsp. *macrocarpus*; subsp. *rubescens* Z. Díaz & Valdés; var. *arrondeaui* (J. Lloyd) Z. Díaz & Valdés), *Asphodelus ramosus* Linnaeus (subsp. *distalis* Z. Díaz & Valdés; subsp. *Ramosus*); *Asphodelus refractus* Boissier, *Asphodelus roseus* Humbert & Maire, *Asphodelus serotinus* Wolley-Dod, *Asphodelus tenuifolius* Cavanilles, and *Asphodelus viscidulus* Boissier [1]. However, on the Missouri Botanical Garden database (Tropicos), more two accepted names (*Asphodelus cerasifer* Gay and *Asphodelus microcarpus* Salzmann & Viviani) were recorded [10]. Considering all the above-mentioned data 18 species and of a total of 27 species, sub-species and varieties must be considered for the *Asphodelus* genus.

Among all the species, *A. aestivus* and *A. fistulosus* are inscribed as "Least Concern" and *A. bento-rainhae* as "Vulnerable" species on International Union for the Conservation of Nature (IUCN) Red List of Threatened Species [11].

Botanical and systematic descriptions of this genus have been discussed by several taxonomists in various flora publications. The plants are hardy herbaceous perennials with narrow tufted radical leaves and an elongated stem bearing a spike of white or yellow flowers. Many have a small rhizomatous crown and thick, fleshy roots [12].

Different ethnomedical uses were described to *Asphodelus* species. Different parts of the plant including leaf, fruit, seed, flower, and root are used as traditional herbal medicines, alone or in mixtures to treat various ailments. In Iberian Peninsula, the following general medicinal uses were described: by rubbing with the cut tubers for the treatment of skin eczema, the ashes of the roots were used against the alopecia, and the leaves and stems decoction was used for the treatment of paralysis and the juice of fresh capsules for earache treatment [2]. Medicinal usage of the *Asphodelus* genus is also common in North African, and West and South Asian countries. Beside its medicinal uses, in Iberian Peninsula the alcohol obtained by fermentation of the tubers is extracted and used as fuel [2] and the local people of Iran, Turkey and Egypt use the root tubers of *A. aestivus* and *A. microcarpus* to produce a strong glue used by shoemakers and cobblers [2,13,14], and as yellow and brown dyes to dye the wool [2].

Root tubers are used as daily food, after being moistened and fried beforehand to eliminate the astringent compounds, and also the young stem, the leaves and the roasted seeds [2,15].

This study aims to present a comprehensive and updated review of documented ethnomedicinal and ethnopharmacological studies including chemical and biological data concerning *Asphodelus* genus.

2. Results and Discussion

Table 1 summarizes the ethnomedicinal data about the *Asphodelus* species including specific information on the plant parts as well as the geographical region where the plant is used. In Table 2 the principal chemical studies and identified compounds of the genus are presented. Tables 3 and 4 summarize the principals of in vitro and in vivo biological activity assays on the total extracts and isolated compounds.

2.1. Ethnomedical Studies

Ethnomedicinal records showed that among the 18 species of the genus *Asphodelus*, only five species namely *A. aestivus*, *A. fistulosus*, *A. microcarpus*, *A. ramosus*, and *A. tenuifolius* have been documented for their traditional uses (Table 1). Most commonly, these species were used as anti-inflammatory and anti-infective agents. In particular, *A. aestivus*, *A. fistulosus* and *A. microcarpus* were reported to be used in dermatomucosal infections in various countries including Cyprus, Egypt, Libya, Palestine, and Spain [16–20]. *A. microcarpus*, *A. ramosus* and *A. tenuifolius* were generally indicted as anti-inflammatory agents specifically for the treatment of psoriasis, eczema, and rheumatism [21–28]. *A. aestivus* and *A. tenuifolius* are also used for ulcer treatment in Turkey, India, and Pakistan [26–29]. *A. ramosus* and *A. tenuifolius* have frequently been reported as diuretics among the inhabitants of Egypt, India, Pakistan, and Turkey [24,25,27–31].

Table 1. Ethnomedicinal uses of the *Asphodelus* species.

Species	Part Used	Country	Traditional Uses/Application	References
A. aestivus	L, R	Turkey	Peptic ulcers	[32]
	R	Turkey	Haemorrhoids, burns, wounds and nephritis	[33]
	NI	Cyprus, Spain	Skin diseases	[16]
A. fistulosus	NI	Egypt, Libya	Fungal infections	[17]
A. luteus *	WP	Palestine	Dermatomucosal infections	[18]
A. microcarpus	FR, L, R	Egypt	Ear-ache, withering and paralysis	[13,14]
	R	Palestine	Dermatomucosal infections	[18]
	R	Egypt	Ectodermal parasites, jaundice, microbial infections and psoriasis	[19–21]
	NI	Algeria	Ear-ache, eczema, colds and rheumatism	[22]
A. ramosus	R	North-Africa	Inflammatory disorders	[23]
	NI	Turkey	Anti-tumoral, diuretic and emmenagogue	[29]
A. tenuifolius	L	India	Diuretic, inflammatory disorders and ulcers	[24]
	L, SE	Egypt	Diuretic	[30]
	R, SE	India	Antipyretic, diuretic, colds and hemorrhoids, inflammatory disorders, rheumatic pain, ulcers and wounds	[25,27]
	SE	Pakistan	ulcers and inflammatory disorders	[26]
	WP	India	Diuretic, inflammatory disorders, bite of bees and wasps, ulcers	[28,34]
	NI	Pakistan	Diuretic	[31]

SE: Seed; L: Leaf; WP: Whole plant; FR: Fruit; R: Root; NI: Not indicated, * *Asphodelus luteus* L.—synonym of *Asphodeline lutea* was formerly included in the family Asphodelaceae.

2.2. Phytochemical Studies

Phytochemical studies as shown in Table 2, revealed the presence of different groups of compounds namely anthraquinones (either in the free or in the glycoside form), phenolic acids, flavonoids, and triterpenoids from *A. acaulis*, *A. albus*, *A. aestivus*, *A. cerasiferus*, *A. fistulosus*, *A. microcarpus*, *A. ramosus*, and *A. tenuifolius*.

Roots were mainly reported to have anthraquinone derivatives such as chrysophanol and aloe-emodin, triterpenoids, and naphthalene derivatives, while aerial parts mostly exhibited the presence of flavonoids such as luteolin, isovitexin and isoorientin, phenolic acids, and few anthraquinones. Fatty acids, namely myristic, palmitic, oleic, linoleic, and linolenic, were found in seeds and roots. Only *A. aestivus* and *A. microcarpus* were studied for essential oil characterization of flowers [22,35].

Table 2. Identified compounds reported from *Asphodelus* genus.

Species	Part Used	Class	Name of Compounds	References
A. acaulis	L	Flavonoids	Luteolin; apigenin	[36]
	R	Anthraquinones	Chrysophanol; asphodelin; 10,7′-bichrysophanol	[37]
	FL	n-alkenes	Hexadecanoic acid (35.6%), pentacosane (17.4%), tricosane (13.4%), heptacosane (8.4%), heneicosane (4.5%), phytol (4.5%), tetracosane (3%), hexacosane (2%), hexahydrofarnesyl acetone (1.7%), tetradecanoic acid (1.4%), docosane (1.3%), nonadecane (1%)	[35]
A. aestivus		Amino acids	Adenosine; tryptophan; phenylalanine	[38]
		Anthraquinones	Aloe-emodin; aloe-emodin acetate; chyrsophanol 1-O-gentiobioside	
	L	Flavonoids	Isovitexin; isoorientin; isoorientin 4′-O-β glucopyranoside; 6′′-O-(malonyl)-isoorientin; 6′′-O-[(S)-3-hydroxy-3-methylglutaroyl]-isoorientin	
		Phenolic acid	Chlorogenic acid	
	SE	Fatty acids	Butyric acid; nervoic acid	[39]
	L	Anthraquinones	Aloe-emodin; chrysophanol	[36,40]
		Flavonoid	Luteolin	[36]
A. albus		Anthraquinones	Chrysophanol; asphodelin; 10,7′-bichrysophanol	[37]
	R	Fatty acids	Myristic (5.3%); palmitic (18.5%); stearic (2.1%); oleic (13.5%); linolenic (9.9%); arachidic (2.7%); behenic (1.2%); lignoceric (2.1%) acids	[41]
		Triterpenoids	β-sitosterol; β-amyrin; campesterol; stigmasterol; fucosterol	
A. albus var. delphinensis	R	Anthraquinones	Asphodeline; microcorpine; aloe-emodine; chrysophanole	[42]
A. cerasifer	L	Anthraquinones	Aloe-emodin	[36]
		Flavonoids	Isoorientin; luteolin; luteolin 7-glucoside	[36,43]
	R	Anthraquinones	Asphodeline; microcorpine; aloe-emodine; chrysophanole	[41]
* A. delphinensis	L	Flavonoids	Isoorientin; luteolin; luteolin 7-glucoside	[43]
	AP	Anthraquinones	Asphodelin; asphodelin 10′-anthrone; aloesaponarin II; aloe-emodin; chrysophanol; desoxyerythrolaccin	[17]
		Flavonoids	Chrysoeriol; luteolin	
A. fistulosus	L	Anthraquinones	Dianhydrorugulosin; aloe-emodin; chrysophanol; 1,8 hydroxy-dianthraquinone	[44]
	R	Anthraquinones	Chrysophanol; asphodelin; 10,7′-Bichrysophanol	[37]
		Anthraquinones	Dianhydrorugulosin; aloe-emodin; chrysophanol; 1,8 hydroxy-dianthraquinone	[44]
		Carbohydrates	Sucrose; raffinose; stachyose	[45]
	SE	Fatty acids	Myristic (0.5%); palmitic (5.7%); stearic (3.6%); oleic (33.1%); linoleic (54.9%)	[45,46]
		Triterpenoids	β-sitosterol; β-amyrin	[45]
** A. luteus	L	Anthraquinones	Aloe-emodin	[36]
*** A. mauritii Semnen	L	Anthraquinones	Aloe-emodin; chrysophanol	[36]
		Flavonoids	Luteolin	

Table 2. Cont.

Species	Part Used	Class	Name of Compounds	References
A. microcarpus	FL	Terpenoids	Germacrene D (78.3%); germacrene B (3.9%); a-elemene (3.8%); caryophyllene (3.3%)	[22]
		Flavonoids	Luteolin; luteolin-6-C-glucoside; luteolin-O-hexoside; luteolin-7-O-glucoside; luteolin-O-acetylglucoside; luteolin-O-deoxyhesosylhexoside; methyl-luteolin, naringenin; apigenin	[47]
		Phenolic acids	3-O caffeoylquinic acid; 5-O caffeoylquinic acid	
	L	Anthraquinone	Chrysophanol, 10 (chrysophanol-7'-yl)-10-Hydroxychrysophanol-9-antrone, asphodoside C, Dianhydrorugulosin; aloe-emodin	[44,48]
		Flavonoids	Luteolin-6-C-glucoside; luteolin-6-C-acetilglucoside; luteolin-C-glucoside; luteolin, isoorientin	[43,49]
		Phenolic acids	5-O caffeoylquinic acid; cichoric acid; cumaril exosa malic acid	[49]
		Anthraquinones	Dianhydrorugulosin; aloe-emodin; chrysophanol; asphodelin; microcarpin, 8 methoxychrysophanol; emodin; 10-(chrysophanol-7'-yl)-10-hydroxychrysophanol-9-anthrone; aloesaponol-III-8-methyl ether; ramosin; aestivin, asphodosides A-E, chrysophanol dianthraquinone; 5,5'-bichrysophanol; chrysophanol-8-mono-β-D-glucoside; Methyl-1,4,5-trihydroxy-7-methyl-9,10-dioxo-9,10-dihydroanthracene-2-carboxylate; 6 methoxychrysophanol	[21,44,50–54]
	R	Arylcoumarins	Asphodelin A 4'-O-β-glucoside; asphodelin A	[19]
		Carbohydrates	Raffinose; sucrose; glucose; fructose	[55]
		Fatty acids	Palmitic; stearic; oleic; linoleic; linolenic; arachidic; behenic; lignoceric; myristic acids	[55,56]
		Naphthalene derivatives	2-acetyl-1,8-dimethoxy-3 methylnaphthalene; 1,6-dimethoxy-3-methyl-2-naphthoic acid	[21]
		Mucilage	Composed of glucose; galactose; arabinose	[55]
		Triterpenoids	β-sitosterol-β-D-glucoside, fucosterol	[13,55]
	SE	Anthraquinones	Aloe-emodin; chrysophanol; chrysophanol-8-mono-β-D-glucoside	[44]
		Carbohydrates	Sucrose; raffinose; stachyose; melibiose	
		Fatty acids	Myristic; palmitic; stearic; oleic; linoleic acids	[45]
		Triterpenoids	β-sitosterol; β-amyrin	
	FL	Flavonoids	Luteolin	[57]
		Phenolic acids	Caffeic acid; chlorogenic acid; p-hydroxy-benzoic acids	
	L	Flavonoids	Luteolin; 7-O-glucosyl luteolin; 7-O-glucosyl apigenin; isoorientin; isoswertiajaponin (7-methyl orientin); isocytisoside (4'-methyl vitexin)	[29]
A. ramosus	R	Anthraquinone	Ramosin; (−)-10'-C-[β-D-xylopyranosyl]-; (−)-10'-C-[β-D-glucopyranosyl]-(1-4)-β-D-glucopyranosyl]-1,1'.8,8,10,10'-hexa hydroxy-3,3'-dimethyl-10,7' bianthracene-9,9'-dione; 10'-deoxy-10-epi-ramosin; 10-(chrysophanol-7'-yl)-10-hydroxychrysophanol-9-anthrone; 7'-(Chrysophanol-4-yl)-chrysophanol-10'anthrone10'-C-α-rhamnopyranosyl; -C-β-xylopyranosyl; -C-β-antiaropyranosyl; -C-α-arabinopyranosyl; -C-β-quinovoopyranosyl	[58–60]
	WP	Flavonoids	Naringin, quercetin, kaemferol	[61]
		Phenolic acids	Gallic acid, chlorogenic acid, vanilic acid, cafeic acid	

Table 2. Cont.

Species	Part Used	Class	Name of Compounds	References
A. tenuifolius	AP	Flavonoids	Luteolin; luteolin-7-O-β-D-glycopyranoside; apigenin, chrysoeriol	[30]
	R	Naphthalene derivatives	1,8-dimethylnaphthalene; 2-acetyl-8-methoxy-3-methyl-1-naphthol; 2-acetyl-1,8-dimethoxy-3-methylnaphthalene	[62]
		Triterpenoids	β-sitosterol; stigmasterol	
	SE	Ester	1-O-17methylstearylmyoinositol	[63]
		Fatty acids	Myristic (3.96%); palmitic (13.84%); oleic (15.60%); linoleic (62.62%); linolenic (2.60%)	[64,65]
		Amino acids	Crystine; serine; glycine; proline; alanine, glycin; serine; alanine and valine in the form of protein	[66]
	WP	Carbohydrates	D-glucose; lactose; D-glucuronic acid; D-arabinose; D-fructose, D-ribose	
		Chromone	2-hentriacontyl-5,7-dihydroxy-8-methyl-4H-1-benzopyran-4-one	[31]
		Triterpenoids	Asphorodin; asphorin A; asphorin B; β-sitosterol; β-amyrin	[26,28,31]

AP: Aerial Part; FL: Flower; FR: Fruit; L: Leaf; R: Root; SE: Seed; WP: Whole plant; NI: Not indicated; * The accepted name is *Asphodelus albus* subsp. *delphinensis* (Gren. & Godr.). ** *Asphodelus luteus* L.—synonym of *Asphodeline lutea* was formerly included in the family *Asphodelaceae*. *** The accepted name is *Asphodelus macrocarpus* subsp. *rubescens*.

2.3. Reported Biological Activities

In vitro and in vivo biological studies concerning *Asphodelus* extracts are presented in Table 3 and those reported from identified pure compounds are shown in Table 4. In some of the studies, no data were obtained concerning the tested doses and/or inhibitory values.

The ethanol and aqueous extracts of *A. aestivus* leaf showed moderate anti-fungal activity against *Aspergillus niger* [33], and whole plant ethanol extracts exhibited weak activity against *Staphylococcus aureus* with minimum inhibitory concentration (MIC) = 42 mg/mL) and *Klebsiella pneumoniae* (MIC = 60 mg/mL) [67]. Both leaf and root extracts showed strong antioxidant activity [15,68]. The root extract also showed significant anti-inflammatory properties, specifically anti-ulcer activity which is one of the documented uses in Turkish traditional medicine [32]. Root and leaf extracts showed antitumoral activity against human cancer cells (lung and prostate) through DNA damage [68,69].

The aerial parts extracts of *A. luteus* showed strong anti-fungal activity against *Trichophyton violaceum* (MIC = 18 μg/mL), *Microsporum canis* (MIC = 25 μg/mL), and *Trichophyton mentagrophytes* (MIC = 30 μg/mL) supporting their traditional use in dermatomucosal infections [18] and weak activity against methicillin-resistant *Staphylococcus aureus* (MRSA) isolates (MIC = 1.25–2.5 mg/mL) [70]. Moreover, the methanol root extracts showed moderate antioxidant activity against 2,2-diphenyl-1-picrylhydrazyl free radicals (DPPH; IC_{50} = 0.54 mg/mL) [61].

The aerial parts and root extracts of *A. microcarpus* showed moderate antioxidant [47,61] and moderate to weak cytotoxic activities [48,49,71]. The ethanol extracts of leaves demonstrated strong antiviral activity against Ebola virus (EBOV) in the concentration of 0.1–0.3 μg/mL [49]. Although the leaf seems to have stronger antimicrobial activity in comparison with roots, in general, both exhibit weak or no antimicrobial/antifungal activity [20,48,49,70]; however, compounds isolated from root tubers extracts showed potent activity such as asphodelin A against S. aureus (MIC = 16 μg/mL), *Escherichia coli* (MIC = 4 μg/mL), *Pseudomonas aeruginosa* (MIC = 8 μg/mL), *Candida albicans* (MIC = 64 μg/mL) [19] and *Botrytis cinerea* (MIC = 128 μg/mL) and asphodoside B against MRSA (IC_{50} = 1.62 μg/mL) [51]. Other isolated compounds from root extracts showed different biological activity; for instance, ramosin showed potent cytotoxic activity against leukemia cell lines [21], aestivin showed potent antimalarial activity against chloroquine-sensitive and resistant strains of *Plasmodium falciparum* with IC_{50} of 0.8–0.7 μg/mL [21] and 3,4-dihydroxy-methyl benzoate exhibited anti-parasitic activity against *Leishmania donovani* promastigotes with IC_{50} of 33.2 μg/mL [54].

Root extracts of *A. ramosus* showed positive in vivo anti-inflammatory activity, confirming the traditional uses of the plant in inflammatory disorders [23].

Several root, seed, aerial parts, fruit, and leaf extracts of *A. tenuifolius* showed strong anti-microbial/antifungal against *K. pneumoniae*, *P. aeruginosa*, *E. coli*, *S. aureus*, *Proteus mirabilis*, *C. albicans*, *Aspergillus fumigatus*, *Vibrio cholerae*, *Salmonella typhi*, and *Candida glabrata*, among other pathogens [24,27,30,34,72,73]. Of note, there is no ethnomedical report of antimicrobial use of *A. tenuifolius*.

The whole plant extract showed in vivo hypotensive and diuretic activity in normotensive rats [74]. The root extract of this species showed anti-oxidant activity (DPPH test, IC_{50} = 2.006 μg/mL) [25] and asphorodin, a compound isolated from the whole plant extract, exhibit a potent inhibition of lipoxygenase enzyme, (IC_{50} = 18.1 μM) [26], which may have an important role as an anti-inflammatory agent. The biological properties of *A. tenuifolius* extracts prove their ethnomedical use mostly as anti-inflammatory or diuretic [24–28,30,31,34].

Table 3. In vitro and in vivo biological studies reported from the *Asphodelus* genus.

Species	Part	Extract	Test/Assay	Result	Reference
A. aestivus	L	Aqueous, Ethanol	In vitro anti-fungal activity (A. niger)—Agar well diffusion method (zone of inhibition in cm^{-1})	Ethanol extract (0.25 and 0.5 mg/mL) showed higher activity than aqueous extract (0.25 and 0.5 mg/mL) and similar activity for concentrations of 1 mg/mL. Both extracts were less active than Fluconazole (100 μ/mL).	[33]
A. aestivus	L	Aqueous, Ethanol	In vitro antioxidant activity—β-carotene bleaching effect, metal chelating, total antioxidant activity, DPPH, ABTS, superoxide radical scavenging activity, hydroxyl radical scavenging activity, DMPD, nitric oxide scavenging activity	Aqueous extract presented higher activity in metal chelating and radical scavenging assays (DPPH, IC$_{50\ aqueous}$ = 4.58 mg/mL and IC$_{50\ methanol}$ = 9.54 mg/mL, superoxide, hydroxyl, DMPD. Ethanol extract presented higher activity in β-carotene bleaching effect and total antioxidant activity. Aqueous and ethanolic extracts presented similar radical scavenging activity in ABTS and NO assays. Both extracts presented significantly inferior results when compared to reference substances	[15]
A. aestivus	L	Acetone, Methanol	In vitro antioxidant activity—β-carotene, reducing power assay, DPPH, ABTS, inhibition of linoleic acid peroxidation, superoxide radical scavenging assays	Reducing power and total antioxidant activity were higher in acetone extract; free radical and superoxide radical scavenging activity were higher in methanol extract (DPPH, IC$_{50\ methanol}$ = 0.16 mg/mL and IC$_{50\ acetone}$ = 0.50 mg/mL). Acetone extract presented higher activity in Reducing power and total antioxidant activity (inhibition of linoleic acid peroxidation). Methanol extract presented higher activity in superoxide radical scavenging and free radical scavenging activity (β-carotene, ABTS and DPPH, IC$_{50\ methanol}$ =0.16 mg/mL, IC$_{50\ acetone}$ = 0.50 mg/mL).	[69]
A. aestivus	L, R	Dichloromethane n-Hexane	In vitro cytotoxic activity—MTT assay against human lung cell cancer (A549) and prostate cell cancer (PC3)	Root: Dichloromethane: A549 (IC$_{50}$ = 16 μg/mL); PC3 (IC$_{50}$ = 19 μg/mL); n-Hexane: PC3 (IC$_{50}$ = 80 μg/mL). Leaves: Dichloromethane: A549 (IC$_{50}$ = 90 μg/mL)	[32]
A. aestivus	R	Aqueous (decoction)	In vivo anti-inflammatory—Ethanol induced gastric ulcer model in rats	Decoction gave significant protection against the lesions	[68]
A. aestivus	R	Diethyl ether, Ethyl acetate, Methanol	In vitro antioxidant activity—DPPH assay	Diethyl ether (IC$_{50}$ = 22.46 μg/mL) have a higher scavenging activity than Ethyl acetate (IC$_{50}$ = 188.90 μg/mL), both have lower activity than reference substance, rutin (7.77 μg/mL). Methanol and aqueous extract had no scavenging activity	
A. aestivus	R	Aqueous (infusion and decoction) Diethyl ether, Ethyl acetate, Methanol	In vitro cytotoxic & apoptotic activity—MCF-7 breast cancer cells–trypan blue exclusion assay, comet assay, Hoechst 33,258, propidium iodide double staining	Methanol and aqueous extracts exhibited strong cytotoxic activities. All extracts showed significant DNA damaging and apoptotic activities.	
A. aestivus	SE	Petroleum ether	In vitro antimicrobial/fungal activity—broth microdilution method	Active against S. aureus (MIC = 512 μg/mL), Enterococcus faecalis (MIC = 512 μg/mL), K. pneumoniae (MIC = 512 μg/mL) and C. albicans (MIC = 512 μg/mL). Not active against Bacillus cereus, Staphylococcus epidermidis, E. coli, P. aeruginosa, S. typhimurium, Salmonella enterica, Candida krusei and Candida parapsilosis	[39]
A. aestivus	WP	n-Butanol, Ethanol	In vitro anti-microbial/fungal activity—well and disk diffusion method	Active against S. aureus (MIC: 42 mg/mL), K. pneumoniae (MIC: 60 mg/mL), E. coli (MIC: 90 mg/mL), C. albicans (MIC: 90 mg/mL)	[67]
A. aestivus	WP	Aqueous	In vitro antioxidant activity—DPPH assay	Inhibition % = 62.5	[75]

Table 3. Cont.

Species	Part	Extract	Test/Assay	Result	Reference
A. fistulosus var. tenuifolius	NI	NI	In vitro anti-microbial/fungal activity	Positive to S. aureus and no activity against E. coli, Proteus vulgaris, Salmonella sp., P. aeruginosa, C. albicans	[76]
A. luteus *	AP	Aqueous	In vitro anti-fungal activity—Agar dilution method	Activity against T. violaceum (MIC = 18 µg/mL), M. canis (MIC = 25 µg/mL) and T. mentagrophytes (MIC = 30 µg/mL)	[18]
A. luteus *	AP R	Methanol, Petroleum Ether	In vitro anti-microbial activity—agar diffusion test; tetrazolium microplate assay (MIC)	Against MRSA isolates Methanol extract: MIC (AP) = 1.25–2.5 mg/mL MIC (R) = 0.65–1.25 mg/mL Petroleum ether extract: Root extract had higher activity than aerial part extract	[70]
A. luteus *	R	Methanol	In vitro antioxidant activity—DPPH assay	IC_{50} (methnol)= 0.54 mg/mL, IC_{50} (reference, BHT) = 0.017 mg/mL	[61]
A. microcarpus	AP	Aqueous	In vitro anti-fungal activity—Agar dilution method	Weak activity against T. violaceum (MIC = 25 µg/mL) and no activity against M. canis and T. mentagrophytes	[18]
A. microcarpus	AP R	Methanol	In vitro anti-microbial activity—agar diffusion test; tetrazolium microplate assay (MIC)	Against MRSA isolates Methanol extract: MIC (AP) = 1.25–5 mg/mL MIC (R) = 1.25–5 mg/mL	[70]
A. microcarpus	FL L R	Aqueous, Ethanol, Methanol	In vitro antimelanogenic activity—tyrosinase inhibition (mushroom tyrosinase assay and mouse melanoma cells viability), kojic acid as positive control	Antimelanogenic activity Ethanol extract (F) had the highest tyrosinase inhibition activity in mushroom assay and melanoma cell assay	[47]
			In vitro antioxidant activity—DPPH and ABTS (reference—Trolox)	Antioxidant activity DPPH (best activity) Ethanol extract (F): IC_{50} = 28.4 µg/mL Ethanol extract (L): IC_{50} = 55.9 µg/mL Trolox: IC_{50} = 3.2 µg/mL	
A. microcarpus	L	Ethanol	In vitro antimicrobial/fungal activity—micro broth dilution method	Active against Bacillus clausii (MIC = 250 µg/mL), S. aureus (MIC = 250 µg/mL), Staphylococcus haemolyticus (MIC = 250 µg/mL) and E. coli (MIC = 500 µg/mL). No activity against Streptococcus spp. and yeasts	[49]
			In vitro antiviral activity (IFN-β induction)—luciferase reporter gene assay	Antiviral activity Active against EBOV in concentration of 0.1-3 µg/mL	
			In vitro cytotoxicity-Cell viability of A549 cells, positive control (camptothecin)	Cytotoxicity IC_{50} (extract) > 100 µg/mL IC_{50} (camptothecin) 0.54 µg/mL	

Table 3. Cont.

Species	Part	Extract	Test/Assay	Result	Reference
A. microcarpus	L	Methanol	In vitro antimicrobial/fungal—two-fold serial dilution technique	Antimicrobial activity: Active against S. aureus (MIC = 78 µg/mL), Bacillus subtilis (MIC = 156 µg/mL), Salmonella spp. (MIC = 313 µg/mL), E. coli (MIC = 125 µg/mL), Aspergillus flavus (MIC = 125 µg/mL), C. albicans (MIC = 78 µg/mL)	[48]
			In vitro antiviral activity—CPE inhibition assay against HSV-1 and HAV-10	Antiviral activity: Moderate activity against Hepatitis A virus (HAV-10) and no activity against Herpes Simplex Virus (HSV-1)	
			In vitro cytotoxicity—viability assay against human tumor cell lines of the lung (A-549), colon (HCT-116), breast (MCF-7) and prostate (PC3). Cisplatin as standard	Cytotoxicity: Highest activity against human lung carcinoma cells (A-549), IC$_{50}$ = 29.3 µg/mL	
A. microcarpus	R	Methnol	In vitro antioxidant activity—DPPH assay	IC$_{50}$ (Methnol) = 0.30 mg/mL, IC$_{50}$ (reference, BHT) = 0.017 mg/mL	[61]
A. microcarpus	R	Methanol	In vitro anti-microbial—Disk diffusion assay	No activity against S. aureus, B. subtilis and E. coli	[20]
A. microcarpus	WP	Aqueous, Ethanol	In vitro antioxidant activity—DPPH assay	Ethanol extract (100 µg/mL) with moderate activity (inhibition percentage—60.3%) higher than aqueous extract (100 µg/mL, inhibition percentage—49.5%)	[71]
A. microcarpus	WP		In vitro cytotoxic activity—Trypan blue technique for Ehrlich Ascites Carcinoma Cells (EACC)	Weak anti-cancer activity of both extracts	
A. ramosus	R	Aqueous, Chloroform, Ethanol, Methanol	In vivo anti-inflammatory—Arachidonic acid test (mouse ear oedema) Carrageenan test (sub-plantar oedema)	Arachidonic acid test: Positive activity from chloroform and ethanol extracts. Carrageenan test: No activity was observed	[23]
A. ramosus	WP	Aqueous, Methanol, Methanol 50%	In vitro antioxidant activity—DPPH assay at 35 °C and 65 °C	Aqueous extract at 65 °C had the highest inhibition percentage	[77]
A. tenuifolius	AP	Butanol, Ethyl acetate, Methylene-chloride	In vitro anti-microbial/fungal activity—Disc diffusion method	All extracts showed antimicrobial activity, the methylene-chloride as the most active against S. aureus (MIC = 1.6 mg/mL), E. faecalis (MIC = 1.0 mg/mL) and P. aeruginosa (MIC = 0.15 mg/mL). All extracts showed antifungal activity against C. albicans, C. parapsilosis, C. glabrata, C. krusei.	[30]
A. tenuifolius	FR	Acetone, Aqueous, Benzene, Chloroform, Methanol, Petroleum ether	In vitro anti-microbial/fungal activity—Kirk-bauer disc diffusion method	Significant activity against S. aureus (acetone, MIC = 125 µg/mL; chloroform and methanol, MIC = 250 µg/mL); S. epidermidis (acetone, MIC = 125 µg/mL; chloroform, MIC = 125 µg/mL); P. vulgaris (methanol, MIC = 250 µg/mL; chloroform, MIC = 125 µg/mL); P. mirabilis (benzene, MIC = 125 µg/mL; acetone and methanol, MIC = 250 µg/mL; chloroform, MIC = 500 µg/mL) E. coli (acetone, chloroform and methanol, MIC = 125 µg/mL); K. pneumoniae (acetone and methanol, MIC = 125 µg/mL; chloroform and benzene, MIC = 500 µg/mL); P. aeruginosa (acetone, MIC = 250 µg/mL; chloroform, MIC = 500 µg/mL); C. albicans (acetone, MIC = 125 µg/mL); A. fumigatus (benzene and chloroform, MIC = 250 µg/mL; acetone, MIC = 500 µg/mL)	[27]
A. tenuifolius	L	Acetone, Methanol	In vitro anti-microbial/fungal activity—Agar disc diffusion method	Methanol extract positive against S. aureus, B. cereus, Citrobacter freundii, Candida tropicalis and acetone extract was positive against K. pneumoniae, C. tropicalis and Cryptococcus luteolus	[24]

Table 3. *Cont.*

Species	Part	Extract	Test/Assay	Result	Reference
A. tenuifolius	R	Methanol	In vitro antioxidant activity—DPPH, ABTS+, NO, OH, O_2^-, ONOO$^-$ assays, Oxidative DNA damage	Positive activity, DPPH (IC$_{50}$ = 2.006 µg/mL), ABTS+ (IC$_{50}$ = 156.94 µg/mL), NO (nd), OH (IC$_{50}$ = 50.13 µg/mL), O_2^- (IC$_{50}$ = 425.92 µg/mL) and ONOO- (IC$_{50}$ = 3.390 µg/mL), oxidative DNA damage: 1.85 µg/mL of extract prevented DNA damage.	[25]
A. tenuifolius	R	Benzene, Chloroform, Ethyl acetate, Methanol, Petroleum ether	In vitro anti-microbial/fungal activity—Disc diffusion method	All extracts were active against *B. subtilis*, *P. vulgaris*, *P. aeruginosa*, *Trichophyton rubrum*, *E. coli*, *K. pneumoniae*, *Shigella sonnei*, *S. aureus*, *C. albicans*, *A. niger* and *A. flavus*	[72]
A. tenuifolius	SE	Aqueous, Ethanol, Methanol, Petroleum ether	In vitro anti-microbial/fungal activity—modified Kirby Bauer disc diffusion method	Petroleum ether: no antibacterial activity Ethanol: activity against *P. aeruginosa*, *Vibrio cholerae* and *S. aureus* (MIC = 16 µg/mL); *P. mirabilis*, *S. typhi*, *Shigella flexneri* and *Serratia marcescens* (MIC = 32 µg/mL). Methanol: activity against *S. aureus* (MIC = 16 µg/mL); *V. cholerae*, *P. aeruginosa*, *S. typhi*, *S. flexneri* and *S. marcescens* (MIC = 16 µg/mL). Aqueous: activity against *V. cholerae*, *S. aureus*, *S. typhi* and *S. flexneri* (MIC = 32 µg/mL); *P. aeruginosa* and *P. mirabilis* (MIC = 16 µg/mL). No antifungal activity against *C. albicans* and *A. niger*	[34]
A. tenuifolius	WP	Methanol	In vitro antimicrobial/fungal activity—disk diffusion method In vitro anti-parasitic activity—trophozoites growth inhibition assay	Good activity against *E. coli* and moderate activity against *S. aureus*, *S. typhi*, *K. pneumoniae*, *P. aeruginosa*, *C. albicans* and *A. niger* Active against *Giardia lamblia* (IC$_{50}$ = 219.82 µg/mL) and *Entamoeba histolytica* (IC$_{50}$ = 344.62 µg/mL)	[73]
A. tenuifolius	WP	Aqueous	In vivo hypotensive activity—blood pressure (BP) measure after parenteral administration of aqueous extracts in rats. Acetylcholine and verapamil as positive controls in co administration with atropine In vivo diuretic activity—measure of rat urine output and urinary electrolytes. After 6 hr administration. Saline solution and furosemide as controls	Hypotensive activity The extract decreased blood pressure in normotensive rats (35.2% decrease with 30 mg/Kg), similar to Verapamil. The response was independent from atropine effect Diuretic activity Significant increase in urinary volume and electrolytes excretion with 300 and 500 mg/Kg	[74]

AP: Aerial Part; FL: Flower; FR: Fruit; L: Leaf; R: Root; SE: Seed; WP: Whole plant; NI: Not indicated; * *Asphodelus luteus* L.—synonym of *Asphodeline lutea* was formerly included in the family *Asphodelaceae*. ABTS+: 2,2′-azinobis-(3-ethylbenzothiazole-6-sulphonate) radical cation, DMPD: *N,N*-dimethyl-*p*-phenylenediamine dihydrochloride, DPPH: 2,2-diphenyl-1-picrylhydrazyl radical, NO: nitric oxide radical, O_2^-: superoxide anion radical , OH: hydroxyl radical, ONOO$^-$: Peroxynitrite radicals, EBOV: Ebola virus.

Table 4. In vitro and in vivo biological studies reported from pure compounds isolated from *Asphodelus* genus.

Species	Pure Compounds	Test/Assay	Result	Reference
	Asphodelin A 4′-O-β-D-glucoside (1), Asphodelin A (2)	In vitro antimicrobial/fungal activity—micro dilution assay	*S. aureus* (MIC$_1$ = 128 µg/mL, MIC$_2$ = 16 µg/mL), *E. coli* (MIC$_1$ = 128 µg/mL, MIC$_2$ = 4 µg/mL), *P. aeruginosa* (MIC$_1$ = 256 µg/mL, MIC$_2$ = 8 µg/mL), *C. albicans* (MIC$_1$ = 512 µg/mL, MIC$_2$ = 64 µg/mL) and *B. cinerea* (MIC$_1$ = 1024 µg/mL, MIC$_2$ = 128 µg/mL.	[19]
	3-methyl anthraline, chrysophanol, and aloe-emodine	Psoriasis	Positive (patent)	[78,79]
		In vitro anti-parasitic activity	Compounds 3 and 4 showed moderate to weak against a culture of *L. donovani* promastigotes (IC$_{50}$ = 14.3 and 35.1 µg/mL, respectively)	
	1,6-dimethoxy-3-methyl-2-naphthoic acid (1), asphodelin (2), chrysophanol (3), 8 methoxychrysophanol (4), emodin (5), 2-acetyl-1,8-dimethoxy-3-methylnaphthalene (6), 10-(chrysophanol-7′-yl)-10-hydroxychrysophanol-9-anthrone (7), aloesaponol-III-8-methyl ether (8), ramosin (9), aestivin (10)	In vitro cytotoxic activity-Human acute leukemia HL60 cells/human chronic leukemia 562 cells	Compounds 7 and 9 exhibited a potent cytotoxic activity against leukemia LH60 and K562 cell lines	[21]
A. microcarpus		In vitro antimalarial activity—chloroquine sensitive & resistant strains of *Plasmodium falciparum* (plasmodial LDH activity)	Compound 10 showed potent antimalarial activities against both chloroquine-sensitive and resistant strains of *P. falciparum* (IC$_{50}$ = 0.8–0.7 µg/mL) without showing any cytotoxicity to mammalian cells	
		In vitro anti-microbial/fungal activity	Compound 4 exhibited moderate antifungal activity against *Cryptococcus neoformans* (IC$_{50}$ = 15.0 µg/mL), compounds 5, 7 and 10 showed good to potent activity against methicillin resistant *S. aureus* (MRSA) (IC$_{50}$ = 6.6, 9.4 µg/mL and 1.4 µg/mL, respectively). Compounds 5, 8 and 9 displayed good activity against *S. aureus* (IC$_{50}$ = 3.2, 7.3 and 8.5 µg/mL, respectively)	
	Methyl-1,4,5-trihydroxy-7-methyl-9,10-dioxo-9,10-dihydroanthracene-2-carboxylate (1), (1R) 3,10-dimethoxy-5-methyl-1H-1,4 epoxybenzo[h]isochromene (2), 3,4-dihydroxy-methyl benzoate (3), 3,4-dihydroxybenzoic acid (4), 6 methoxychrysophanol (6)	In vitro anti-parasitic activity	Compound 3 showed activity against a culture of *L. donovani* promastigotes (IC$_{50}$ = 33.2 µg/mL)	[54]
		In vitro anti-microbial/activity	Compound 1 showed a potent activity against methicillin resistant *S. aureus* (MRSA) and *S. aureus* (IC$_{50}$: 1.5 and 1.2 µg/mL, Respectively)	
	5 Compounds, Asphodosides A–E	In vitro anti-microbial activity	Compounds 2–4 showed activity against methicillin resistant *S. aureus* (MRSA) (IC$_{50}$: 1.62, 7.0 and 9.0 µg/mL, respectively). activity against *S. aureus* (non-MRSA), IC$_{50}$ = 1.0, 3.4 and 2.2 µg/mL, respectively	[51]
A. tenuifolius	Asphorodin	In vitro anti-inflammatory-inhibition of lipoxygenase enzyme	Potent inhibitory activity (IC$_{50}$ = 18.1 µM), Reference: baicalein (22.6 µM)	[26]

3. Materials and Methods

Ethnobotanical data was collected by our team in Portugal and relevant literature was reviewed until December 2017, by probing scientific databases (PubMed, Scopus, Google Scholar, b-on, Web of knowledge) and other web sources such as records from WCSP, IUCN, APG and the Missouri Botanical Garden database. Various keywords were used during the bibliographic research including: ASPHODELUS SPECIES; TRADITIONAL USES; ETHNOMEDICINAL EVIDENCE; BIOLOGICAL ACTIVITIES; ISOLATED MOLECULES; PHYTOCHEMISTRY. Information was gathered and summarized in table form where appropriate.

4. Conclusions

In conclusion, among the 18 species of the genus *Asphodelus*, only 30 percent of the species namely *A. aestivus, A. fistulosus, A. microcarpus, A. ramosus,* and *A. tenuifolius* have been documented for their traditional uses. In phytochemical studies 50 percent of the species (*A. acaulis, A. aestivus, A. albus, A. cerasifer, A. fistulosus, A. macrocarpus, A. microcarpus, A. ramosus, A. tenuifolius*) have been evaluated for their constituents however there is no documented data related to traditional uses of *A. acaulis, A. albus* and *A. cerasiferus*.

All the species with ethnomedical documented data were submitted to biological activity tests, showing a total or partial correlation with their traditional use as anti-microbial, anti-fungal, anti-parasitic, cytotoxic, anti-inflammatory, or antioxidant agents.

Root tubers plant part were mainly reported to have anthraquinone derivatives, triterpenoids, and naphthalene derivatives, while aerial parts mostly exhibited the presence of flavonoids, phenolic acids, and few anthraquinones.

Considering the previous phytochemical studies, 1,8 dihydroxyanthracene derivatives (e.g., aloe-emodin and chrysophanol) were the most common reported anthraquinones of *A. aestivus, A. luteus* and *A. microcarpus* extracts which could be responsible for the reported antimicrobial/fungal activities [78,80]. Aloe-emodin as a potent cytotoxic compound might be related to the reported anti-tumoral activity of *A. aestivus* [68,78].

Flavonoids namely luteolin and apigenin derivatives were frequently reported from the aerial parts of all studied *Aphodelus* species, which according to their known antioxidant and anti-inflammatory properties [81,82], could be correlated to their traditional uses in inflammatory diseases in agreement with the reported biological studies. Phenolic acids, namely caffeic acid and chlorogenic acid reported from aerial parts and root tubers might be responsible for the general antioxidant activity presented in the biological studies.

Phytosterols (e.g., fucosterol, β-sitosterol, and stigmasterol) and β-amyrin were the most common found triterpenoids from roots and seeds. According to the literature, β-amyrin possess antibacterial/antifungal properties [83] which complement the reported biological activities of *A. tenuifolius*.

The present study allowed the importance and potential of the genus *Asphodelus* as a source of new compounds to be ascertained, with biological activity and new herbal products based on *Asphodelus* genus used in traditional medicine being ascertained, as well as its quality, mode of action, and safety of use. It should be pointed out that, to the best of our knowledge, the latter aspect (the safety of *Asphodelus* species) has not yet been the object of in-depth studies.

Acknowledgments: The authors wish to thank the Fundação para a Ciência e Tecnologia (FCT) for financial support of iMed.ULisboa project (UID/DTP/04138/2013) as well as a doctoral fellowship granted to the first author (SFRH/BD/125310/2016). This work was supported by the National Research Foundation (FCT), in Portugal, having no role in the data collection and analysis, interpretation of the findings, preparation of the manuscript, or the decision to submit the manuscript for publication. None of the authors has a conflict of interest.

Author Contributions: Maryam Malmir and Olga Silva did the literature review and wrote the first draft of the manuscript; Beatriz Silva-Lima and Manuela Caniça gave scientific involvement and evaluated critically the literature review; Rita Serrano was associated with the edition of the paper.

Conflicts of Interest: The authors declare no conflict of interest.

References

1. World Checklist of Selected Plant Families (WCSP). *Facilitated by the Royal Botanic Gardens, Kew*. Available online: http://apps.kew.org/wcsp/ (accessed on 20 December 2017).
2. Díaz Linfante, Z. *Asphodelus L. Flora Iberica*; Talavera, S., Andrés, C., Arista, M., Piedra, M.P.F., Rico, E., Crespo, M.B., Quintanar, A., Herrero, A., Aedo, C., Eds.; Consejo Superior de Investigaciones Científicas (CSIC), Real Jardin Botánico: Madrid, Spain, 2013; Volume 20, ISBN 276-308-152.
3. Cronquist, A. *An Integrated System of Classification of Flowering Plants*; Columbia University Press: New York, NY, USA, 1981; ISBN 9780231038805.
4. Beadle, N.C.W. *The Vegetation of Australia*; Cambridge University Press: New York, NY, USA, 1981; ISBN 0521241952.
5. Rudall, P.J. Unique floral structures and iterative evolutionary themes in Asparagales: Insights from a morphological cladistic analysis. *Bot. Rev.* **2002**, *68*, 488–509. [CrossRef]
6. The Angiosperm Phylogeny Group. An update of the Angiosperm Phylogeny Group classification for the orders and families of flowering plants: APG II. *Bot. J. Linn. Soc.* **2003**, *141*, 399–436. [CrossRef]
7. The Angiosperm Phylogeny Group. An update of the Angiosperm Phylogeny Group classification for the orders and families of floweringplants: APG III. *Bot. J. Linn. Soc.* **2009**, *161*, 105–121. [CrossRef]
8. The Angiosperm Phylogeny Group. An Ordinal Classification for the Families of Flowering Plants. *Ann. Mo. Bot. Gard.* **1998**, *85*, 531–553.
9. The Angiosperm Phylogeny Group. An update of the Angiosperm Phylogeny Group classification for the orders and families of flowering plants: APG IV. *Bot. J. Linn. Soc.* **2016**, *181*, 1–20. [CrossRef]
10. Tropicos. Missouri Botanical Garden. 2017. Available online: http://www.tropicos.org/Name/18400528 (accessed on 20 December 2017).
11. Caldas, F.B.; Moreno Saiz, J.C. *The International Union for the Conservation of Nature (IUCN) Red List of Threatened Species*; The International Union for the Conservation of Nature: Grand, Switzerland, 2011.
12. Talavera, S.; Andras, C.; Arista, M.; Piedra, M.P.F.; Rico, E.; Crespo, M.B.; Quintanar, A.; Herrero, A.; Aedo, C. (Eds.) Asphodelus. In *Flora Iberica. Plantas Vasc. la Península Ibérica e Islas Balear*; Consejo Superior de Investigaciones Científicas (CSIC), Real Jardin Botánico: Madrid, Spain, 2014.
13. Hammouda, F.M.; Rizk, A.M.; Ghaleb, H.; Abdel-Gawad, M.M. Chemical and Pharmacological Studies of *Asphodelus microcarpus*. *Planta Medica* **1972**, *22*, 188–195. [CrossRef] [PubMed]
14. Boulos, L. *Medicinal Plants of North Africa*; Reference Publications, Inc.: Algonac, MI, USA, 1983.
15. Peksel, A.; Imamoglu, S.; Altas Kiymaz, N.; Orhan, N. Antioxidant and radical scavenging activities of *Asphodelus aestivus* Brot. extracts. *Int. J. Food Prop.* **2013**, *16*, 1339–1350. [CrossRef]
16. González-Tejero, M.R.; Casares-Porcel, M.; Sánchez-Rojas, C.P.; Ramiro-Gutiérrez, J.M.; Molero-Mesa, J.; Pieroni, A.; Giusti, M.E.; Censorii, E.; de Pasquale, C.; Della, A.; et al. Medicinal plants in the Mediterranean area: Synthesis of the results of the project Rubia. *J. Ethnopharmacol.* **2008**, *116*, 341–357. [CrossRef] [PubMed]
17. Abd El-Fattah, H. Chemistry of *Asphodelus fistulosus*. *Int. J. Pharmacogn.* **1997**, *35*, 274–277. [CrossRef]
18. Ali-Shtayeh, M.S.; Abu Ghdeib, S.I. Antifungal activity of plant extracts against dermatophytes. *Mycoses* **1999**, *42*, 665–672. [CrossRef] [PubMed]
19. El-Seedi, H.R. Antimicrobial arylcoumarins from *Asphodelus microcarpus*. *J. Nat. Prod.* **2007**, *70*, 118–120. [CrossRef] [PubMed]
20. Abuhamdah, S.; Abuhamdah, R.; Al-Olimat, S.; Paul, C. Phytochemical investigations and antibacterial activity of selected medicinal plants from Jordan. *Eur. J. Med. Plants* **2013**, *3*, 394–404. [CrossRef]
21. Ghoneim, M.M.; Ma, G.; El-Hela, A.; Mohammad, A.; Kottob, S.; El-Ghaly, S.; Cutler, S.J.; Ross, S. Biologically active secondary metabolites from *Asphodelus microcarpus*. *Nat. Prod. Commun.* **2013**, *8*, 1117–1119. [PubMed]
22. Amar, Z.; Noureddine, G.; Salah, R. A Germacrene—D, caracteristic essential oil from *A. microcarpus* Salzm and Viv. flowers growing in Algeria. *Glob. J. Biodivers. Sci. Manag.* **2013**, *3*, 108–110.
23. Rimbau, V.; Risco, E.; Canigueral, S.; Iglesias, J. Antiinflammatory activity of some extracts from plants used in the traditional medicine of north-African countries. *Phytother. Res.* **1996**, *10*, 421–423. [CrossRef]
24. Vaghasiya, Y.; Chanda, S.V. Screening of methanol and acetone extracts of fourteen Indian medicinal plants for antimicrobial activity. *Turk. J. Biol.* **2007**, *31*, 243–248.

25. Kalim, M.D.; Bhattacharyya, D.; Banerjee, A.; Chattopadhyay, S. Oxidative DNA damage preventive activity and antioxidant potential of plants used in Unani system of medicine. *BMC Complement. Altern. Med.* **2010**, *10*, 77. [CrossRef] [PubMed]
26. Safder, M.; Imran, M.; Mehmood, R.; Malik, A.; Afza, N.; Iqbal, L.; Latif, M. Asphorodin, a potent lipoxygenase inhibitory triterpene diglycoside from *Asphodelus tenuifolius*. *J. Asian Nat. Prod. Res.* **2009**, *11*, 945–950. [CrossRef] [PubMed]
27. Panghal, M.; Kaushal, V.; Yadav, J.P. In vitro antimicrobial activity of ten medicinal plants against clinical isolates of oral cancer cases. *Ann. Clin. Microbiol. Antimicrob.* **2011**, *10*, 21. [CrossRef] [PubMed]
28. Saxena, V.K.; Singh, R.B. Unsaponified Matter of *Asphodelus tenuifolius* Fat. *Curr. Sci.* **1975**, *44*, 723.
29. Reynaud, J.; Flament, M.M.; Lussignol, M.; Becchi, M. Flavonoid content of *Asphodelus ramosus* (Liliaceae). *Can. J. Bot.* **1997**, *75*, 2105–2107. [CrossRef]
30. Faidi, K.; Hammami, S.; Salem, A.B.; El Mokni, R.; Mastouri, M.; Gorcii, M.; Ayedi, M.T. Polyphenol derivatives from bioactive butanol phase of the Tunisian narrow-leaved asphodel (*Asphodelus tenuifolius* Cav., Asphodelaceae). *J. Med. Plant Res.* **2014**, *8*, 550–557. [CrossRef]
31. Safder, M.; Mehmood, R.; Ali, B.; Mughal, U.R.; Malik, A.; Jabbar, A. New secondary metabolites from *Asphodelus tenuifolius*. *Helv. Chim. Acta* **2012**, *95*, 144–151. [CrossRef]
32. Gürbüz, I.; Üstün, O.; Yeşilada, E.; Sezik, E.; Akyürek, N. In vivo gastroprotective effects of five Turkish folk remedies against ethanol-induced lesions. *J. Ethnopharmacol.* **2002**, *83*, 241–244. [CrossRef]
33. Peksel, A.; Altas-Kiymaz, N.; Imamoglu, S. Evaluation of antioxidant and antifungal potential of *Asphodelus aestivus* Brot. growing in Turkey. *J. Med. Plants Res.* **2012**, *6*, 253–265. [CrossRef]
34. Dangi, A.S.; Aparna, S.M.; Yadav, J.P.; Arora, D.R.; Chaudhary, U. Antimicrobial Potential of *Asphodelus tenuifolius*. *J. Evol. Med. Dent. Sci.* **2013**, *2*, 5663–5667.
35. Polatoğlu, K.; Demirci, B.; Can Başer, K.H. High Amounts of n-Alkanes in the Composition of *Asphodelus aestivus* Brot. Flower Essential Oil from Cyprus. *J. Oleo Sci.* **2016**, *65*, 867–870. [CrossRef] [PubMed]
36. Williams, C.A. Biosystematics of the Monocotyledoneac—Flavonoid Patterns in Leaves of the Liliaceae. *Biochem. Syst. Ecol.* **1975**, *3*, 229–244. [CrossRef]
37. Van Wyk, B.E.; Yenesew, A.; Dagne, E. Chemotaxonomic significance of anthraquinones in the roots of asphodeloideae (Asphodelaceae). *Biochem. Syst. Ecol.* **1995**, *23*, 277–281. [CrossRef]
38. Çalış, I.; Birincioğlu, S.S.; Kırmızıbekmez, H.; Pfeiffer, B.; Heilmann, J. Secondary Metabolites from *Asphodelus aestivus*. *Z. Naturforsch.* **2006**, *61*, 1304–1310. [CrossRef]
39. Fafal, T.; Yilmaz, F.F.; Birincioğlu, S.S.; Hoşgör-Limoncu, M.; Kivçak, B. Fatty acid composition and antimicrobial activity of *Asphodelus aestivus* seeds. *Hum. Vet. Med.* **2016**, *8*, 103–107.
40. Van Oudtshoorn, M.V.R. Chemotaxonomic Investigations in Asphodeleae and Aloineae (Liliaceae). *Phytochemistry* **1964**, *3*, 383–390. [CrossRef]
41. Abdel-Gawad, M.; Hasan, A.; Raynaud, J. Estude de l'insaponifiable et des acides gras des tuberculus d' *Asphodelus albus*. *Fitoterapia* **1976**, *47*, 111–112.
42. Abdel-Gawad, M.M.; Raynaud, J.; Netien, G. Les Anthraquinones Libres d'*Asphodelus albus* var. Delphinensze et d'*Asphodelus cerasifer*. *Planta Med.* **1976**, *30*, 232–236. [CrossRef] [PubMed]
43. Raynaud Par, J.; Abdel-Gawad, M.M. *Contribution à L'étude Chimiotaxinomique du Genre Asphodelus (Liliaceae)*; Publications of the Lyon Linnéenne Society: Lyon, France, 1974.
44. Hammouda, F.M.; Rizk, A.M.; Seif El-Nasr, M.M. Anthraquinones of Certain Egyptian *Asphodelus* Species. *Z. Naturforsch. C* **1974**, *29*, 351–354. [CrossRef]
45. Fell, K.; Hammouda, F.; Rizk, A. The Constituents of the Seeds of *Asphodelus microcarpus* Viviani and *A. fistulosus* L. *J. Pharm. Pharmacol.* **1968**, *20*, 646–649. [CrossRef] [PubMed]
46. Khan, S.A.; Qureshi, M.I.; Bhatty, M.K. Composition of the Oil of *Asphodclus fistulosus* (Piazi) Seeds. *J. Am. Oil Chem. Soc.* **1961**, *38*, 452–453. [CrossRef]
47. Di Petrillo, A.; González-Paramás, A.M.; Era, B.; Medda, R.; Pintus, F.; Santos-Buelga, C.; Fais, A. Tyrosinase inhibition and antioxidant properties of *Asphodelus microcarpus* extracts. *BMC Complement. Altern. Med.* **2016**, *16*, 453. [CrossRef] [PubMed]
48. El-Ghaly, E.-S. Phytochemical and biological activities of *Asphodelus microcarpus* leaves. *J. Pharmacogn. Phytochem.* **2017**, *6*, 259–264.

49. Di Petrillo, A.; Fais, A.; Pintus, F.; Santos-Buelga, C.; González-Paramás, A.M.; Piras, V.; Orrù, G.; Mameli, A.; Tramontano, E.; Frau, A. Broad-range potential of *Asphodelus microcarpus* leaves extract for drug development. *BMC Microbiol.* **2017**, *17*, 159. [CrossRef] [PubMed]
50. González, A.G.; Freire, R.; Hernández, R.; Salazar, J.A.; Suárez, E. Asphodelin and Microcarpin, Two New Bianthraquinones from *Asphodelus microcarpus*. *Chem. Ind.* **1973**, *4*, 851.
51. Ghoneim, M.M.; Elokely, K.M.; El-Hela, A.A.; Mohammad, A.; Jacob, M.; Radwan, M.M.; Doerksen, R.J.; Cutler, S.J.; Ross, S.A. Asphodosides A-E, anti-MRSA metabolites from *Asphodelus microcarpus*. *Phytochemistry* **2014**, *105*, 79–84. [CrossRef] [PubMed]
52. Rizk, A.M.; Hammouda, F.M.; Abdel-Gawad, M.M. Anthraquinones of *Asphodelus microcarpus*. *Phytochemistry* **1972**, *11*, 2122–2125. [CrossRef]
53. Ghaleb, H.; Rizk, A.M.; Hammouda, F.M.; Abdel-Gawad, M.M. The Active Constituentes of *Asphodelus microcarpus* Salzm et Vivi. *Qual. Plant. Mater. Veg.* **1972**, *21*, 237–251. [CrossRef]
54. Ghoneim, M.M.; Elokely, K.M.; El-Hela, A.A.; Mohammad, A.; Jacob, M.; Cutler, S.J.; Doerksen, R.J.; Ross, S.A. Isolation and characterization of new secondary metabolites from *Asphodelus microcarpus*. *Med. Chem. Res.* **2014**, *23*, 3510–3515. [CrossRef] [PubMed]
55. Rizk, A.M.; Hammouda, F.M. Phytochemical Studies of *Asphodelus microcarpus* (Lipids and Carbohydrates). *Planta Med.* **1970**, *18*, 168–172. [CrossRef] [PubMed]
56. Picci, V.; Cerri, R.; Ladu, M. Note di biologia e fitochimica sul genere *Asphodelus* (Liliaceae) di Sardegna, I *Asphodelus microcarpus* Salzmann e Viviani. *Riv. Ital. Essenze Profum. Piante Off.* **1978**, *60*, 647–650.
57. Chimona, C.; Karioti, A.; Skaltsa, H.; Rhizopoulou, S. Occurrence of secondary metabolites in tepals of *Asphodelus ramosus* L. *Plant Biosyst.* **2013**, *148*, 31–34. [CrossRef]
58. Adinolfi, M.; Corsaro, M.M.; Lanzetta, R.; Parrilli, M.; Scopa, A. A Bianthrone C-Glycoside from *Asphodelus Ramosus* Tubers. *Phytochemistry* **1989**, *28*, 284–288. [CrossRef]
59. Lanzetta, R.; Parrilli, M.; Adinolfi, M.; Aquila, T.; Corsaro, M. Bianthrone C-Glucosides. 2. Three New Compounds from *Asphodelus ramosus* Tubers. *Tetrahedron* **1990**, *46*, 1287–1294. [CrossRef]
60. Adinolfi, M.; Lanzelta, R.; Marciano, C.E.; Parrilli, M.; Giulio, A.D.E. A New Class of Anthraquinone-Anthrone-C-Glycosides from *Asphodelus ramosus* Tubers. *Tetrahedron* **1991**, *47*, 4435–4440. [CrossRef]
61. Adawi, K. Comparison of the Total Phenol, Flavonoid Contents and Antioxidant Activity of Methanolic Roots Extracts of *Asphodelus microcarpus* and *Asphodeline lutea* Growing in Syria. *Int. J. Pharmacogn. Phytochem. Res.* **2017**, *9*, 159–164. [CrossRef]
62. Abdel-Mogib, M.; Basaif, S. Two New Naphthalene and Anthraquinone Derivatives from *Asphodelus tenuifolius*. *Pharmazie* **2002**, *57*, 286–287. [PubMed]
63. Singh, N.; Jaswel, S.S. Structure of an Ester from the Seeds of *Asphodelus tenuifolius* Cav. *Indian J. Chem.* **1974**, *12*, 1325–1327.
64. Madaan, T.R.; Bhatia, I.S. Lipids of *Asphodelus tenuifolius* Cav. Seeds. *Indian J. Bichem. Biophys.* **1973**, *10*, 55–58.
65. Ahmad, F.; Ahmad, M.U.; Alam, A.; Sinha, S.; Osman, S.M. Studies on Herbaceous Seed Oils-I. *J. Oil Technol. Assoc. India* **1976**, *8*, 3–4. [CrossRef]
66. Saxena, V.K.; Singh, R.B. Chemical Examination of *Asphodelus tenuifolius*. *J. Inst. Chem.* **1976**, *48*, 18–20.
67. Oskay, M.; Aktaş, K.; Sari, D.; Azeri, C. *Asphodelus aestivus* (Liliaceae) un Antimikrobiyal Etkisinin çukur ve Disk Diffüzyon Yöntemiyle Karşilaştirmali Olarak Belirlenmesi. *Ekoloji* **2007**, *16*, 62–65.
68. Aslantürk, S.Ö.; Çelik, T.A. Investigation of antioxidant, cytotoxic and apoptotic activities of the extracts from tubers of *Asphodelus aestivus* Brot. *Afr. J. Pharm. Pharmacol.* **2013**, *7*, 610–621. [CrossRef]
69. Al Groshi, A.; Nahar, L.; Andrew, E.; Auzi, A.; Sarker, S.D.; Ismail, F.M.D. Cytotoxicity of *Asphodelus aestivus* against two human cancer cell lines. *Nat. Prod. Chem. Res.* **2017**, *5*. [CrossRef]
70. Al-kayali, R.; Kitaz, A.; Haroun, M. Antibacterial Activity of *Asphodelin lutea* and *Asphodelus microcarpus* Against Methicillin Resistant *Staphylococcus aureus* Isolates. *Int. J. Pharmacogn. Phytochem. Res.* **2016**, *8*, 1964–1968.
71. Aboul-Enein, A.M.; El-Ela, F.A.; Shalaby, E.A.; El-Shemy, H.A. Traditional Medicinal Plants Research in Egypt: Studies of Antioxidant and Anticancer Activities. *J. Med. Plants Res.* **2012**, *6*, 689–703. [CrossRef]
72. Menghani, E. Isolation and Characterization of Bioactives from Arid Zone Plants. *Int. J. Pharm. Res. Dev.* **2012**, *4*, 113–118.

73. Ahmed, A.; Howladar, S.; Mohamed, H.; Al-Robai, S. Phytochemistry, Antimicrobial, Antigiardial and Antiamoebic Activities of Selected Plants from Albaha Area, Saudi Arabia. *Br. J. Med. Med. Res.* **2016**, *18*. [CrossRef]
74. Aslam, N.; Janbaz, K.H.; Jabeen, Q. Hypotensive and diuretic activities of aqueous-ethanol extract of *Asphodelus tenuifolius*. *Bangladesh J. Pharmacol.* **2016**, *11*, 830–837. [CrossRef]
75. Unal, I.; Ince, O.K. Characterization of Antioxidant activity, Vitamins and Elemental Composition of Ciris (*Asphodelus aestivus* L.) from Tunceli, Turkey. *Instrum. Sci. Technol.* **2017**, *45*, 469–478. [CrossRef]
76. Al-Yahya, M.A.; Al-Meshal, I.A.; Mossa, J.S.; Khatibi, A.; Hammouda, Y. Phytochemical and Biological Screening of Saudi Medicinal Plants: Part II. *Fitoterapia* **1983**, *1*, 21–24.
77. Apaydin, E.; Arabaci, G. Antioxidant Capacity and Phenolic Compounds with HPLC of *Asphodelus ramosus* and Comparison of the Results with *Allium cepa* L. and *Allium porrum* L. Extracts. *Turk. J. Agric. Nat. Sci.* **2017**, *4*, 499–505.
78. Abu-darwish, S.M.; Ateyyat, M. The Pharmacological and Pesticidal Actions of Naturally Occurring 1, 8-dihydroxyanthraquinones Derivatives. *World J. Agric. Sci.* **2008**, *4*, 495–505.
79. Moady, M.; Petereit, H.U. Anti-Psoriatic Composition, Method of Making and Method of Using. U.S. Patent 5,955,081A, 21 of 1999.
80. García-sosa, K.; Villarreal-alvarez, N.; Lübben, P.; Peña-rodríguez, L.M. Chrysophanol, an Antimicrobial Anthraquinone from the Root Extract of Colubrina greggii. *J. Mex. Chem. Soc.* **2006**, *50*, 76–78.
81. Kumar, S.; Pandey, A.K. Chemistry and Biological Activities of Flavonoids: An Overview. *Sci. World J.* **2013**, *2013*, 1–16. [CrossRef] [PubMed]
82. Küpeli, E.; Aslan, M.; Gürbüż, I.; Yesilada, E. Evaluation of in vivo Biological Activity profile of Isoorientin. *Z. Naturforsch.* **2004**, *59*, 787–790. [CrossRef]
83. Hernndez, L.; Palazon, J.; Navarro-Oca, A. The Pentacyclic Triterpenes alpha, beta-amyrins: A Review of Sources and Biological Activities. *Phytochem. Glob. Perspect. Role Nutr. Health* **2012**, *426*. [CrossRef]

© 2018 by the authors. Licensee MDPI, Basel, Switzerland. This article is an open access article distributed under the terms and conditions of the Creative Commons Attribution (CC BY) license (http://creativecommons.org/licenses/by/4.0/).

Article

Immunomodulatory and Inhibitory Effect of Immulina®, and Immunloges® in the Ig-E Mediated Activation of RBL-2H3 Cells. A New Role in Allergic Inflammatory Responses

Kurt Appel [1], Eduardo Munoz [2], Carmen Navarrete [2], Cristina Cruz-Teno [2], Andreas Biller [3] and Eva Thiemann [3,*]

1. VivaCell Biotechnology GmbH, Denzlingen 79211, Germany; appel@vivacell.de
2. VivaCell Biotechnology España, Córdoba 14004, Spain; e.munoz@uco.es (E.M.); c.navarrete@vivacellspain.com (C.N.); c.cruz@vivacellspain.com (C.C.-T.)
3. Dr. Loges + Co. GmbH, Winsen 21423, Germany; biller@loges.de
* Correspondence: thiemann@loges.de; Tel.: +49-4171-707-180

Received: 5 December 2017; Accepted: 23 February 2018; Published: 26 February 2018

Abstract: Immulina®, a high-molecular-weight polysaccharide extract from the cyanobacterium *Arthrospira platensis* (*Spirulina*) is a potent activator of innate immune cells. On the other hand, it is well documented that *Spirulina* exerts anti-inflammatory effects and showed promising effects with respect to the relief of allergic rhinitis symptoms. Taking into account these findings, we decided to elucidate whether Immulina®, and immunLoges® (a commercial available multicomponent nutraceutical with Immulina® as a main ingredient) beyond immune-enhancing effects, might also exert inhibitory effects in the induced allergic inflammatory response and on histamine release from RBL-2H3 mast cells. Our findings show that Immulina® and immunLoges® inhibited the IgE-antigen complex-induced production of TNF-α, IL-4, leukotrienes and histamine. The compound 48/80 stimulated histamine release in RBL-2H3 cells was also inhibited. Taken together, our results showed that Immulina® and immunLoges® exhibit anti-inflammatory properties and inhibited the release of histamine from mast cells.

Keywords: allergy; inflammation; mast cells; Immulina®; immunLoges®

1. Introduction

Arthrospira platensis (*Spirulina*) is a microscopic and filamentous cyanobacterium. It has a long tradition of use as food and has been used by humans for many years as food supplement in the form of health drinks, pills, or tablets. *Spirulina* became famous after it was used by NASA as a dietary supplement for astronauts. The safety of Spirulina has been well established. The Food and Drug Administration (FDA) has categorized *Arthrospira* products "Generally Recognized as Safe" (GRAS). *Spirulina* has high nutritional values due to its high content in proteins (up to 70%), essential amino acids, minerals, essential fatty acids, vitamins, and antioxidants [1]. Apart from high nutritional values *Spirulina* showed hypolipidemic, hypoglycemic, antihypertensive properties, microbial-modulating, antioxidant and anti-inflammatory activities [2]. In addition, *Spirulina* may be useful for the treatment of allergic conditions [3–5]. A prospective study found a high prevalence of *Spirulina* usage for the relief of allergic rhinitis symptoms in Turkey [6]. On the other hand, polysaccharides contained in *Spirulina* also have immunostimulatory activity. The main active compounds within Immulina®, a commercial extract of *Spirulina*, are bacterial Braun-type lipoproteins that activate innate immune cells through a toll-like receptor (TLR) 2-dependent pathway [7]. Previous studies indicate that Immulina® is a potent in vitro [8,9] and in vivo [10,11] immune cell activator. In addition, Immulina® can exhibit a protective effect against influenza A (H1N1) viral infection [12]. Immulina® which has been commercialized as a dietary supplement for modulating immune function, as well as ameliorating various diseases, is also an ingredient of multicomponent-nutraceuticals e.g., of immunLoges®.

Macrophages are an important component of the innate immune system and clear organisms through their phagocytic function. Toll-like receptors expressed by macrophages play an important role in their activation. Macrophages at tissue sites of allergic inflammation originate from mononuclear cells in the bone marrow. In addition to their phagocytic and antigen presenting role, macrophages have the potential to be proinflammatory or anti-inflammatory based on the spectrum of mediators they can release. There are two major pathways of the macrophage activation required for their main functional activities: the classical M1 pathway and the alternative M2 pathway [13,14]. The classical pathway gives rise to the M1 macrophages during an immune response to infection and is mediated by the Th1 cytokine interferon gamma (IFN-γ) and the toll-like receptor agonists, such as lipopolysaccharide (LPS). IL-17 also modulates the differentiation of macrophages to the pro-inflammatory M1 phenotype [15]. The classically activated M1 macrophages express and secrete large amounts of potent pro-inflammatory mediators such as tumor necrosis factor (TNF) and reactive oxygen species. Moreover, the M1 macrophages express high levels of major histocompatibility complex (MHC) class II and, co-stimulatory molecules, such as CD86, which are important for the activation and stimulation of CD4 T cells. The alternatively activated M2 macrophages are induced by the Th2-derived cytokines IL-4 or IL-13 (IL-4/IL-13-induced phenotype is also called the "M2a" subset of M2 population). The M2 macrophages produce a number of factors that are responsible for the anti-parasitic response and promote tissue repair such as Fizz1, Ym1, Arg1, Mannose Receptor (MR or CD206) as well as anti-inflammatory, regulatory, and B cell-stimulating factors such as TGF-β1, IL-4, and IL-10 [14,16].

Allergic airway diseases, such as asthma and allergic rhinitis, are common diseases caused by hypersensitivity of the immune system. Approximately 10–20% of the world population is affected by allergies, with the number of allergy patients increasing annually [17,18]. Most allergy patients are genetically predisposed to produce IgE. Mast cells play an important role in innate and adaptive immunity, especially in allergic and inflammatory responses. Mast cells express a high-affinity IgE receptor on membranes, and the binding of IgE-antigen complexes to the FcεRI (high-affinity receptor for the Fc region of immunoglobulin E) receptor triggers complex biological reactions. Activation of rat basophilic leukemia cells (RBL-2H3) cells by an IgE-antigen complex induced the production of TNF-alpha and IL-4. Mast cells are a key player in early allergic response, which typically occurs within minutes of exposure to an appropriate antigen, and other biological responses, including inflammatory disorders [19]. These cells are critical effector cells in IgE-dependent immediate hypersensitivity reactions [20]. Mast cell degranulation can initiate an acute inflammatory response and contribute to the progression of chronic diseases [21]. When an IgE-antigen binds with FcεRI, the receptor is activated, and a variety of biologically active mediators are released, causing allergic reactions, including the release of histamine, arachidonic acid metabolites, and inflammatory cytokines [22]. Importantly, arachidonic acid metabolites, including prostaglandins and leukotrienes, mediate acute and chronic allergic reactions [23,24]. RBL-2H3 cells are a mast cell line that originated from rat basophilic leukemia and have been widely used to study IgE-FcεRI interactions and degranulation. Furthermore, RBL-2H3 cells are a useful model for in vitro screening of antiallergy drug candidates.

The aim of the present study was to determine whether Immulina® despite its well documented immunostimulatory activity and/or immunLoges®, also have anti-inflammatory activities and may be useful for the treatment of allergic conditions.

2. Results

2.1. Cytotoxicity Effects of Immunloges® and Immulina® in Raw264.7 Cells

To analyze the potential cytotoxic activities of the test substances Raw264.7 cells were incubated with increasing concentrations of immunLoges® and Immulina® for 24 and the cytotoxicity was measured by YOYO-1 staining and fluorescence microscopy using an Incucyte System. Only immunLoges® at the highest concentration tested (1000 µg/mL) showed cytotoxicity at 24 h (Figure 1). After analyzing the potential cytotoxicity in primary cells, it was decided to use the concentrations described in Material and Methods.

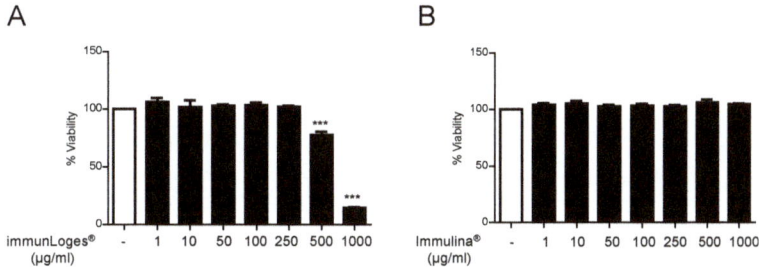

Figure 1. Cytotoxic activity of immunLoges® and Immulina® in Raw264.7 cells. The cells were treated with immunLoges® (**A**) or Immulina® (**B**) at the indicated concentrations for 24 h. Cytotoxicity was measured by fluorescence microscopy using an Incucyte System. Data represent the mean ± SD (n = 3). *** $p < 0.001$ vs. untreated. (one-way analysis of variance (ANOVA) followed Newman-Keuls test).

2.2. NF-kB Activation in Raw264.7 Cells

Although previous studies indicate that Immulina® is a potent in vitro and in vivo immune cell activator, we wanted to know the activity of the commercial extract immunLoges®. In this sense we evaluated the effect of the active component and the commercial extract on the NF-κB transcriptional activity. This activity was evaluated by using the luciferase reporter construct KBF-Luc [25]. Activation by LPS clearly increased (13-fold induction) the luciferase gene expression driven by the NF-κB dependent promoter in stably transfected Raw264.7 cells. We found that, immunLoges® and Immulina® increase this activity (Figure 2A,B).

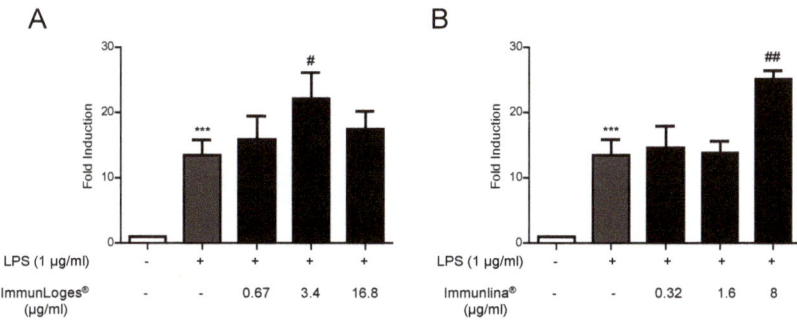

Figure 2. NF-kB activity of immunLoges® (**A**) and Immulina® (**B**) in Raw264.7-KBF-Luc cells. Cells were pre-incubated with the test substances at the indicated concentrations for 30 min and then stimulated with LPS for 6 h. The results of the specific transactivation are expressed as a fold induction over untreated cells. Data represent the mean ± SD (n = 3). *** $p < 0.001$ vs. untreated, # $p < 0.05$, ## $p < 0.01$ vs. LPS treatment. (one-way ANOVA followed Newman-Keuls test).

2.3. Effect of Immunloges® and Immulina® on M1 and M2 Polarization

Macrophages are also important effector cells that mediate the immune responses. They act as antigen presenting cells (APC), thereby activating an antigen-specific T cell response in the periphery and central nervous system (CNS). Macrophages detect the endogenous danger signals that are present in the debris of necrotic cells through Toll-like receptors (TLRs) 2,6,intracellular pattern-recognition receptors and the interleukin-1 receptor (IL-1R), most of which signal through the adaptor molecule myeloid differentiation primary-response gene 88 (MyD88). This function makes macrophages one of the primary sensors of danger in the host.

Treatment of RAW264.7 macrophages with IL-17 promoted their polarization towards a pro-inflammatory M1 phenotype, as shown by increased expression of M1 markers such as TNF-α, CCL2 or IL-1β. Raw264.7 cells were pre-treated for 18 h with the test substances and then were exposed to recombinant murine IL-17 to induce M1 polarization and M1 markers were analyzed by qPCR. The treatment with immunLoges® and Immulina® clearly enhanced the expression of IL17-induced M1 markers TNF-α, CCL2 and IL-1ß as shown in Figure 3.

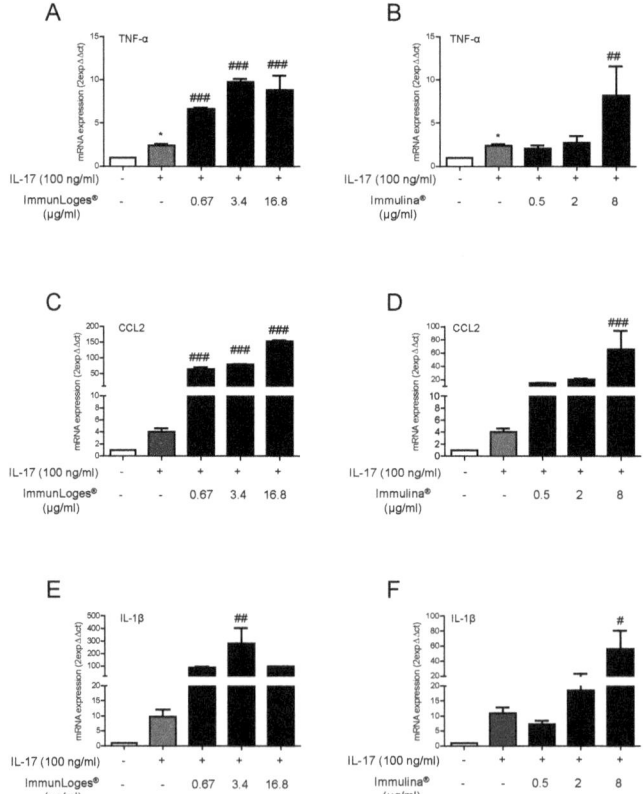

Figure 3. Effect of immunLoges®, Immulina® on IL-17-induced M1 markers expression. Cells were pre-incubated with immunLoges® or Immulina® at the indicated concentrations for 18 h and then stimulated with IL-17 for 24 h. TNF-α (**A,B**), CCL2 (**C,D**) and IL-1β (**E,F**) expression was determined. The GAPDH gene was used to standardize mRNA expression in each sample and gene expression was quantified using the 2-ΔΔCt method. Data represent the mean ± SD (n = 3). * p < 0.05 vs. untreated, # p < 0.05, ## p < 0.01, ### p < 0.001 vs. IL-17 treatment. (one-way ANOVA followed Newman-Keuls test.

To study the effect of immunLoges® and Immulina® on M2 polarization, Raw264.7 cells were treated for 24 h with the test substances at the indicated concentrations. As positive control we used the recombinant mouse IL-4 to induce M2 polarization. M2 markers such as Arg1, Mrc1 and IL-10 were determined. The treatment with immunLoges® and Immulina® clearly induced the expression of the M2 markers Arg1 (Figure 4A,B), and Mrc1 (Figure 4C,D). In the case of IL10 gene expression, the treatment did not impair the gene expression (Figure 4E,F).

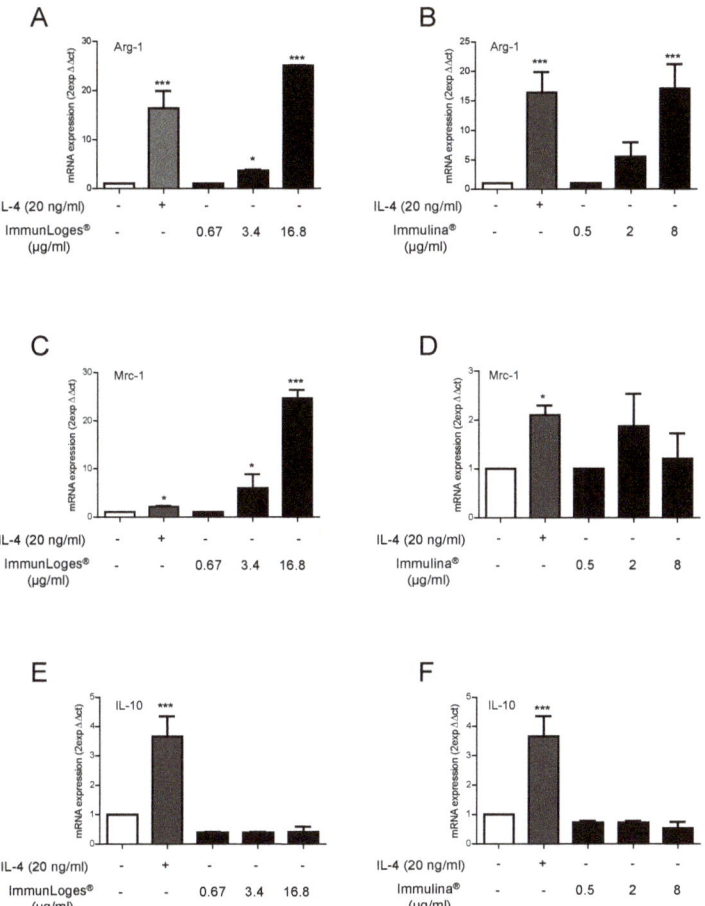

Figure 4. Effect of immunLoges®, Immulina® on M2 polarization. Cells were pre-incubated with immunLoges® or Immulina® at the indicated concentrations for 24 h. Arg-1 (**A**,**B**), Mrc-1 (**C**,**D**) and IL-10 (**E**,**F**) expression was determined. The GAPDH gene was used to standardize mRNA expression in each sample and gene expression was quantified using the 2-ΔΔCt method. Data represent the mean ± SD (n = 3). * $p < 0.05$, *** $p < 0.001$ vs. untreated. (one-way ANOVA followed Newman-Keuls test).

2.4. Cytotoxicity Effects of Immunloges® and Immulina® in RBL-2H3 Cells

To evaluate the cytotoxicity of immunLoges® and Immulina® on RBL-2H3, the cells were treated with different concentrations for 24 h and the cytotoxicity was measured by fluorescence microscopy using an Incucyte System. Neither Immulina® nor immunLoges® showed cytotoxicity at the concentrations investigated in this study (Figure 5).

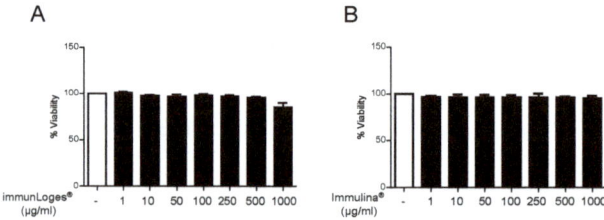

Figure 5. Cytotoxic activity of immunLoges® (**A**) and Immulina® (**B**) in RBL-2H3 cells. The cells were treated with immunLoges® or Immulina® at the indicated concentrations for 24 h. Cytotoxicity was measured by fluorescence microscopy using an Incucyte System. Data represent the mean ± SD ($n = 3$).

2.5. Effect on Cytokines Release in IgE-Antigen Complex Stimulated RBL-2H3 Cells

Mast cells express a high-affinity IgE receptor (FcεRI) on membranes, and the binding of IgE-antigen complexes to FcεRI triggers complex biological reactions. RBL-2H3 cells were sensitized with anti-DNP-IgE (monoclonal anti-dinotrophenyl antibody produced in mouse, IgE isotype) and stimulated with DNP-HSA (dinitrophenyl-human serum albumin). Inflammatory cytokines mediate the pathological reaction in inflammatory diseases. Besides the secretion of stored cytokines during degranulation, the production of newly generated inflammatory cytokines (TNF-α, IL-4, IL-6 etc.) is induced.

The effects of the test substances on pro-inflammatory cytokine production were assessed by ELISA. Activation of RBL-2H3 cells by an IgE-antigen complex induced the production of TNF-α and IL-4. The treatment with immunLoges® (Figure 6A) or Immulina® (Figure 6B) decreased the levels of TNF-α and IL-4 (Figure 6C,D) in FcεRI-activated RBL-2H3 cells.

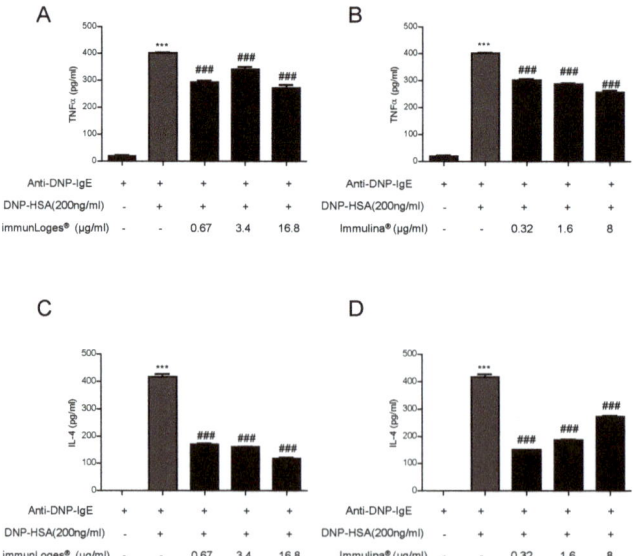

Figure 6. Effect of immunLoges® and Immulina® on TNF-α and IL-4 production in IgE-antigen complex-stimulated RBL-2H3 cells. The cells were treated with immunLoges® (**A,C**) or Immulina® (**B,D**) at the indicated concentrations and the levels of TNF-α and IL-4 were determined in the supernatants. Data represent the mean ± SD ($n = 3$). *** $p < 0.001$ vs. Anti-DNP-IgE, ### $p < 0.001$ vs. Anti-DNP-IgE+DNP-HAS treatment. (one-way ANOVA followed Newman-Keuls test).

2.6. Effect on Leukotrienes Production in IgE-Antigen Complex-Stimulated RBL-2H3 Cells

In parallel to the analysis of cytokine production we determined the formation of leukotrienes LTC4/D4/E4 in supernatants of IgE-antigen complex-stimulated RBL-2H3 cells. The treatment with immunLoges® (Figure 7A) and Immulina® (Figure 7B) reduced the levels of leukotrienes.

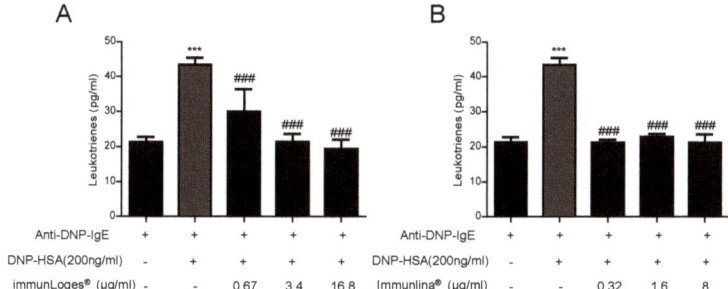

Figure 7. Effect of immunLoges® and Immulina® on leukotrienes production in IgE-antigen complex-stimulated RBL-2H3 cells. The cells were treated with immunLoges® (**A**) and Immulina® (**B**) at the indicated concentrations and the levels of leukotrienes were determined in the supernatants. Data represent the mean ± SD (n = 3). *** $p < 0.001$ vs. Anti-DNP-IgE, ### $p < 0.001$ vs. Anti-DNP-IgE+DNP-HSA treatment. (one-way ANOVA followed Newman-Keuls test).

2.7. Effect on the Release of Histamine in IgE-Antigen Complex-Stimulated RBL-2H3 Cells

Optimum conditions for IgE mediated degranulation in RBL-2H3 cells were sensitizing with anti-DNP-IgE and stimulating with 2,4-Dinitrophenyl hapten conjugated to human serum albumin (DNP-HSA). The release of histamine in IgE-antigen complex-stimulated RBL-2H3 cells were significantly increased compared with untreated cells. Pre-treatment for 2 h with the test substances at the indicated concentrations clearly suppressed the degranulation in IgE-antigen complex-stimulated RBL-2H3 cells. In addition, we tested the mast cell stabilizing agent Disodium cromoglycate (DSCG). As it is shown in Figure 8 the treatment with this compound inhibited the degranulation in a similar manner to that observed for immunLoges® (Figure 8A) and Immulina® (Figure 8B).

The effects of synthetic compound 48/80 were also examined, since earlier studies have shown that this compound is a very potent inducer of mast cell degranulation. Compound 48/80 is a G protein stimulant compound originally described as a mast cell secretagogue. Pre-treatment for 2 h with the compounds at the indicated concentrations inhibited the release of histamine from RBL-2H3 stimulated by compound 48/80 (Figure 8 C,D).

Finally, we determined the effect of immunLoges® and Immulina® on the basal release of histamine in the mast cell line. The treatment with the test substances at the concentrations investigated did not induce histamine secretion (Figure 9).

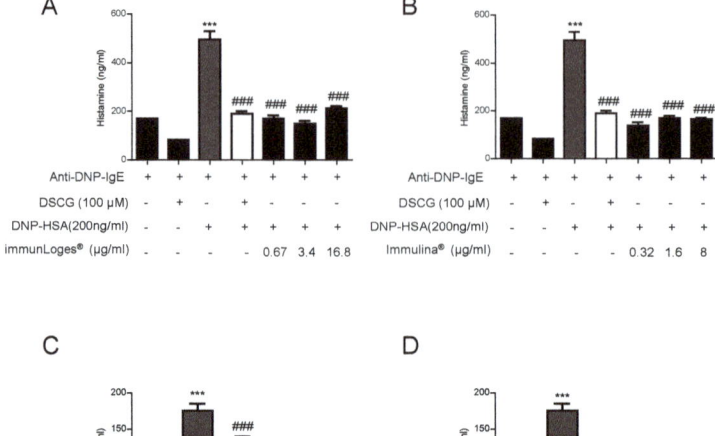

Figure 8. Effect of immunLoges® and Immulina® on mast stabilization. The cells were pretreated for 2 h with immunLoges® (**A,C**) and Immulina® (**B,D**) at the indicated concentrations and stimulated as indicated. The levels of histamine were determined in the supernatants. Data represent the mean ± SD ($n = 3$). *** $p < 0.001$ vs. Anti-DNP-IgE or untreated cells, ### $p < 0.001$ vs. Anti-DNP-IgE+DNP-HAS or 48/80 treatment. (one-way ANOVA followed Newman-Keuls test).

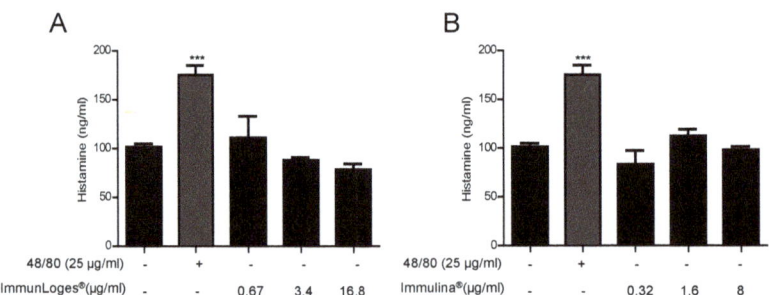

Figure 9. Effect of immunLoges® and Immulina® on mast stabilization. The cells were treated for 2 h with immunLoges® (**A**) and Immulina® (**B**) at the indicated concentrations. The levels of histamine were determined in the supernatants. Data represent the mean ± SD ($n = 3$). *** $p < 0.001$ vs. untreated cells (one-way ANOVA followed Newman-Keuls test).

3. Discussion

Previous reports showed that Immulina® is a potent immune cell activator and activates NF-kB through TLR2 receptor [7,8]. Here we describe the activation of NF-kB in macrophages treated with immunLoges® and Immulina® in the presence of LPS confirming those previous studies. Herein we show that immunLoges® preserves the immunomodulatory activity.

Traditionally, macrophages have been seen only as immune cells, and their homeostatic responsibilities have been overlooked. Mosser et al. [26], proposed that macrophages exist in three main categories: host defense, wound healing, and immune regulation. M1 macrophages (classically activated) clears tissue of cellular debris and pathogens via aggressive phagocytosis and release of pro-inflammatory cytokines including: IL-1, IL-6, IL-12, IL-23, and TNF-α. The M2 macrophage acts as a support cell by promoting the healing of damaged cells by angiogenesis and tissue remodeling via secretion of anti-inflammatory cytokines. While these two operationally useful polarities exist, it is widely believed that macrophages exist mainly in the continuum between these two phenotypes [27]. Regulation of this continuum involves complex processes that have been extensively studied, yet remain relatively unclear. Recently it was shown that macrophages will respond to injury or infection by upregulating pro-inflammatory (M1) responses, but later switch to an anti-inflammatory (M2) response once the infection is under control [28,29]. This M1 activation program is typically associated with protection against disease, and M1 polarization has been shown to aid host control of several bacteria, including *Listeria* monocytogenes, *Salmonella typhimurium*, *Mycobacterium tuberculosis* and *Chlamydia* infections [30–32].

Here we describe a dual role for immunLoges® and Immulina® in the macrophage polarization. On the one hand, we observed an increase of M1 markers such as TNF-α, CCL2 and IL-1β in the presence of LPS. This effect on M1 polarization is important from the point of view of the common response of macrophages to bacterial infections which mainly involve the up-regulation of genes involved in M1 polarization. Moreover, other M1 associated up-regulated genes encode the enzymes indoleamine-pyrrole 2,3 dioxygenase and NO synthase 2 (NOS2), which are related with macrophage microbicidal activity [30]. The effect of immunLoges® and Immulina® on M2 polarization could be related with immunoregulatory activity since M2 macrophages dampen inflammation, suppress antitumor immunity and promote wound healing and tissue remodeling.

Although more sophisticated mechanisms are still unknown, data from animal studies suggest that mast cells act similarly to macrophages and other inflammatory cells and contribute to human diseases through cell–cell interactions and the release of proinflammatory cytokines, chemokines, and proteases to induce inflammatory cell recruitment. There is growing interest in understanding how mast cells participate in positive immunoregulation particularly in the innate aspects of an immune response [33–35].

Mast cells are important in innate immune system. They have been appreciated as potent contributors to allergic reaction. Mast cells have a widespread distribution and are found predominantly at the interface between the host and the external environment and acted as both first responders in harmful situations as well as to respond to changes in their environment by communicating with a variety of other cells implicated in immunological responses. Therefore, the critical role of mast cells in both innate and adaptive immunity, including immune tolerance, has gained increased prominence [36–38]. Conversely, mast cell dysfunction has pointed to these cells as the main offenders in several chronic allergic/inflammatory disorders, cancer, and autoimmune diseases [39–41].

Moreover, it has been well documented that *Spirulina* exhibits anti-inflammatory properties by inhibiting the release of histamine from mast cells [3]. In a randomized double-blind placebo-controlled trial individuals with allergic rhinitis *Spirulina* significantly reduced the levels of IL-4 that is important in regulating immunoglobulin (Ig)E-mediated allergy [5].

Since high molecular weight polysaccharides are present in Immulina® it is possible that different preparations of these algae may exert different immunological activities. Immulina® and immunLoges® decreased the levels of TNF-α and IL-4 in FcεRI-activated RBL-2H3 cells. These results indicated the anti-allergic property of both substances in allergic inflammation therapy.

In parallel to the analysis of cytokine production we tested the release of leukotrienes, which are important mediators both in host defense mechanisms and in inflammatory disease states. Leukotrienes can be formed in mast cells, granulocytes, and monocytes/macrophages in response to extracellular stimulation. In this study, we determined the formation of leukotrienes LTC4/D4/E4 in supernatants of IgE-antigen complex-stimulated RBL-2H3 cells. The treatment with Immulina® reduced the levels of leukotrienes.

Apart from the anti-inflammatory parameters we also investigated the effects on mast cell stabilization and histamine release. Immulina® inhibited the release of histamine in IgE-antigen complex-stimulated RBL-2H3 cells in a similar manner to that observed for the mast cell stabilizing agent Disodium cromoglycate (DSCG). In addition, we determined the effect of the test substances on the secretion of histamine induced by compound 48/80. This G protein stimulant compound was originally described as a mast cell secretagogue. Again, Immulina® inhibited the compound 48/80 stimulated histamine release in RBL-2H3 cells.

In parallel we tested immunLoges® in the above-mentioned assays and immunLoges® showed similar effects compared to Immulina®. Therefore, neither additional compounds (vitamins etc.) nor excipients in immunLoges® interfered with the described activities.

Immulina® is a patented, high molecular weight polysaccharide extract from *spirulina* and the predominant active compounds are Braun-type lipoproteins. Although Immulina® is a potent NF-kB and immune cell activator, our results clearly showed that Immulina® has anti-inflammatory and anti-histaminic effects in RBL-2H3 cells.

In general, the results obtained for both products, Immulina® and immunLoges®, are dose-response independent, which can be explained by the fact that most of the studies for both compounds had strong activity at the lowest concentrations studied.

Overall, our results with Immulina® and immunLoges® agree with the concept that *Spirulina* exerts anti-inflammatory effects and may be useful for the treatment of allergic conditions, but further in vivo studies should be performed to clarify these effects for Immulina® and immunLoges®.

4. Materials and Methods

4.1. Test items

Immulina® is a patented extract [22] derived from *Arthrospira platensis* (Nordstedt) Gomont, Phormidiaceae (previously called *Spirulina platensis* and commonly known as *Spirulina*). This extract represents a 10–15% yield from spray dried *Arthrospira platensis* and is standardized by biological activity to ensure consistent activity of the extract from batch to batch. Biological standardization is performed using a humanTHP-1 monocyte activation assay that measures activity as indicated by increased expression of an NF-kB-driven luciferase reporter as previously described (Pugh et al. 2001). The Immulina® extract (batch number 600137) used in this research was supplied by ChromaDex Inc. (Irvine, CA, USA). immunLoges® (batch 500875, Dr. Loges + Co. GmbH, Winsen, Germany) is a multi-complex formulation of Immulina®, Betox 93®, Selenium, Zink, Vitamin C, D and resveratrol. One tablet immunLoges® (420 mg) contains among others 200 mg Immulina®. The composition of commercial extract for ImmunLoges® is described in the Table 1 below. For a better comparison of the results, we tested the immunLoges® concentration which correspond to the Immulina® concentration in immunLoges®. The stock solutions were prepared in DMSO and then diluted in cell culture media. The final DMSO concentration never exceeded 0.1% (v/v).

Table 1. Composition of commercial extract immunLoges®. The composition of immunLoges® per capsule is indicated in the table. Excipients (54.5 mg) such as cellulose, talcum and magnesium.

Composition	Per Capsule
Immulina®	200 mg
Betox 93®	60 mg
Vineatrol®	0.5 mg
Vitamin C	40 mg
Vitamin D	10 µg
Selenite	50 µg
Zinc	5 mg

4.2. Cell Culture

RAW264.7 mouse macrophage cell line was obtained from ATCC (TIB-71) and maintained at 37 °C in a humidified atmosphere containing 5% CO_2 in (Dulbecco's Modified Eagle Medium) (DMEM) supplemented with 10% fetal calf serum (FCS), 2 mM L-glutamine and 1% (v/v) penicillin/streptomycin. Raw264.7-kappaB Factor (KBF)-Luc cell line is stably transfected cell line with the KBF-Luc plasmid that contains the luciferase gene driven by an artificial promoter made by 3 multimerized copies of the NF-kB binding site from the MHCI Class I promoter. These cells were maintained in DMEM medium supplemented with antibiotics and 10% heat-inactivated FBS. The rat basophilic leukemia cell line RBL-2H3 was obtained from ATCC (CRL-2256) and grown in Dulbecco's modified Eagle medium with 100 IU/mL penicillin, 100 mg/mL streptomycin, and 10% heat-inactivated fetal calf serum (FCS) in a humidified 5% CO_2 atmosphere at 37 °C.

4.3. Cytotoxicity

To identify non-cytotoxic concentrations of the test items RAW264.7 or RBL-2H3 cells were treated with increasing concentrations of the test compounds for 24 h. Cytotoxicity was measured by fluorescence (Essen BioScience, Inc., Ann Arbor, MI, USA) using YOYO-1, purchased from Thermo Fisher Scientific (Waltham, MA, USA), is a green fluorescent dye used in DNA staining. The dye was diluted in cell cultured medium and added to a final concentration of 0.1 µM.

After 3–4 h the total number of DNA containing objects was measured by treating cells with 0.0625% triton X-100 to permeabilize the cell membrane. Object counting analysis was performed using the Incucyte FLR software (Essen BioScience, Inc., Ann Arbor, MI, USA) to calculate the total number of YOYO-1 fluorescence positive cells and total DNA containing objects (end point). The cytotoxicity index was calculated by dividing the number of YOYO-1 fluorescence positive objects by the total number of DNA containing objects for each treatment group.

4.4. NF-kB Activation in Raw264.7-KBF-Luc Cells

To study the potential NF-kB modulatory activity of the test items we used the stably transfected Raw264.7-KBF-Luc cell line. Cells were pretreated with the test substances for 30 min and then were stimulated with LPS for 6 h. In addition, cells were treated with the test compounds in the absence of LPS. Then the cells were washed and lysed in 25 mM Tris-phosphate pH 7.8, 8 mM $MgCl_2$, 1 mM DTT, 1% Triton X-100, and 7% glycerol. Luciferase activity was measured using an Autolumat LB 953 (EG&G Berthold, Oak Ridge, TN, USA) following the instructions of the luciferase assay kit (Promega, Madison, WI, USA). The results of the specific transactivation are expressed as a fold induction over untreated cells.

4.5. IL-17-Induced M1 Polarization

Serum-starved Raw264.7 macrophages were pre-incubated with increasing concentrations of the test compounds for 18 h, before an additional 24-h treatment with the recombinant mouse IL-17 (100 ng/mL). Then cells were collected and total RNA extracted using the High Pure RNA Isolation kit (Roche Diagnostics, Hong Kong, China). Total RNA (1 µg) was retrotranscribed using the iScriptTM cDNA Synthesis Kit (Bio-Rad; Hercules, CA, USA), and the cDNA generated analyzed by real-time PCR, using the iQTM SYBR Green Supermix (Bio-Rad; Hercules, CA, USA). Real-time PCR was performed using a CFX96 Real-time PCR Detection System (Bio-Rad; Hercules, CA, USA). The GAPDH gene was used to standardize mRNA expression in each sample and gene expression was quantified using the 2-ΔΔCt method. The markers analyzed for M1 polarization were TNF-α, CCL2 and IL-1ß.

4.6. IL-4-Induced M2 Polarization

Serum-starved Raw264.7 macrophages were pre-incubated with increasing concentrations of the test compounds for 24 h. As positive control for M2 polarization we treated the cells with the recombinant mouse IL-4 (20 ng/mL) for 24 h. Then cells were collected and total RNA extracted using the High Pure RNA Isolation kit (Roche Diagnostics, Hong Kong, China). Total RNA (1 µg) was retrotranscribed using the iScriptTM cDNA Synthesis Kit (Bio-Rad; Hercules, CA, USA), and the cDNA generated analyzed by real-time PCR, using the iQTM SYBR Green Supermix (Bio-Rad; Hercules, CA, USA). Real-time PCR was performed using a CFX96 Real-time PCR Detection System (Bio-Rad; Hercules, CA, USA). The GAPDH gene was used to standardize mRNA expression in each sample and gene expression was quantified using the 2-$\Delta\Delta$Ct method. The markers analyzed for M2 polarization were Arg1, IL-10 and Mrc1.

4.7. Detection of Leukotrienes and Cytokines in RBL-2H3 Cells

RBL-2H3 cells were seeded at 2×10^5 cells in 24-well plates and after 24 h cells were incubated overnight with or without anti-DNP IgE mAb (Santa Cruz Biotechnology, Dallas, TX, USA) (0.5 µg/mL) for sensitization. Next, cells were pre-incubated with the test substances for 1 h and then stimulated with DNP-HSA (Santa Cruz Biotechnology) (200 ng/mL) for 3 h. Supernatants were collected, centrifuged for 10 min at $400 \times g$ (4 °C) and the leukotrienes (LTC4/D4/E4), IL-4 and TNF-α were analyzed. Leukotrienes kit was obtained from Enzo Life Sciences (New York, NY, USA) IL-4 and TNF-α ELISA kits were purchased from R&D Systems (Minneapolis, MN, USA). The levels of leukotrienes and cytokines were analyzed with plate-bound ELISA kits according to the manufacturer's recommendations.

4.8. Detection of Histamine Release in RBL-2H3 Cells

RBL-2H3 cells were seeded at 2×10^5 cells in 24-well plates and after 24 h the medium in each well was aspirated and replaced by fresh medium. Next cells were pre-incubated with the compounds for 2 h and then were stimulated with the compound 48/80 (25 µg/mL) (Sigma Aldrich, St. Louis, MO, USA) in the presence or absence of non-cytotoxic concentrations of the test substances for 15 min. After treatment, supernatants were collected and centrifuged for 10 min at $400 \times g$ (4 °C). Histamine Elisa Fast Track was purchased from LDN GmbH, Germany. Histamine in the supernatant was determined by an ELISA assay according to the manufacturer's recommendations.

4.9. RBL-2H3 Mast Cells Stabilization

RBL-2H3 cells were seeded at 2×10^5 cells in 24-well plates and after 24 h cells were incubated overnight with or without anti-DNP IgE mAb (0.5 µg/mL) for sensitization. Next, cells were pre-incubated with the test substances or the well-known mast stabilizing agent Disodium cromoglycate (DSCG) for 2 h and then stimulated with DNP-HSA (1 µg/mL) for 30 min. Supernatants were collected, centrifuged for 10 min at $400 \times g$ (4 °C) and histamine release was determined as above described.

4.10. Statistical Analysis

Data were analyzed by one-way analysis of variance (ANOVA) followed by Newman Keuls post-hoc test using GraphPad Prism version 6.00 (GraphPad, San Diego, CA, USA). Results were presented as means ± SD and differences were considered significant when $p < 0.05$.

Acknowledgments: Financial support for the analysis was provided by Dr. Loges + Co. GmbH, the manufacturer of immunLoges®. The founding sponsors had no role in the design of the study; in the collection, analyses, and in interpretation of data.

Author Contributions: Kurt Appel and Eduardo Munoz conceived and designed the experiments; Carmen Navarrete and Cristina Cruz-Teno performed the experiments; Kurt Appel, Andreas Biller, Eva Thiemann and Eduardo Munoz analyzed the data; Andreas Biller contributed reagents/materials/analysis tools; Eva Thiemann and Kurt Appel wrote the paper.

Conflicts of Interest: The authors declare no conflict of interest.

References

1. Karkos, P.D.; Leong, S.C.; Karkos, C.D.; Sivaji, N.; Assimakopoulos, D.A. Spirulina in clinical practice: Evidence-based human applications. *Evid. Based Complement. Altern. Med.* **2011**, *2011*, 531053. [CrossRef] [PubMed]
2. Finamore, A.; Palmery, M.; Bensehaila, S.; Peluso, I. Antioxidant, immunomodulating, and microbial-modulating activities of the sustainable and ecofriendly spirulina. *Oxid. Med. Cell. Longev.* **2017**, *2017*, 3247528. [CrossRef] [PubMed]
3. Yang, H.N.; Lee, E.H.; Kim, H.M. Spirulina platensis inhibits anaphylactic reaction. *Life Sci.* **1997**, *61*, 1237–1244. [CrossRef]
4. Kim, H.M.; Lee, E.H.; Cho, H.H.; Moon, Y.H. Inhibitory effect of mast cell-mediated immediate-type allergic reactions in rats by spirulina. *Biochem. Pharmacol.* **1998**, *55*, 1071–1076. [CrossRef]
5. Mao, T.K.; Van de Water, J.; Gershwin, M.E. Effects of a spirulina-based dietary supplement on cytokine production from allergic rhinitis patients. *J. Med. Food* **2005**, *8*, 27–30. [CrossRef] [PubMed]
6. Sayin, I.; Cingi, C.; Oghan, F.; Baykal, B.; Ulusoy, S. Complementary therapies in allergic rhinitis. *ISRN Allergy* **2013**, *2013*, 938751. [CrossRef] [PubMed]
7. Balachandran, P.; Pugh, N.D.; Ma, G.; Pasco, D.S. Toll-like receptor 2-dependent activation of monocytes by spirulina polysaccharide and its immune enhancing action in mice. *Int. Immunopharmacol.* **2006**, *6*, 1808–1814. [CrossRef] [PubMed]
8. Pugh, N.; Ross, S.A.; ElSohly, H.N.; ElSohly, M.A.; Pasco, D.S. Isolation of three high molecular weight polysaccharide preparations with potent immunostimulatory activity from spirulina platensis, aphanizomenon flos-aquae and chlorella pyrenoidosa. *Planta Med.* **2001**, *67*, 737–742. [CrossRef] [PubMed]
9. Grzanna, R.; Polotsky, A.; Phan, P.V.; Pugh, N.; Pasco, D.; Frondoza, C.G. Immolina, a high-molecular-weight polysaccharide fraction of spirulina, enhances chemokine expression in human monocytic thp-1 cells. *J. Altern. Complement. Med.* **2006**, *12*, 429–435. [CrossRef] [PubMed]
10. Lobner, M.; Walsted, A.; Larsen, R.; Bendtzen, K.; Nielsen, C.H. Enhancement of human adaptive immune responses by administration of a high-molecular-weight polysaccharide extract from the cyanobacterium arthrospira platensis. *J. Med. Food* **2008**, *11*, 313–322. [CrossRef] [PubMed]
11. Nielsen, C.H.; Balachandran, P.; Christensen, O.; Pugh, N.D.; Tamta, H.; Sufka, K.J.; Wu, X.; Walsted, A.; Schjorring-Thyssen, M.; Enevold, C.; et al. Enhancement of natural killer cell activity in healthy subjects by immulina(r), a spirulina extract enriched for braun-type lipoproteins. *Planta Med.* **2010**, *76*, 1802–1808. [CrossRef] [PubMed]
12. Pugh, N.D.; Edwall, D.; Lindmark, L.; Kousoulas, K.G.; Iyer, A.V.; Haron, M.H.; Pasco, D.S. Oral administration of a spirulina extract enriched for braun-type lipoproteins protects mice against influenza a (h1n1) virus infection. *Phytomedicine* **2015**, *22*, 271–276. [CrossRef] [PubMed]
13. Wynn, T.A.; Chawla, A.; Pollard, J.W. Macrophage biology in development, homeostasis and disease. *Nature* **2013**, *496*, 445–455. [CrossRef] [PubMed]
14. Mills, C.D. M1 and m2 macrophages: Oracles of health and disease. *Crit. Rev. Immunol.* **2012**, *32*, 463–488. [CrossRef] [PubMed]
15. Guillot, A.; Hamdaoui, N.; Bizy, A.; Zoltani, K.; Souktani, R.; Zafrani, E.S.; Mallat, A.; Lotersztajn, S.; Lafdil, F. Cannabinoid receptor 2 counteracts interleukin-17-induced immune and fibrogenic responses in mouse liver. *Hepatology* **2014**, *59*, 296–306. [CrossRef] [PubMed]
16. Tugal, D.; Liao, X.; Jain, M.K. Transcriptional control of macrophage polarization. *Arterioscler. Thromb. Vasc. Biol.* **2013**, *33*, 1135–1144. [CrossRef] [PubMed]
17. Flohr, C.; Quinnell, R.J.; Britton, J. Do helminth parasites protect against atopy and allergic disease? *Clin. Exp. Allergy* **2009**, *39*, 20–32. [CrossRef] [PubMed]
18. Erb, K.J. Can helminths or helminth-derived products be used in humans to prevent or treat allergic diseases? *Trends Immunol.* **2009**, *30*, 75–82. [CrossRef] [PubMed]
19. Bochner, B.S.; Schleimer, R.P. Mast cells, basophils, and eosinophils: Distinct but overlapping pathways for recruitment. *Immunol. Rev.* **2001**, *179*, 5–15. [CrossRef] [PubMed]
20. Galli, S.J. New concepts about the mast cell. *N. Engl. J. Med.* **1993**, *328*, 257–265. [CrossRef]
21. Nguyen, M.; Pace, A.J.; Koller, B.H. Age-induced reprogramming of mast cell degranulation. *J. Immunol.* **2005**, *175*, 5701–5707. [CrossRef] [PubMed]

22. Gilfillan, A.M.; Tkaczyk, C. Integrated signalling pathways for mast-cell activation. *Nat. Rev. Immunol.* **2006**, *6*, 218–230. [CrossRef] [PubMed]
23. Church, M.K.; Levi-Schaffer, F. The human mast cell. *J. Allergy Clin. Immunol.* **1997**, *99*, 155–160. [CrossRef]
24. Metcalfe, D.D.; Kaliner, M.; Donlon, M.A. The mast cell. *Crit. Rev. Immunol.* **1981**, *3*, 23–74. [PubMed]
25. Yano, O.; Kanellopoulos, J.; Kieran, M.; Le Bail, O.; Israel, A.; Kourilsky, P. Purification of kbf1, a common factor binding to both h-2 and beta 2-microglobulin enhancers. *EMBO J.* **1987**, *6*, 3317–3324. [PubMed]
26. Mosser, D.M.; Edwards, J.P. Exploring the full spectrum of macrophage activation. *Nat. Rev. Immunol.* **2008**, *8*, 958–969. [CrossRef] [PubMed]
27. Mantovani, A.; Sica, A.; Locati, M. Macrophage polarization comes of age. *Immunity* **2005**, *23*, 344–346. [CrossRef] [PubMed]
28. Csoka, B.; Nemeth, Z.H.; Selmeczy, Z.; Koscso, B.; Pacher, P.; Vizi, E.S.; Deitch, E.A.; Hasko, G. Role of A_{2A} adenosine receptors in regulation of opsonized *E. coli*-induced macrophage function. *Purinergic Signal.* **2007**, *3*, 447–452. [CrossRef] [PubMed]
29. Martinez, F.O.; Gordon, S.; Locati, M.; Mantovani, A. Transcriptional profiling of the human monocyte-to-macrophage differentiation and polarization: New molecules and patterns of gene expression. *J. Immunol.* **2006**, *177*, 7303–7311. [CrossRef] [PubMed]
30. Benoit, M.; Desnues, B.; Mege, J.L. Macrophage polarization in bacterial infections. *J. Immunol.* **2008**, *181*, 3733–3739. [CrossRef] [PubMed]
31. Chacon-Salinas, R.; Serafin-Lopez, J.; Ramos-Payan, R.; Mendez-Aragon, P.; Hernandez-Pando, R.; Van Soolingen, D.; Flores-Romo, L.; Estrada-Parra, S.; Estrada-Garcia, I. Differential pattern of cytokine expression by macrophages infected in vitro with different mycobacterium tuberculosis genotypes. *Clin. Exp. Immunol.* **2005**, *140*, 443–449. [CrossRef] [PubMed]
32. Rottenberg, M.E.; Gigliotti-Rothfuchs, A.; Wigzell, H. The role of ifn-gamma in the outcome of chlamydial infection. *Curr. Opin. Immunol.* **2002**, *14*, 444–451. [CrossRef]
33. Galli, S.J.; Grimbaldeston, M.; Tsai, M. Immunomodulatory mast cells: Negative, as well as positive, regulators of immunity. *Nat. Rev. Immunol.* **2008**, *8*, 478–486. [CrossRef] [PubMed]
34. Metz, M.; Maurer, M. Mast cells–key effector cells in immune responses. *Trends Immunol.* **2007**, *28*, 234–241. [CrossRef] [PubMed]
35. Abraham, S.N.; St John, A.L. Mast cell-orchestrated immunity to pathogens. *Nat. Rev. Immunol.* **2010**, *10*, 440–452. [CrossRef] [PubMed]
36. Da Silva, E.Z.; Jamur, M.C.; Oliver, C. Mast cell function: A new vision of an old cell. *J. Histochem. Cytochem.* **2014**, *62*, 698–738. [CrossRef] [PubMed]
37. Gangwar, R.S.; Friedman, S.; Seaf, M.; Levi-Schaffer, F. Mast cells and eosinophils in allergy: Close friends or just neighbors. *Eur. J. Pharmacol.* **2016**, *778*, 77–83. [CrossRef] [PubMed]
38. Saluja, R.; Khan, M.; Church, M.K.; Maurer, M. The role of IL-33 and mast cells in allergy and inflammation. *Clin. Transl. Allergy* **2015**, *5*, 33. [CrossRef] [PubMed]
39. Fang, Y.; Xiang, Z. Roles and relevance of mast cells in infection and vaccination. *J. Biomed. Res.* **2016**, *30*, 253–263. [PubMed]
40. Ferrer, M. Immunological events in chronic spontaneous urticaria. *Clin. Transl. Allergy* **2015**, *5*, 30. [CrossRef] [PubMed]
41. Shi, G.P.; Bot, I.; Kovanen, P.T. Mast cells in human and experimental cardiometabolic diseases. *Nat. Rev. Cardiol.* **2015**, *12*, 643–658. [CrossRef] [PubMed]

© 2018 by the authors. Licensee MDPI, Basel, Switzerland. This article is an open access article distributed under the terms and conditions of the Creative Commons Attribution (CC BY) license (http://creativecommons.org/licenses/by/4.0/).

Review

Ethnopharmacology, Chemistry and Biological Properties of Four Malian Medicinal Plants

Karl Egil Malterud

Section Pharmacognosy, Department of Pharmaceutical Chemistry, School of Pharmacy, University of Oslo, P.O. Box 1068 Blindern, Oslo 0316, Norway; k.e.malterud@farmasi.uio.no; Tel.: +47-22-856563

Academic Editor: Milan S. Stankovic
Received: 15 December 2016; Accepted: 14 February 2017; Published: 21 February 2017

Abstract: The ethnopharmacology, chemistry and pharmacology of four Malian medicinal plants, *Biophytum umbraculum*, *Burkea africana*, *Lannea velutina* and *Terminalia macroptera* are reviewed. These plants are used by traditional healers against numerous ailments: malaria, gastrointestinal diseases, wounds, sexually transmitted diseases, insect bites and snake bites, etc. The scientific evidence for these uses is, however, limited. From the chemical and pharmacological evidence presented here, it seems possible that the use in traditional medicine of these plants may have a rational basis, although more clinical studies are needed.

Keywords: Malian medicinal plants; *Biophytum umbraculum*; *Burkea africana*; *Lannea velutina*; *Terminalia macroptera*

1. Introduction

Africa has a very varied flora, and the study of African medicinal plants has engaged many scientists for a long time. The oldest written documents on this are found in the Papyrus Ebers, written ca. 3500 years ago [1,2]. One early study by the Norwegian medical doctor Henrik Greve Blessing, carried out in 1901–1904, but only existing as a handwritten manuscript, was recently discovered and has now been published [3].

Studies on African medicinal plants have in nearly all cases been limited to geographically limited areas—this is necessary, due to the very wide floral variation throughout this huge continent. Some examples of the large number of books dealing with African medicinal plants are the classical work by Watt and Breyer-Brandwijk [4] and Burkill's multivolume treatise [5]. A few more recent examples include the books by Iwu [1], Kuete [6], Neuwinger [7] and van Wyk [8]—this list is very far from complete! Journals such as *African Journal of Traditional, Complementary and Alternative Medicines*, *Journal of Ethnopharmacology*, *Journal of Ethnobiology and Ethnomedicine*, as well as many others, are also rich sources of knowledge of African medicinal plants and their properties and use.

For many years, scientists in Mali and Norway have collaborated in investigating Malian medicinal plants, and this has resulted in more than sixty publications in peer-reviewed international journals, numerous contributions at national and international conferences, nine Ph.D. degrees and more than 100 M.Sc. degrees awarded. A description of this project is available on the web [9].

In this brief survey, I intend to review the ethnopharmacology, chemistry and pharmacology of some Malian medicinal plants. The plants discussed have been studied in my research group in the Section of Pharmacognosy, School of Pharmacy, University of Oslo. Our research has in the main been directed towards isolation and identification of secondary metabolites of plant origin and their properties as antioxidants, radical scavengers and inhibitors of enzymes involved in peroxidative processes.

Four of the less known species are chosen for this treatise, viz. *Biophytum umbraculum*, *Burkea africana*, *Lannea velutina* and *Terminalia macroptera* (Table 1). References to previous work were found through the SciFinder database, which covers Medline and Chemical Abstracts. Titles in other languages than English, French or German have been used in translation.

Table 1. Data for herbarium voucher samples of plants discussed in this review.

Plant	Deposited in	Registry Number	Article Reference
Biophytum umbraculum	DMT	2653	[10]
Burkea africana	DMT	no number (registered under plant name)	[11]
Lannea velutina	DMT	1014	[12]
Terminalia macroptera	DMT	2468	[13,14]

DMT: Department of Traditional Medicine, Bamako, Mali.

A recent review [15] covers medicinal plants from Mali. That review is, however, differently angled, and the plants covered in detail in the present article are only briefly mentioned.

2. *Biophytum umbraculum*

Biophytum umbraculum Welw. (Oxalidaceae) is a small herb, up to 15 cm in height. It is widespread in the tropical parts of Africa and Asia [16]. Several synonyms exist for this plant, including *B. petersianum* Klotzsch (the name most commonly found in the scientific literature) [17].

2.1. Ethnopharmacology

Few systematic surveys on the traditional use of this plant have been done. Diallo et al. [18] reported that leaves are used in wound healing, using the dried and powdered plant. Another survey [19] in Mali found that while the plant was used against pain, insect bites and snake bites, treatment of cerebral malaria was by far the most common indication.

2.2. Chemistry

The polysaccharides of this plant are fairly well investigated. Pectic polysaccharides with several biological activities have been isolated and identified [20–22]. These polysaccharides contain rhamnogalacturonan, xylogalacturonan, and arabinogalactan regions. Very little was known about the low molecular weight constituents prior to our studies. Saponins were known to be present in the plant [23,24], but their structures still remain unknown.

From a methanol extract of above ground parts of *B. umbraculum*, we isolated the C-glycosylflavones cassiaoccidentalin A (**1**), isovitexin (**2**) and isoorientin (**3**) (Figure 1) [10]. Isovitexin and isoorientin are not rare substances, but this appears to the first report on their occurrence in *B. umbraculum*. They have, however, been reported from another *Biophytum* species, *B. sensitivum* [25].

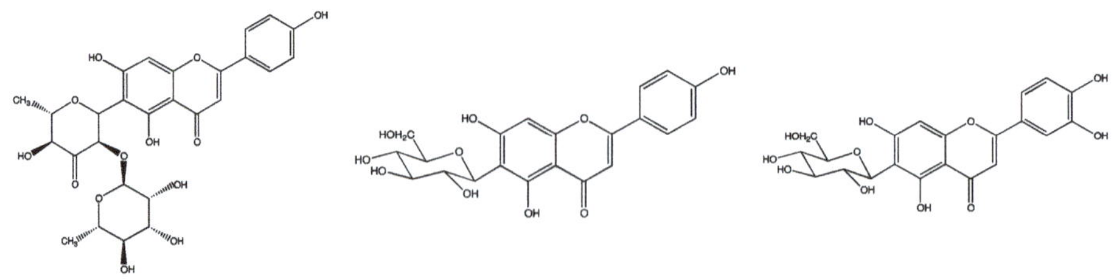

Figure 1. Cassiaoccidentalin A (**1**); isovitexin (**2**) and isoorientin (**3**) from *B. umbraculum*.

Cassiaoccidentalin A was first reported from *Cassia occidentalis* (Leguminosae) [26] along with substances (**2**) and (**3**). After our report on this, cassiaoccidentalin A was found in *Serjania marginata* (Sapindaceae) leaves [27]. To our knowledge, no other sources for this compound are known.

2.3. Biological Activity

Crude extracts of the plant show complement fixing activity [18], hypotensive effects [28], calcium antagonism [29], and increased corticosteroid secretion [30]. Decreased methane production was observed in cattle fed *B. petersianum* [24].

The polysaccharides referred to in Section 2.2 have been investigated for immunological effects. Complement fixing, macrophage activation, dendritic cell activation and immunomodulating activity against Peyer's patch cells has been reported [20–22]. The pectic polysaccharides from this plant exhibited protective ability against *Streptococcus pneumoniae* in a mouse model [31]. Cassiaoccidentalin A was investigated for suppression of the HIV promoter [32], but was found to be without effect. The plant where it was first found, *C. occidentalis*, is a known antimalarial plant and is used in Mali as one of the components of the "Improved Traditional Medicine" Malarial [33], which has been subjected to clinical studies. In view of the traditional use in Mali against cerebral malaria and the related chemistry between *B. umbraculum* and *C. occidentalis*, we investigated the malaria-related and antiinflammatory properties of *B. umbraculum* extracts and pure substances [34]. The ethyl acetate extract showed antiplasmodial, anti-complement and antiinflammatory activity and inhibited lipopolysaccharide stimulation in macrophages, but this appears to be due to other constituents than the isolated flavonoids. These findings may be relevant for the ethnopharmacological use of the plant against cerebral malaria. No clinical studies have been done, however.

3. *Burkea africana*

The tree *Burkea africana* Hook. (Leguminosae), usually less than 10 m in height, grows over large parts of tropical and subtropical Africa [35]. No synonyms are given for this plant [17].

3.1. Ethnopharmacology

In traditional medicine, the powdered stem bark is used in Mali on wounds in the mouth [18]. Other sources [4,35–38] cite a number of uses in different African countries.

3.2. Chemistry

Like many African plants, *B. africana* has only been subjected to a limited number of scientific studies. It has been reported [39] that the activity of the bark, which was stated as being used in folk medicine against gastrointestinal symptoms and headache, was due to its content of tyramine. The finding of tyramine was later corroborated [40], and the occurrence of harman-type alkaloids was reported, as well [41,42]. 5-Hydroxypipecolic acid, an unusual amino acid, was reported from the seeds of this plant [43].

Work in Ferreira's group in the late 1980s [44,45] led to the identification of several new oligomeric flavonoids (proanthocyanidins) based on the flavan-3-ol fisetinidol (5-deoxycatechin) in the heartwood of *B. africana*. The leaf lipids [46] and the stem bark polysaccharides [18] have been investigated, as well.

In view of the widespread medicinal use and the limited knowledge of the constituents of this plant, we decided to investigate the chemistry and the antioxidative properties of the stem bark, which appears to be the part of the plant that is most commonly used for medicinal purposes in Mali. The putative wound-healing and antiinflammatory activities ascribed to preparations of *B. africana* stem bark might well be correlated to antioxidative effects.

Using stem bark collected in Blendio, about 320 km south-east of the capital Bamako, we made an ethanolic extract which was subjected to liquid-liquid partition and repeated chromatography. This led to two major constituents (Figure 2), one of which was identified as fisetinidol-(4β→8)-catechin-3-O-gallate (**4**), previously known only from the heartwood of this tree [45]. The other substance was given the structure bis-fisetinidol-(4→6, 4→8)-catechin-3-O-gallate (**5**). Compound (**5**) was a new natural product [11].

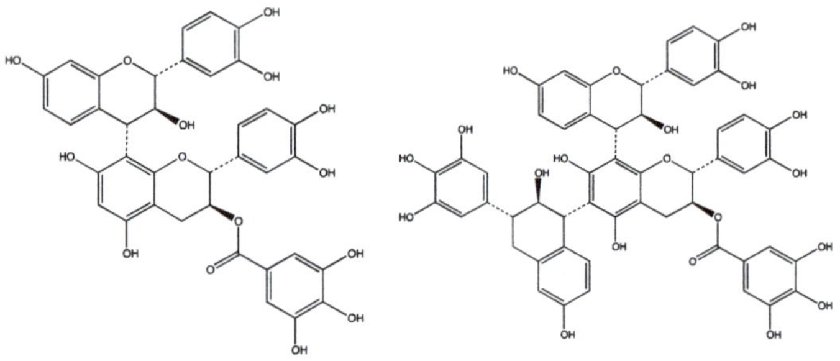

(**4**)   (**5**)

Figure 2. Structures of compounds (**4**) and (**5**) from *Burkea africana* stem bark.

3.3. Biological Activity

A pharmacological screening of Malian medicinal plants [47] revealed that a methanolic bark extract of *B. africana* had considerable radical scavenging activity as well as molluscicidal properties, and that both dichloromethane and methanol extracts of the bark were fungicidal. The methanolic bark extract has later been shown to attenuate oxidative stress in cells [48]. Fungicidal effects have been reported for heartwood extracts, as well [49], and moderate activity of water and methanol extracts of the bark against *Candida albicans* was reported [50]. An ethanol extract of the stem bark was shown to be antidiarrhoeal in mice [51]. Moderate inhibition of hyaluronidase and phospholipase A_2 as well as antiproteolytic activity of methanol and water extracts of *B. africana* was reported [52]. An acetone leaf extract was tested for some antioxidant and antiinflammatory activities, finding inhibition of 15-lipoxygenase, acetylcholinesterase and nitric oxide production, as well as radical scavenging activity in different assay systems [53].

We found [11,54] that the crude aqueous ethanolic extract (80% ethanol) had very high radical scavenging activity, good 15-lipoxygenase inhibitory activity, and was also active as an inhibitor of iron-induced peroxidation of phospholipids. Of the subfractions from liquid-liquid partition, the ethyl acetate part was the most active, indicating that the active constituents were semipolar.

Substances **4** and **5** were about equiactive as radical scavengers and 15-lipoxygenase inhibitors, as might be expected from their polyphenolic structures, but differed in their inhibitory activity against phospholipid peroxidation.

We concluded that the hydroethanolic extract of *Burkea africana* stem bark showed strong activity against several peroxidative processes, probably due to its high content of polyphenolic substances, most of which appear to be proanthocyanidin-type tannins [11,54]. Wound-healing properties of tannins are well described in the literature (e.g., [55,56]), so our findings may be related to the ethnopharmacological use of the plant. Although animal experiments have been reported, it appears that no clinical studies have been carried out.

4. *Lannea velutina*

Lannea velutina A. Rich. (Anacardiaceae), syn. *Calesiam velutinum* (A. Rich.) Kuntze, *Odina velutina* (A. Rich.) Oliv. [17] is a shrub or tree growing from Senegal to Ghana.

4.1. Ethnopharmacology

Several sources report on the medicinal use of this plant, but without details [57,58]. More specifically, decoctions are said to be used against diarrhoea, rachitis, muscle pains, gastric pains, and as a tonic. The bark is applied to wounds and leprosy [59]. In Mali, the stem bark is used in wound healing [18], and the leaves are sold in markets for making a decoction used as an antidote to poisoning [60]. The fruits are edible [61]. According to one source, the smoke from the twigs is used in witchcraft [62]. In our field work, we found that in Dioila and Kolokani, leaf decoctions, stem bark decoctions and root decoctions of *L. velutina* are used against a multitude of diseases, including skin diseases, fever, gastrointestinal diseases, and in wound treatment [12,63].

4.2. Chemistry

Prior to our studies, this plant appeared to be nearly uninvestigated. Unidentified tannins are reported to occur in all parts of the plant, mainly in the bark [64]. We found that the root bark of this plant was a rich source of procyanidins [63]. These had a linear 4→8 structure and commonly had catechin as starter unit and epicatechin as extender unit. Monomeric (catechin), dimeric, trimeric, tetrameric, hexameric, heptameric, nonameric, decameric and dodecameric procyanidin fractions were isolated (Figure 3). Most of the procyanidins were decameric or higher.

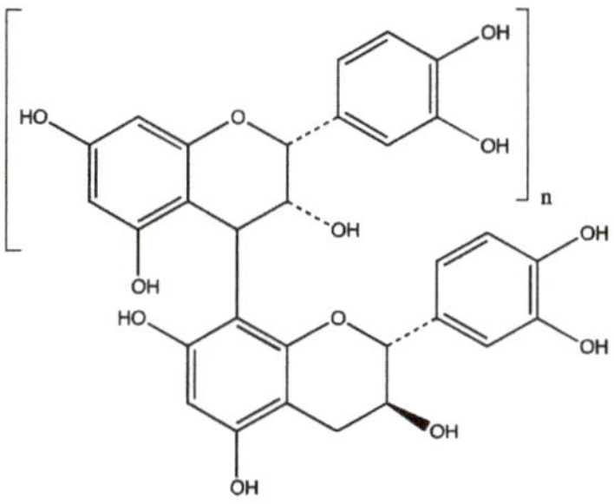

Figure 3. General structure of procyanidins from *L. velutina* (lower part: catechin, upper part: epicatechin, $n = 0$ (catechin monomer) to 11 (procyanidin dodecamer).

Mass spectroscopy revealed that *O*-methylation, *O*-galloylation and substitution of epiafzelechin (deoxyepicatechin) for epicatechin occurred, although to a minor extent.

4.3. Biological Activity

In a survey of Malian medicinal plants [47], extracts (dichloromethane, methanol, water) of the leaves, bark and roots of *L. velutina* were investigated for antifungal, larvicidal, molluscicidal, antioxidant and radical scavenging activities. Although activities varied between plant parts and extracts, positive results were obtained for antioxidant activity (bark and root methanol extracts), antifungal activity (leaf dichloromethane extract active against both *Candida albicans* and *Cladosporium cucumerinum*, the other extracts had more selective activity), larvicidal activity against the malaria mosquito *Anopheles gambiae* (dichloromethane bark and leaf extract, methanol leaf extract), and molluscicidal activity towards the schistomiasis-transmitting snail *Biomphalaria pfeifferi*. Antioxidant and antibacterial activity of an ethanol extract of *L. velutina* bark were reported [65], with the highest activity towards *Bacillus subtilis*, *Staphylococcus aureus* (Gram-positive), *Pseudomonas aeruginosa* and *Salmonella typhimurium* (Gram-negative).

We investigated the radical scavenger and 15-lipoxygenase inhibitory activity of different extracts of root bark and stem bark [12]. Lipophilic extracts (petrol ether, dichloromethane, chloroform) were inactive as scavengers of the DPPH radical, water extracts had moderate activity, while semipolar extracts (methanol, 80% aqueous ethanol) of both root bark and stem bark were highly active. Similar observations (high activity of semipolar extracts) were made for inhibition of the peroxidative enzyme 15-lipoxygenase, although lipophilic extracts were more active than aqueous extracts in this assay. In a continuation of these experiments [63], radical scavenging activity and 15-lipoxygenase inhibition were investigated for the purified proanthocyanidins. All of them were highly active, although there was a slight trend towards higher activity on a weight basis for the higher molecular weight substances in both assays, translating to a clear molecular weight—activity correlation on a molecular basis. Two of the smaller compounds, epicatechin and trimeric proanthocyanidin (mostly catechin-→epicatechin→epicatechin) were tested for ability to counteract pro-oxidant toxicity induced by glutamate in cerebellar neurons. The trimer gave a significant decrease in cell death. Although a decrease was observed for epicatechin, as well, this did not reach statistical significance.

As mentioned above (Section 3.3), the contents of proanthocyanidins may be of relevance to the ethnopharmacological use of the plants against wounds. Proanthocyanidins may also be involved in the use of the plant against gastric pains and diarrhea [66]. As for many ethnopharmacologically used plants, clinical data are lacking.

5. *Terminalia macroptera*

Terminalia macroptera Guill. & Perr. (Combretaceae) (syn. *Terminalia chevalieri* Diels, *Myrobalanus macroptera* (Guill. & Perr.) Kuntze) [17] is a tree up to 20 m in height. It is widespread in West Africa from Senegal to Cameroon, often on wet land [67].

5.1. Ethnopharmacology

Different parts of *T. macroptera* have been used for numerous ailments, including malaria [68–70], GI tract ailments (e.g., diarrhoea, gastritis, colitis, piles) [67,71,72], and infectious diseases including sexually transmitted diseases [73,74].

We carried out a systematic survey on the medicinal use of *T. macroptera* by traditional healers in three districts in Mali; Siby, Dioila and Dogonland [75]. Although there were regional differences, major areas of use were against pain and rheumatism (all areas), wounds (mainly in Siby), and hepatitis (mainly in Dogonland). Cough, diarrhoea and fever/malaria were also treated with *T. macroptera* preparations. Root bark, stem bark and leaves were used, but *Loranthus* parasitic plants on this tree were often employed in Dogonland.

5.2. Chemistry

The first published research on the constituents of *T. macroptera* appears to be by Prista et al. [76], reporting on the identification of chlorogenic acid and quercetin, two very common phenolic natural products. Other flavonoids isolated from *T. macroptera* flowers include the flavone C-glucosides orientin, isoorientin [77], vitexin, isovitexin [78,79] and the flavonol glycosides isorhamnetin 3-O-(6-O-α-L-rhamnosyl)-β-D-glucoside, quercetin 3-O-(6-O-β-D-glucosyl)-α-L-rhamnoside, quercetin 3-O-β-D-glucoside and quercetin 3-O-(6-O-α-L-rhamnosyl)-β-D-glucoside [80]. Unidentified proanthocyanidins have been reported to occur in the leaf extract [81].

Other polyphenols have been found, as well. Conrad et al [82] reported a new phenolic glucoside, vanillic acid 4-O-β-D-(6'-O-galloyl) glucopyranoside from the bark, and Kone et al [79] found a series of benzoic and cinnamic acids in the stem and root bark extracts of the plant. Ellagitannins, some of them new natural products, have been reported [82–84]. Ellagic acid and methylellagic acids were reported from the heartwood in an early investigation [85].

The other major group of natural products reported from *T. macroptera* is the terpenoids, more specifically triterpenoids. Idemudia [85] reported the presence of terminolic acid; more recently, 23-galloylarjunolic acid and its glucosyl ester [86] and glucosides of 24-deoxysericoside and chebuloside II [78] have been identified.

Research on the polysaccharides of *T. macroptera* has recently been carried out in our department [87–89]. Pectins are important in this plant, and their biological activities have been discussed below (Biological activity, Section 5.3).

We investigated leaf constituents in *T. macroptera*, since this part of the plant was not well known. A series of compounds was isolated and identified [13] from the methanolic extract (Figure 4): The flavonoids rutin (**6**) and narcissin (**7**), the hydrolyzable tannins corilagin (**8**), chebulagic acid (**9**), chebulinic acid (**10**) and chebulic acid trimethyl ester (**11**), methyl gallate (**12**) and shikimic acid (**13**). In the dichloromethane extract, poly-*cis*-isoprene (**14**; calculated average chain length ca 25 units) was the main constituent. All of these compounds are new to *T. macroptera*, and chebulic acid trimethyl ester is a new compound. It may, however, be an artifact, formed from chebulic acid during methanol extraction.

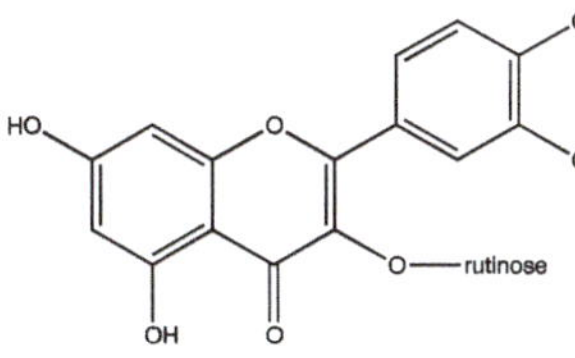

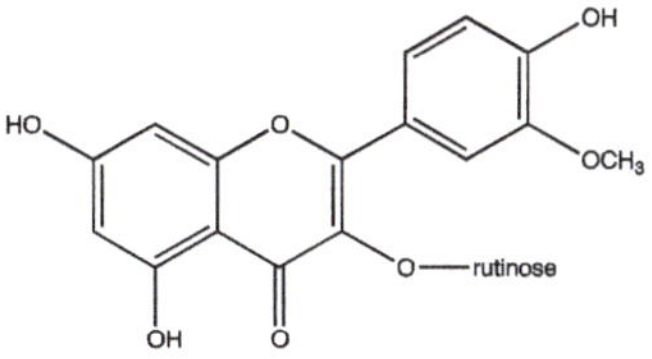

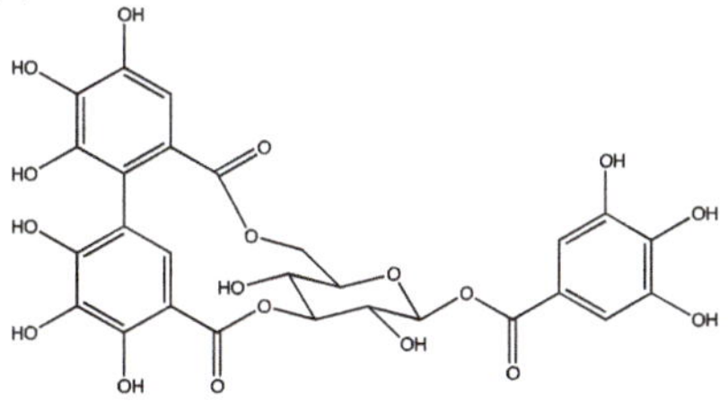

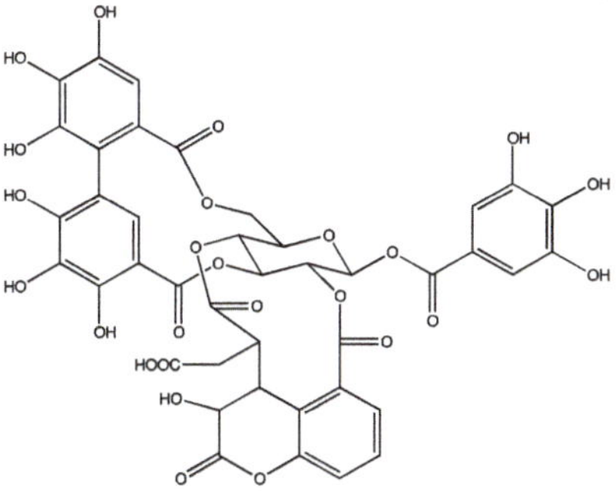

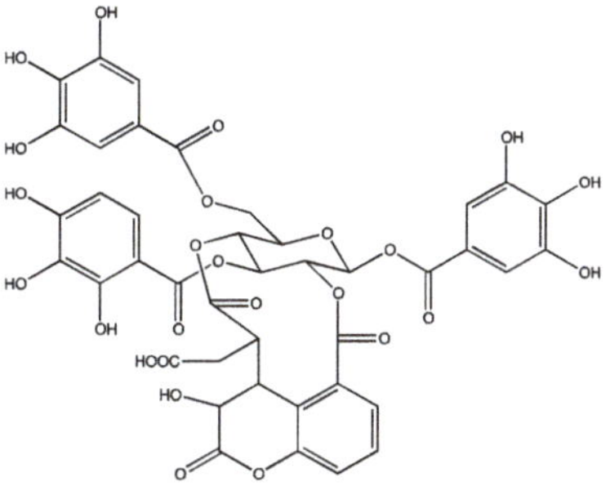

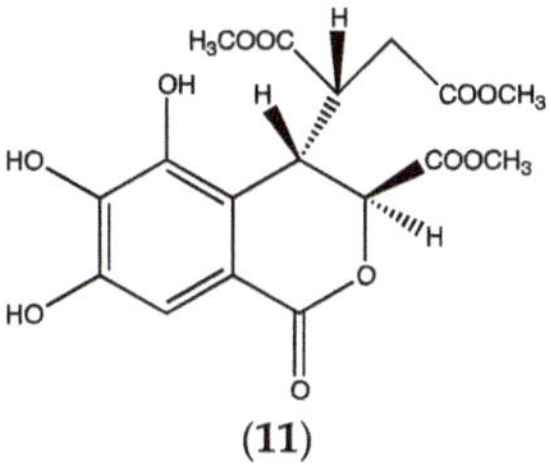

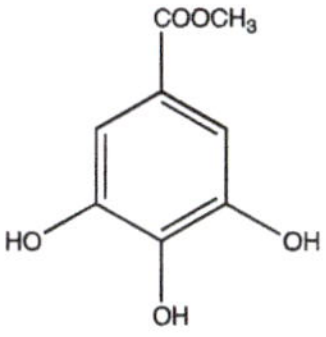

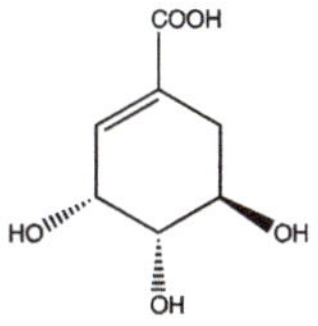

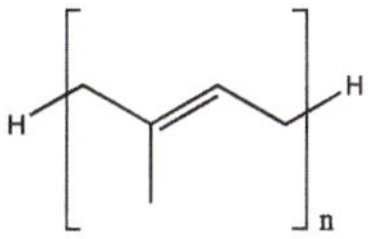

Figure 4. Constituents of *Terminalia macroptera* leaves: Flavonoids (**6–7**), ellagitannins (**8–10**), other constituents (**11–14**).

5.3. Biological Activity

Pharmacological studies of *T. macroptera* were started in the 1990s by Silva and co-workers. In a series of papers [74,90–93], extracts of roots and leaves were investigated for antimicrobial activity towards a series of pathogenic microorganisms, including *Neisseria gonorrheae* and *Helicobacter pylori*. Other groups have investigated antimicrobial effects, as well: Antibacterial and/or antifungal effects were reported [81,82,94–96]. Effects on the malaria protozoa *Plasmodium falciparum* [68,97] as well as on another pathogenic protozoa, *Trypanosoma brucei* [97] might be of special importance in view of the serious nature of the diseases caused by these organisms.

The pharmacology of the pectic polysaccharides from *T. macroptera* was recently investigated [87–89]. These polysaccharides were found to have immunomodulatory and complement fixing properties. Interestingly, such activities were present in preparations made in the same way as traditional healers do [88].

We found [13] that the methanol crude extract had high radical scavenger activity (6.2 ± 0.4 µg/mL) and showed moderate inhibition (52 ± 5 µg/mL) of xanthine oxidase, an enzyme involved in production of superoxide radical anion. Xanthine oxidase inhibition is of medicinal importance in the treatment of gout. While poly-*cis*-isoprene was inactive in all our assays, the other isolated compounds showed various activities. Corilagin and chebulagic acid were very good radical scavengers, with IC_{50} values less than half of the positive control quercetin. Due to lack of material, chebulinic acid could not be tested.

Rutin and chebulagic acid inhibited xanthine oxidase. They were, however, less active than quercetin (positive control). Shikimic acid was inactive [13]. Toxicity towards brine shrimp was low (LD_{50} > 100 µg/mL for all extracts, >200 µM for all pure compounds) [14]. This test is commonly used as an indication of general toxicity [98]. The crude methanol extract had good activity as a 15-lipoxygenase inhibitor (IC_{50} 27.9 ± 1.5 µg/mL, comparable to the positive control quercetin) and an α-glucosidase inhibitor (IC_{50} 0.47 ± 0.03 µg/mL, much more active than the positive control acarbose with an IC_{50} value of 130 ± 18 µg/mL). In 15-lipoxygenase inhibition, both chebulagic acid, corilagin and narcissin were considerably more active than the positive control quercetin, while chebulagic acid showed remarkable inhibitory activity towards α-glucosidase (IC_{50} < 0.1 µM). Corilagin was less active, but being present in much higher concentration in the extract, it may well be the most important component in this respect [14,15]. In sum, we have found that *T. macroptera* leaves constitute a rich source of bioactive compounds, with good activity both as an antioxidant and radical scavenger and as an inhibitor of α-glucosidase. This might be of importance for their use by traditional healers.

Biological activities in vitro (radical scavenging; enzyme inhibition (α-glucosidase, 15-lipoxygenase, xanthine oxidase); complement fixation) and in vivo (toxicity towards brine shrimp) of ethanol and water extracts of root bark, stem bark and leaves of *T. macroptera* were investigated in a separate set of experiments [99]. Ethanol extracts of root bark and stem bark showed the highest activity. Radical scavenging and enzyme inhibition was correlated to total phenolic content, while complement fixation was not. The extracts were non-toxic towards brine shrimp.

The ethnopharmacological use of the plant against GI tract ailments might be related to its content of gallotannins and ellagitannins. The polysaccharides of the plant have anti-inflammatory properties in vitro, but no clinical studies have been carried out.

6. Conclusions

The plants *Biophytum umbraculum*, *Burkea africana*, *Lannea velutina* and *Terminalia macroptera* are used in traditional medicine in Mali against diverse ailments. Extracts of these plants show a variety of biological effects in vitro and in vivo. These effects may be related to the medicinal use of these plants, and may therefore indicate that their use in traditional medicine could have a rational basis. Clinical studies are, however, needed to draw conclusions on this.

Acknowledgments: No funding was received for preparing and publishing this review. References to external funding of our work cited in this review are given in each paper. A special acknowledgement goes to the healers of Mali for generously sharing their knowledge with us.

Conflicts of Interest: The author declares no conflict of interest.

References

1. Iwu, M.M. *Handbook of African Medicinal Plants*, 2nd ed.; CRC Press: Boca Raton, FL, USA, 2014.
2. Petrovska, B.B. Historical review of medicinal plants' usage. *Pharmacogn. Rev.* **2012**, *6*, 1–5.
3. Paulsen, B.S.; Ekeli, H.; Johnson, Q.; Norum, K.R. *South African Traditional Medicinal Plants from Kwazulu-Natal*; Fagbokforlaget: Oslo, Norway, 2012.
4. Watt, J.M.; Breyer-Brandwijk, M.G. *The Medicinal and Poisonous Plants of Southern and Eastern Africa: Being an Account of Their Medicinal and Other Uses, Chemical Composition, Pharmacological Effects and Toxicology in Man and Animal*; Livingstone: Edinburgh, UK, 1962.
5. Burkill, H.M. *The Useful Plants of West Tropical Africa*; Royal Botanic Gardens: Kew, UK, 1985–2000; Volumes 1–5.
6. Kuete, V. *Medicinal Plant Research in Africa: Pharmacology and Chemistry*; Elsevier: Amsterdam, The Netherlands, 2013.
7. Neuwinger, H.D. *African Traditional Medicine*; Medpharm: Stuttgart, Germany, 2000.
8. Van Wyk, B.E.; Van Oudtshoorn, B.; Gericke, N. *Medicinal Plants of South Africa*, 2nd ed.; Briza: Pretoria, South Africa, 2009.
9. Universitetet I Oslo. The Malian Medicinal Plant Project. Available online: http://www.mn.uio.no/farmasi/english/research/projects/maliplants/ (accessed on 14 November 2016).
10. Pham, A.T.; Nguyen, C.; Malterud, K.E.; Diallo, D.; Wangensteen, H. Bioactive flavone-C-glycosides of the African medicinal plant *Biophytum umbraculum*. *Molecules* **2013**, *18*, 10312–10319.
11. Mathisen, E.; Diallo, D.; Andersen, Ø.M.; Malterud, K.E. Antioxidants from the bark of *Burkea africana*, an African medicinal plant. *Phytother. Res.* **2002**, *16*, 148–153.
12. Maiga, A.; Malterud, K.E.; Diallo, D.; Paulsen, B.S. Antioxidant and 15-lipoxygenase inhibitory activities of the Malian medicinal plants *Diospyros abyssinica* (Hiern) F. White (Ebenaceae), *Lannea velutina* A. Rich. (Anacardiaceae) and *Crossopteryx febrifuga* (Afzel) Benth. (Rubiaceae). *J. Ethnopharmacol.* **2006**, *104*, 132–137.
13. Pham, A.T.; Malterud, K.E.; Paulsen, B.S.; Diallo, D.; Wangensteen, H. DPPH radical scavenging and xanthine oxidase inhibitory activity of *Terminalia macroptera* leaves. *Nat. Prod. Commun.* **2011**, *6*, 1125–1128.
14. Pham, A.T.; Malterud, K.E.; Paulsen, B.S.; Diallo, D.; Wangensteen, H. α-Glucosidase inhibition, 15-lipoxygenase inhibition, and brine shrimp toxicity of extracts and isolated compounds from *Terminalia macroptera* leaves. *Pharm. Biol.* **2014**, *52*, 1166–1169.
15. Wangensteen, H.; Diallo, D.; Paulsen, B.S. Medicinal plants from Mali: Chemistry and biology. *J. Ethnopharmacol.* **2015**, *176*, 429–437.
16. Pham, A.T. Chemical, Biological and Ethnopharmacological Studies of Two Malian Medicinal Plants: Terminalia Macroptera and Biophytum Umbraculum. Ph.D. Thesis, University of Oslo, Oslo, Norway, 2014. Available online: https://www.duo.uio.no/handle/10852/41479 (accessed on 5 December 2016).
17. The Plant List. Available online: www.theplantlist.org (accessed on 13 November 2016).
18. Diallo, D.; Sogn, C.; Samaké, F.B.; Paulsen, B.S.; Michaelsen, T.E.; Keita, A. Wound healing plants in Mali, the Bamako region. An ethnobotanical survey and complement fixation of water extracts from selected plants. *Pharm. Biol.* **2002**, *40*, 117–128.
19. Grønhaug, T.E.; Glæserud, S.; Skogsrud, M.; Ballo, N.; Bah, S.; Diallo, D.; Paulsen, B.S. Ethnopharmacological survey of six medicinal plants from Mali, West-Africa. *J. Ethnobiol. Ethnomed.* **2008**, *4*, 26.
20. Inngjerdingen, K.T.; Coulibaly, A.; Diallo, D.; Michaelsen, T.E.; Paulsen, B.S. A complement fixing polysaccharide from *Biophytum petersianum* Klotzsch, a medicinal plant from Mali, West Africa. *Biomacromolecules* **2006**, *7*, 48–53.
21. Inngjerdingen, M.; Inngjerdingen, K.T.; Patel, T.R.; Allen, S.; Chen, X.; Rolstad, B.; Morris, G.A.; Harding, S.E.; Michaelsen, T.E.; Diallo, D.; et al. Pectic polysaccharides from *Biophytum petersianum* Klotzsch, and their activation of macrophages and dendritic cells. *Glycobiology* **2008**, *18*, 1074–1084.

22. Grønhaug, T.E.; Kiyohara, H.; Sveaass, A.; Diallo, D.; Yamada, H.; Paulsen, B.S. Beta-D-(1→4)-galactan-containing side chains in RG-I regions of pectic polysaccharides from *Biophytum petersianum* Klotzsch. contribute to expression of immunomodulating activity against intestinal Peyer's patch cells and macrophages. *Phytochemistry* **2011**, *72*, 2139–2147.
23. Santoso, B.; Kilmaskossu, A.; Sambodo, P. Effects of saponin from *Biophytum petersianum* Klotzsch on ruminal fermentation, microbial protein synthesis and nitrogen utilization in goats. *Anim. Feed Sci. Technol.* **2007**, *137*, 58–68.
24. Hariadi, B.T.; Santoso, B. Evaluation of tropical plants containing tannin on in vitro methanogenesis and fermentation parameters using rumen fluid. *J. Sci. Food Agric.* **2010**, *90*, 456–461.
25. Bucar, F.; Jachak, S.M.; Kartnig, T.; Schubert-Zsilavecz, M. Phenolic compounds from *Biophytum sensitivum*. *Pharmazie* **1998**, *53*, 651–653.
26. Hatano, T.; Mizuta, S.; Ito, H.; Yoshida, T. C-glycosidic flavonoids from *Cassia occidentalis*. *Phytochemistry* **1999**, *52*, 1379–1383.
27. Heredia-Vieira, S.C.; Simonet, A.M.; Vilegas, W.; Macías, F.A. Unusual C,O-fused glucosylapigenins from *Serjania marginata* leaves. *J. Nat. Prod.* **2015**, *78*, 77–84.
28. Titrikou, S.; Aklikokou, A.K.; Gbeassor, M. Effet de l'extrait de *Biophytum petersianum* (Oxalidaceae) Klotzsch sur le systeme cardiovasculaire de cobaye. *Pharm. Med. Tradit. Afr.* **1998**, *10*, 32–41. (In French)
29. Titrikou, S.; Eklu-Gadegbeku, K.; Mouzou, A.; Aklikokou, K.; Gbeassor, M. Calcium antagonistic activity of *Biophytum petersianum* on vascular smooth muscles of Wistar rats. *Iran. J. Pharmacol. Ther.* **2007**, *6*, 185–189.
30. Kodjo, K.M.; Contesse, V.; Do Rego, J.L.; Aklikokou, K.; Titrikou, S.; Gbeassor, M.; Vaudry, H. In vitro effects of crude extracts of *Parkia biglobosa* (Mimosaceae), *Stereospermum kunthianum* (Bignoniaceae) and *Biophytum petersianum* (Oxalidaceae) on corticosteroid secretion in rat. *J. Steroid Biochem. Mol. Biol.* **2006**, *100*, 202–208.
31. Inngjerdingen, K.T.; Langerud, B.K.; Rasmussen, H.; Olsen, T.K.; Austarheim, I.; Grønhaug, T.E.; Aaberge, I.S.; Diallo, D.; Paulsen, B.S.; Michaelsen, T.E. Pectic polysaccharides isolated from Malian medicinal plants protect against *Streptococcus pneumoniae* in a mouse pneumococcal pneumonia infection model. *Scand. J. Immunol.* **2013**, *77*, 372–388.
32. Uchiumi, F.; Hatano, T.; Ito, H.; Yoshida, T.; Tanuma, S.I. Transcriptional suppression of the HIV promoter by natural compounds. *Antivir. Res.* **2003**, *58*, 89–98.
33. Willcox, M.; Sanogo, R.; Diakite, C.; Giani, S.; Paulsen, B.S.; Diallo, D. Improved traditional medicines in Mali. *J. Altern. Complement. Med.* **2012**, *18*, 212–220.
34. Austarheim, I.; Pham, A.T.; Nguyen, C.; Zou, Y.F.; Diallo, D.; Malterud, K.E.; Wangensteen, H. Antiplasmodial, anti-complement and anti-inflammatory in vitro effects of *Biophytum umbraculum* Welw. traditionally used against cerebral malaria in Mali. *J. Ethnopharmacol.* **2016**, *190*, 159–164.
35. Nonyane, F.; Masupa, T. *Burkea africana* Hook. Available online: http://www.plantzafrica.com/plantab/burkeaafricana.htm (accessed on 14 November 2016).
36. Delaveau, P.A.; Desvignes, E.; Adoux, E.; Tessier, A.M. Baguettes frotte-dents d'Afrique occidentale. Examen chimique et microbiologique. *Ann. Pharm. Fr.* **1979**, *37*, 185–190. (In French)
37. Pedersen, M.E.; Vestergaard, H.T.; Hansen, S.L.; Bah, S.; Diallo, D.; Jäger, A.K. Pharmacological screening of Malian medicinal plants used against epilepsy and convulsions. *J. Ethnopharmacol.* **2009**, *121*, 472–475.
38. Maroyi, A. *Burkea africana* Hook. Available online: http://uses.plantnet-project.org/en/Burkea_africana_ (PROTA) (accessed on 16 February 2017).
39. Correira da Silva, A.; Paiva, M.Q. Comparison of the pharmacodynamic activity of bark extracts from *Burkea africana* with one of its alkaloid constituents. *An. Fac. Farm. Porto.* **1971**, *31*, 107–118.
40. Ferreira, M.A. Indole alkaloids from *Burkea africana*. *An. Fac. Farm. Porto.* **1972**, *32*, 37–41.
41. Ferreira, M.A. Chemical study of *Burkea africana*. I. Identification of β-sitosterol and tetrahydroharman. *Garcia de Orta. Ser. Farmacogn.* **1973**, *2*, 7–21.
42. Ferreira, M.A. Indolic alkaloids from *Burkea africana*. II. Characterization of harmane and dihydroharmane. *Garcia de Orta Ser. Farmacogn.* **1973**, *2*, 23–32.
43. Watson, R.; Fowden, L. Amino acids of *Caesalpinia tinctoria* and some allied species. *Phytochemistry* **1973**, *12*, 617–622.
44. Malan, J.C.S.; Young, D.A.; Steenkamp, J.A.; Ferreira, D. Oligomeric flavonoids. Part 2. The first profisetinidins with dihydroflavonol constituent units. *J. Chem. Soc. Perkin Trans. 1* **1988**, *9*, 2567–2572.

45. Bam, M.; Malan, J.C.S.; Young, D.A.; Brandt, E.V.; Ferreira, D. Profisetinidin-type 4-arylflavan-3-ols and related δ-lactones. *Phytochemistry* **1990**, *29*, 283–287.
46. Davidson, B.C. Seasonal changes in leaf lipid and fatty acid composition of nine plants consumed by two African herbivores. *Lipids* **1998**, *33*, 109–113.
47. Diallo, D.; Marston, A.; Terreaux, C.; Touré, Y.; Paulsen, B.S.; Hostettmann, K. Screening of Malian medicinal plants for antifungal, larvicidal, molluscicidal, antioxidant and radical scavenging activities. *Phytother. Res.* **2001**, *15*, 401–406.
48. Cordier, W.; Gulumian, M.; Cromarty, A.D.; Steenkamp, V. Attenuation of oxidative stress in U937 cells by polyphenolic-rich bark fractions of *Burkea africana* and *Syzygium cordatum*. *BMC Complement. Altern. Med.* **2013**, *13*, 116.
49. Neya, B.; Hakkou, M.; Pétrissans, M.; Gérardin, P. On the durability of *Burkea africana* heartwood: Evidence of biocidal and hydrophobic properties responsible for durability. *Ann. For. Sci.* **2004**, *61*, 277–282.
50. Steenkamp, V.; Fernandes, A.C.; Van Rensburg, C.E.J. Screening of Venda medicinal plants for antifungal activity against *Candida albicans*. *S. Afr. J. Bot.* **2007**, *73*, 256–258.
51. Tanko, Y.; Iliya, B.; Mohammed, A.; Mahdi, M.A.; Musa, K.Y. Modulatory effect of ethanol stem bark extract of *Burkea africana* on castor oil induced diarrhoea on experimental animals. *Arch. Appl. Sci. Res.* **2011**, *3*, 122–130.
52. Molander, M.; Nielsen, L.; Søgaard, S.; Staerk, D.; Rønsted, N.; Diallo, D.; Chifundera, K.Z.; Van Staden, J.; Jäger, A.K. Hyaluronidase, phospholipase A_2 and protease inhibitory activity of plants used in traditional treatment of snakebite-induced tissue necrosis in Mali, DR Congo and South Africa. *J. Ethnopharmacol.* **2014**, *157*, 171–180.
53. Dzoyem, J.P.; Eloff, J.N. Anti-inflammatory, anticholinesterase and antioxidant activity of leaf extract of twelve plants used traditionally to alleviate pain and inflammation in South Africa. *J. Ethnopharmacol.* **2015**, *160*, 194–201.
54. Mathisen, E. Radical Scavengers and Antioxidants from Burkea Africana, an African Medicinal Plant. Master's Thesis, University of Oslo, Oslo, Norway, 1999. (In Norwegian)
55. Rane, M.M.; Mengi, S.A. Comparative effect of oral administration and topical application of alcoholic extract of *Terminalia arjuna* bark on incision and excision wounds in rats. *Fitoterapia* **2003**, *74*, 553–558.
56. Kisseih, E.; Lechtenberg, M.; Petereit, F.; Sendker, J.; Zacharski, D.; Brandt, S.; Agyare, C.; Hensel, A. Phytochemical characterization and in vitro wound healing activity of leaf extracts from *Combretum mucronatum* Schum. & Thonn.: Oligomeric procyanidins as strong inductors of cellular differentiation. *J. Ethnopharmacol.* **2015**, *174*, 628–636.
57. Taïta, P. Use of woody plants by locals in Mare aux Hippopotames biosphere reserve in western Burkina Faso. *Biodivers. Conserv.* **2003**, *12*, 1205–1217.
58. Belem, B.; Nacoulma, B.M.I.; Gbangou, R.; Kambou, S.; Hansen, H.H.; Gausset, Q.; Lund, S.; Raebild, A.; Lompo, D.; Ouedraogo, M.; et al. Use of non wood forest products by local people bordering the "Parc national Kaboré Tambi", Burkina Faso. *J. Transdiscipl. Environ. Stud.* **2007**, *6*, 1–21.
59. Jansen, P.C.M. *Lannea velutina* A. Rich. PROTA. Available online: http://uses.plantnet-project.org/en/Lannea_velutina_(PROTA) (accessed on 16 February 2017).
60. Maiga, A.; Diallo, D.; Fane, S.; Sanogo, R.; Paulsen, B.S.; Cisse, B. A survey of toxic plants on the market in the district of Bamako, Mali: Traditional knowledge compared with a literature search of modern pharmacology and toxicology. *J. Ethnopharmacol.* **2005**, *96*, 183–193.
61. Gueye, M.; Ayessou, N.C.; Koma, S.; Diop, S.; Akpo, L.E.; Samb, P.I. Wild fruits traditionally gathered by the Malinke ethnic group in the edge of Niokolo Koba park (Senegal). *Am. J. Plant Sci.* **2014**, *5*, 1306–1317.
62. Pageard, R. Plantes à brûler chez les Bambara. *J. Soc. Afr.* **1967**, *37*, 87–130.
63. Maiga, A.; Malterud, K.E.; Mathisen, G.H.; Paulsen, R.E.; Thomas-Oates, J.; Bergström, E.; Reubsaet, L.; Diallo, D.; Paulsen, B.S. Cell protective antioxidants from the root bark of *Lannea velutina* A. Rich., a Malian medicinal plant. *J. Med. Plants Res.* **2007**, *1*, 66–79.
64. Séréné, A.; Millogo Rasolodimby, J.; Guinko, S.; Nacro, M. Concentration en tanins des organes de plantes tannifères du Burkina Faso. *J. Soc. Ouest.-Afr. Chim.* **2008**, *25*, 55–61. (In French)
65. Ouattara, L.; Koudou, J.; Zongo, C.; Barro, N.; Savadogo, A.; Bassole, I.H.N.; Ouattara, A.S.; Traore, A.S. Antioxidant and antibacterial activities of three species of *Lannea* from Burkina Faso. *J. Appl. Sci.* **2011**, *11*, 157–162.

66. Cires, M.J.; Wong, X.; Carrasco-Pozo, C.; Gotteland, M. The gastrointestinal tract as a key target organ for the health-promoting effects of dietary proanthocyanidins. *Front. Nutr.* **2017**, *3*, 57.
67. Arbonnier, M. *Trees, Shrubs and Lianas of West African Dry Zones*; Margraf: Weikersheim, Germany, 2004.
68. Sanon, S.; Ollivier, E.; Azas, N.; Mahiou, V.; Gasquet, M.; Ouattara, C.T.; Nebie, I.; Traore, A.S.; Esposito, F.; Balansard, G.; et al. Ethnobotanical survey and in vitro antiplasmodial activity of plants used in traditional medicine in Burkina Faso. *J. Ethnopharmacol.* **2003**, *86*, 143–147.
69. Benoit-Vical, F.; Soh, P.N.; Saléry, M.; Harguem, L.; Poupat, C.; Nongonierma, R. Evaluation of Senegalese plants used in malaria treatment: Focus on *Chrozophora senegalensis*. *J. Ethnopharmacol.* **2008**, *116*, 43–48.
70. Traore, M.S.; Baldé, M.A.; Diallo, M.S.T.; Baldé, E.S.; Diané, S.; Camara, A.; Diallo, A.; Balde, A.; Keïta, A.; Keita, S.M.; et al. Ethnobotanical survey on medicinal plants used by Guinean traditional healers in the treatment of malaria. *J. Ethnopharmacol.* **2013**, *150*, 1145–1153.
71. Etuk, E.U.; Uhwah, M.O.; Ajagbonna, O.P.; Onyeyili, P.A. Ethnobotanical survey and preliminary evaluation of medicinal plants with antidiarrhoea properties in Sokoto state, Nigeria. *J. Med. Plants Res.* **2009**, *3*, 763–766.
72. Kayode, J.; Ige, O.E.; Adetogo, T.A.; Igbakin, A.P. Conservation and biodiversity erosion in Ondo state, Nigeria: (3). Survey of plant barks used in native pharmaceutical extraction in Akoko region. *Ethnobot. Leafl.* **2009**, *13*, 665–667.
73. Kayode, J.; Jose, R.A.; Ige, O.E. Conservation and biodiversity erosion in Ondo state, Nigeria: (4). Assessing botanicals used in the cure of sexually transmitted diseases in Owo region. *Ethnobot. Leafl.* **2009**, *13*, 734–738.
74. Silva, O.; Ferreira, E.; Pato, M.V.; Canica, M.; Gomes, E.T. In vitro anti-*Neisseria gonorrhoeae* activity of *Terminalia macroptera* leaves. *FEMS Microbiol. Lett.* **2002**, *217*, 271–274.
75. Pham, A.T.; Dvergsnes, C.; Togola, A.; Wangensteen, H.; Diallo, D.; Paulsen, B.S.; Malterud, K.E. *Terminalia macroptera*, its current medicinal use and future perspectives. *J. Ethnopharmacol.* **2011**, *137*, 1486–1491.
76. Prista, L.N.; De Almeida e Silva, L.; Alves, A.C. Phytochemical study of the barks and leaves of *Terminalia macroptera*. *Garcia de Orta* **1962**, *10*, 501–509.
77. Nongonierma, R.; Proliac, A.; Raynaud, J. Two mono-C-glycosyl flavonoids from the flowers of *Terminalia macroptera* Guill. et Perr. (Combretaceae). *Pharmazie* **1987**, *42*, 871–872.
78. Nongonierma, R.; Proliac, A.; Raynaud, J. Vitexin and isovitexin in the flowers of *Terminalia macroptera* Guill. et Perr. (Combretaceae). *Pharmazie* **1988**, *43*, 293.
79. Kone, D.; Diop, B.; Diallo, D.; Djilani, A.; Dicko, A. Identification, quantitative determination, and antioxidant properties of polyphenols of some Malian medicinal plant parts used in folk medicine. In *Macro to Nano Spectroscopy*; Uddin, J., Ed.; Intech: Rijeka, Croatia, 2012; pp. 131–142.
80. Nongonierma, R.; Proliac, A.; Raynaud, J. O-glycosyl flavonoids of flowers from *Terminalia macroptera* Guill. et Perr. (Combretaceae). *Pharm. Acta Helv.* **1990**, *65*, 233–235.
81. Karou, S.D.; Tchacondo, T.; Tchibozo, M.A.D.; Anani, K.; Ouattara, L.; Simpore, J.; De Souza, C. Screening Togolese medicinal plants for few pharmacological properties. *Pharmacogn. Res.* **2012**, *4*, 116–122.
82. Conrad, J.; Vogler, B.; Klaiber, I.; Reeb, S.; Guse, J.H.; Roos, G.; Kraus, W. Vanillic acid 4-O-β-D-(6'-O-galloyl) glucopyranoside and other constituents from the bark of *Terminalia macroptera* Guill. et Perr. *Nat. Prod. Lett.* **2001**, *15*, 35–42.
83. Silva, O.; Gomes, E.T.; Wolfender, J.L.; Marston, A.; Hostettmann, K. Application of high performance liquid chromatography coupled with ultraviolet spectroscopy and electrospray mass spectrometry to the characterization of ellagitannins from *Terminalia macroptera* roots. *Pharm. Res.* **2000**, *17*, 1396–1401.
84. Conrad, J.; Vogler, B.; Reeb, S.; Klaiber, I.; Papajewski, S.; Roos, G.; Vasquez, E.; Setzer, M.C.; Kraus, W. Isoterchebulin and 4,6-O-isoterchebuloyl-D-glucose, novel hydrolyzable tannins from *Terminalia macroptera*. *J. Nat. Prod.* **2001**, *64*, 294–299.
85. Idemudia, O.G. Terpenoids of Nigerian *Terminalia* species. *Phytochemistry* **1970**, *9*, 2401–2402.
86. Conrad, J.; Vogler, B.; Klaiber, I.; Roos, G.; Walter, U.; Kraus, W. Two triterpene esters from *Terminalia macroptera* bark. *Phytochemistry* **1998**, *48*, 647–650.
87. Zou, Y.F.; Zhang, B.Z.; Barsett, H.; Inngjerdingen, K.T.; Diallo, D.; Michaelsen, T.E.; Paulsen, B.S. Complement fixing polysaccharides from *Terminalia macroptera* root bark, stem bark and leaves. *Molecules* **2014**, *19*, 7440–7458.
88. Zou, Y.F.; Zhang, B.Z.; Inngjerdingen, K.T.; Barsett, H.; Diallo, D.; Michaelsen, T.E.; Paulsen, B.S. Complement activity of polysaccharides from three different plant parts of *Terminalia macroptera* extracted as healers do. *J. Ethnopharmacol.* **2014**, *155*, 672–678.

89. Zou, Y.F.; Barsett, H.; Ho, G.T.T.; Inngjerdingen, K.T.; Diallo, D.; Michaelsen, T.E.; Paulsen, B.S. Immunomodulating pectins from root bark, stem bark and leaves of the Malian medicinal plant *Terminalia macroptera*: Structure activity relations. *Carbohydr. Res.* **2015**, *403*, 167–173.
90. Silva, O.; Duarte, A.; Cabrita, J.; Pimentel, M.; Diniz, A.; Gomes, E. Antimicrobial activity of Guinea-Bissau traditional remedies. *J. Ethnopharmacol.* **1996**, *50*, 55–59.
91. Silva, O.; Duarte, A.; Pimentel, M.; Viegas, S.; Barroso, H.; Machado, J.; Pires, I.; Cabrita, J.; Gomes, E. Antimicrobial activity of *Terminalia macroptera* root. *J. Ethnopharmacol.* **1997**, *57*, 203–207.
92. Silva, O.; Ferreira, E.; Vaz Pato, M.; Gomes, E. Guinea-Bissau's plants: In vitro susceptibility studies on *Neisseria gonorrheae*. *Int. J. Pharmacogn.* **1997**, *35*, 323–328.
93. Silva, O.; Viegas, S.; De Mello-Sampayo, C.; Costa, M.J.P.; Serrano, R.; Cabrita, J.; Gomes, E.T. Anti-*Helicobacter pylori* activity of *Terminalia macroptera* root. *Fitoterapia* **2012**, *83*, 872–876.
94. Batawila, K.; Kokou, K.; Koumaglo, K.; Gbéassor, M.; De Foucault, B.; Bouchet, P.; Akpagana, K. Antifungal activities of five Combretaceae used in Togolese traditional medicine. *Fitoterapia* **2005**, *76*, 264–268.
95. Traoré, M.S.; Baldé, M.A.; Camara, A.; Baldé, E.S.; Diané, S.; Diallo, M.S.T.; Keita, A.; Cos, P.; Maes, L.; Pieters, L.; et al. The malaria co-infection challenge: An investigation into the antimicrobial activity of selected Guinean medicinal plants. *J. Ethnopharmacol.* **2015**, *174*, 576–581.
96. Traore, Y.; Ouattara, K.; Ouattara, A.; Méité, S.; Bagré, I.; Konan, K.F.; Coulibaly, A.; Nathalie, K.G. Evaluation of the antistaphylococcic activity of *Terminalia macroptera* Guill et Perr (*Combretaceae*) stem bark extracts. *Am. J. Biosci.* **2015**, *3*, 221–225.
97. Traore, M.S.; Diane, S.; Diallo, M.S.T.; Balde, E.S.; Balde, M.A.; Camara, A.; Diallo, A.; Keita, A.; Cos, P.; Maes, L.; et al. In vitro antiprotozoal and cytotoxic activity of ethnopharmacologically selected Guinean plants. *Planta Med.* **2014**, *80*, 1340–1344.
98. McLaughlin, J.L. Crown gall tumours on potato discs and brine shrimp lethality: Two simple bioassays for higher plant screening and fractionation. In *Methods in Plant Biochemistry*; Hostettmann, K., Ed.; Academic Press: London, UK, 1991; Volume 6, pp. 1–32.
99. Zou, Y.F.; Ho, G.T.T.; Malterud, K.E.; Le, N.H.T.; Inngjerdingen, K.T.; Barsett, H.; Diallo, D.; Michaelsen, T.E.; Paulsen, B.S. Enzyme inhibition, antioxidant and immunomodulatory activities, and brine shrimp toxicity of extracts from the root bark, stem bark and leaves of *Terminalia macroptera*. *J. Ethnopharmacol.* **2014**, *155*, 1219–1226.

© 2017 by the author. Licensee MDPI, Basel, Switzerland. This article is an open access article distributed under the terms and conditions of the Creative Commons Attribution (CC BY) license (http://creativecommons.org/licenses/by/4.0/).

Review

Phytochemicals: Extraction, Isolation, and Identification of Bioactive Compounds from Plant Extracts

Ammar Altemimi [1,*], Naoufal Lakhssassi [2], Azam Baharlouei [2], Dennis G. Watson [2] and David A. Lightfoot [2]

1. Department of Food Science, College of Agriculture, University of Al-Basrah, Basrah 61004, Iraq
2. Department of Plant, Soil and Agricultural Systems, Plant Biotechnology and Genome Core-Facility, Southern Illinois University at Carbondale, Carbondale, IL 62901, USA; naoufal.lakhssassi@siu.edu (N.L.); baharlouei@siu.edu (A.B.); dwatson@siu.edu (D.G.W.); ga4082@siu.edu (D.A.L.)
* Correspondence: ammaragr@siu.edu; Tel.: +964-773-564-0090

Academic Editor: Ulrike Mathesius
Received: 19 July 2017; Accepted: 19 September 2017; Published: 22 September 2017

Abstract: There are concerns about using synthetic phenolic antioxidants such as butylated hydroxytoluene (BHT) and butylated hydroxyanisole (BHA) as food additives because of the reported negative effects on human health. Thus, a replacement of these synthetics by antioxidant extractions from various foods has been proposed. More than 8000 different phenolic compounds have been characterized; fruits and vegetables are the prime sources of natural antioxidants. In order to extract, measure, and identify bioactive compounds from a wide variety of fruits and vegetables, researchers use multiple techniques and methods. This review includes a brief description of a wide range of different assays. The antioxidant, antimicrobial, and anticancer properties of phenolic natural products from fruits and vegetables are also discussed.

Keywords: antimicrobial; antioxidants; medicinal plants; BHT

1. Introduction

Many antioxidant compounds can be found in fruits and vegetables including phenolics, carotenoids, anthocyanins, and tocopherols [1]. Approximately 20% of known plants have been used in pharmaceutical studies, impacting the healthcare system in positive ways such as treating cancer and harmful diseases [2]. Plants are able to produce a large number of diverse bioactive compounds. High concentrations of phytochemicals, which may protect against free radical damage, accumulate in fruits and vegetables [3]. Plants containing beneficial phytochemicals may supplement the needs of the human body by acting as natural antioxidants [4]. Various studies have shown that many plants are rich source of antioxidants. For instance, vitamins A, C, E, and phenolic compounds such as flavonoids, tannins, and lignins, found in plants, all act as antioxidants [3]. The consumption of fruits and vegetables has been linked with several health benefits, a result of medicinal properties and high nutritional value [5]. Antioxidants control and reduce the oxidative damage in foods by delaying or inhibiting oxidation caused by reactive oxygen species (ROS), ultimately increasing the shelf-life and quality of these foods [6]. Beta carotene, ascorbic acid, and many phenolics play dynamic roles in delaying aging, reducing inflammation, and preventing certain cancers [7]. Increasing the consumption of fruits and vegetables has been recommended by many agencies and health care systems throughout the world [8].

The objective of this paper is to provide a review of phytochemical studies that have addressed extracting, measuring and identifying bioactive compounds of plants. This review includes an overview of the lipid oxidation process, details of plants known to be antioxidant and antimicrobial sources, phenolic compounds, antioxidants from vegetables and fruits, cancer prevention, extraction techniques for phenolic compounds, isolation and purification of bioactive molecules, and techniques for structural classification of bioactive molecules.

2. Methods Used for Bioactive Compound Extraction, Isolation, and Purification

2.1. Extraction of Phenolic Compounds Using Solvents

Scientists have studied and analyzed the impact of different types of solvents, such as methanol, hexane, and ethyl alcohol, for the purpose of antioxidant extraction from various plants parts, such as leaves and seeds. In order to extract different phenolic compounds from plants with a high degree of accuracy, various solvents of differing polarities must be used [9]. Moreover, scientists have discovered that highly polar solvents, such as methanol, have a high effectiveness as antioxidants.

Anokwuru et al. reported that acetone and N,N dimethylformamide (DMF) are highly effective at extracting antioxidants, while Koffi et al. found that methanol was more effective in at a large amount of phenolic contents from walnut fruits when compared to ethanol [10–12].

It has been reported that ethanolic extracts of Ivorian plants extracted higher concentrations/amount of phenolics compared to acetone, water, and methanol [11]. Multiple solvents have been commonly used to extract phytochemicals, and scientists usually employed a dried powder of plants to extract bioactive compounds and eliminate the interference of water at the same time.

Solvents used for the extraction of biomolecules from plants are chosen based on the polarity of the solute of interest. A solvent of similar polarity to the solute will properly dissolve the solute. Multiple solvents can be used sequentially in order to limit the amount of analogous compounds in the desired yield. The polarity, from least polar to most polar, of a few common solvents is as follows: Hexane < Chloroform < Ethylacetate < Acetone < Methanol < Water.

2.2. Microwave-Assisted Extraction (MAE)

MAE has attracted the attention of researchers as a technique to extract bioactive compounds from a wide variety of plants and natural residues [12]. Microwaves have electromagnectic radiation that occurs at frequencies between 300 MHz to 300 GHz, and wavelengths between 1 cm and 1 m. These electromagnetic waves consist of both an electrical field and a magnetic field. These are described as two perpendicular fields. The first application of microwaves was to heat up objects that can absorb a part of the electromagnetic energy and convert it into heat. Commercial microwave instruments commonly use the frequency 2450 MHz, which corresponds to an energy output of 600–700 Watts [13].

Recently, advanced techniques have become available to reduce the loss of bioactive compound without increasing the extraction time. Therefore, microwave-assisted extraction is demonstrated to be a good technique in multiple fields, especially in the medicinal plant area. Moreover, this technique reduced the losses of the biochemical compounds being extracted [14]. Microwave-assisted extraction (MAE) has been used as an alternative to conventional techniques for the extraction of antioxidants because of its ability to reduce both time and extraction solvent volume [15]. In fact, the main objective of using MAE is to heat the solvent and extract antioxidants from plants with a lesser amount of these solvents [13].

Li et al. reported that conventional methods using various solvents presented less antioxidant activity and phenolic content than MAE [16]. Therefore, the finding confirmed that MAE was more effective at increasing antioxidant activity by measuring ferric reducing antioxidant power (FRAP), oxygen radical absorbance capacity (ORAC), and total phenolic content (TPC). The efficiency of the microwave extraction can be changed through some factors such as extraction temperature, solvent composition, and extraction time. The extraction temperature was usually studied more than other factors due to its ability to increase the efficiency of the microwave extraction. Tsubaki et al. reported that 170 °C was the most effective temperature for extracting phenolic compounds from Chinese tea. In addition, increasing the extraction temperature beyond this point resulted in a reduced extraction yield [17]. Recently, Christophoridou et al. used a new microwave-assisted extraction (MAE) process, which converts energy to heat, thereby cooperating with solvents in order to extract a specific compound [18]. Williams et al. showed many advantages of MAE, including lower solvent consumption, shorter extraction times, and higher sensitivity towards target molecules [19]. A comparison of some antioxidant methods used has been provided in Table 1.

Table 1. Comparison of methods for assessing antioxidant capacity based upon mechanism, endpoint, quantitation method, and whether the assay is adaptable to measure lipophilic and hydrophilic antioxidants.

Antioxidant Assay	Mechanism	Endpoint	Quantification	Lipophilic and Hydrophilic AOC
ORAC	HAT	Fixed time	AUC	Yes
TRAP	HAT	Lag phase	IC50 lag time	No
FRAP	SET	Time varies	ΔOD fixed time	No
TEAC	SET	Time varies	ΔOD fixed time	Yes
DPPH	SET	IC50	ΔOD fixed time	No
LDL oxidation	SET	Lag phase	Lag time	No

2.3. Ultrasonic-Assisted Extraction

Ultrasound-assisted extraction (UAE) has been used in diverse applications of food-processing technology to extract bioactive compounds from plant materials [19]. Ultrasound, with levels greater than 20 kHz, is used to disrupt plant cell walls, which helps improve the solvent's ability to penetrate the cells and obtain a higher extraction yield. UAE can use a low operating temperature through processing, maintaining a high extract quality for compounds. UAE is known to be one of the easiest extraction techniques because it uses common laboratory equipment such as an ultrasonic bath. In this technique, a smashed sample is mixed with the suitable solvent and placed into the ultrasonic bath, while temperature and extraction time are controlled [20].

UAE of various organic and inorganic samples can use a wide range of solvents. Common equipment used in ultrasound-assisted extraction includes an ultrasonic bath and an ultrasonic probe system. Unfortunately, ultrasonic probe has two main negative properties mainly related to experimental repeatability and reproducibility [21].

Tabaraki et al. noted that green technology is necessary to protect the environment from toxic substances [22]. Therefore, extraction of phenolic compounds by ultrasound has grown during recent years due to its role in reducing the amount of solvent and energy used. Corrales et al. have shown that UAE can break down plant tissue and work properly during the production process and release of active compounds in solvents with a high efficiency [21]. Results showed an increase in antioxidant activity from 187.13 µmol TE g^{-1}DM to 308 µmol TE g^{-1}DM by using UAE as an effective method to extract antioxidants from different sources. Recently, Albu et al. studied and applied the use of ultrasound to extract phenolic compounds from rosemary [23]. Multiple criteria have been compared including ultrasonic bath extractions, ultrasonic probe system, a shaking water bath at various temperatures, and different solvents to select the most efficient method. In all situations, the operation time was dramatically decreased by applying and using the ultrasonic bath and probe systems.

Similar behavior was reported by Cho et al. when extracting resveratrol from grapes [24]. In another study, Barbero et al. suggested the use of ultrasound in different industries because of its positive effects in the extraction of capsaicinoids of hot peppers [25]. Moreover, the ultrasonic method had the ability to decrease the degradation of phenolics [26]. Mulinacci et al. compared the extraction time of phenolic compounds from strawberries with other extraction methods such as solid–liquid, subcritical water, and microwave-assisted method [27]. The results confirmed that UAE was the most effective method.

2.4. Techniques of Isolation and Purification of Bioactive Molecules from Plants

Purification and isolation of bioactive compounds from plants is a technique that has undergone new development in recent years [28,29]. This modern technique offers the ability to parallel the development and availability of many advanced bioassays on the one hand, and provided precise techniques of isolation, separation, and purification on the other. The goal when searching for bioactive compounds is to find an appropriate method that can screen the source material for bioactivity such as antioxidant, antibacterial, or cytotoxicity, combined with simplicity, specificity, and speed [27].

In vitro methods are usually more desirable than in vivo assays because animal experiments are expensive, take more time, and are prone to ethical controversies. There are some factors that make it impossible to find final procedures or protocols to isolate and characterize certain bioactive molecules. This could be due to different parts (tissues) in a plant, many of which will produce quite different compounds, in addition to the diverse chemical structures and physicochemical properties of the bioactive phytochemicals [30]. Both the selection and the collection of plant materials are considered primary steps to isolate and characterize a bioactive phytochemical. The next step involves a retrieval of ethno-botanical information to discern possible bioactive molecules. Extracts can then be made with various solvents to isolate and purify the active compounds that are responsible for the bioactivity. Column chromatographic techniques can be used for the isolation and purification of the bioactive compounds. Developed instruments such as High Pressure Liquid Chromatography (HPLC) accelerate the process of purification of the bioactive molecule. Different varieties of spectroscopic techniques like UV-visible, Infrared (IR), Nuclear Magnetic Resonance (NMR), and mass spectroscopy can identify the purified compounds [31].

2.5. Purification of the Bioactive Molecule

Many bioactive molecules have been isolated and purified by using paper thin-layer and column chromatographic methods. Column chromatography and thin-layer chromatography (TLC) are still mostly used due to their convenience, economy, and availability in various stationary phases [32]. Silica, alumina, cellulose, and polyamide exhibit the most value for separating the phytochemicals. Plant materials include high amounts of complex phytochemicals, which make a good separation difficult [32]. Therefore, increasing polarity using multiple mobile phases is useful for highly valued separations. Thin-layer chromatography has always been used to analyze the fractions of compounds by column chromatography. Silica gel column chromatography and thin-layer chromatography (TLC) have been used for separation of bioactive molecules with some analytical tools [32].

2.6. Structural Clarification of the Bioactive Molecules

Determination of the structure of certain molecules uses data from a wide range of spectroscopic techniques such as UV-visible, Infrared (IR), Nuclear Magnetic Resonance (NMR), and mass spectroscopy. The basic principle of spectroscopy is passing electromagnetic radiation through an organic molecule that absorbs some of the radiation, but not all. By measuring the amount of absorption of electromagnetic radiation, a spectrum can be produced. The spectra are specific to certain bonds in a molecule. Depending on these spectra, the structure of the organic molecule can be identified. Scientists mainly use spectra produced from either three or four regions—Ultraviolet (UV), Visible, Infrared (IR), radio frequency, and electron beam [31]—for structural clarification.

2.7. UV-Visible Spectroscopy

UV-visible spectroscopy can be performed for qualitative analysis and for identification of certain classes of compounds in both pure and biological mixtures. Preferentially, UV-visible spectroscopy can be used for quantitative analysis because aromatic molecules are powerful chromophores in the UV range. Natural compounds can be determined by using UV-visible spectroscopy [33]. Phenolic compounds including anthocyanins, tannins, polymer dyes, and phenols form complexes with iron that have been detected by the ultraviolet/visible (UV-Vis) spectroscopy [34]. Moreover, spectroscopic UV-Vis techniques were found to be less selective and give information on the composition of the total polyphenol content. The UV-Vis spectroscopy was used to determine the total phenolic extract (280 nm), flavones (320 nm), phenolic acids (360 nm), and the total anthokyanids (520 nm). This technique is not time-consuming, and presents reduced cost compared to other techniques [35].

2.8. Infrared Spectroscopy

Some of the frequencies will be absorbed when infrared light passes through a sample of an organic compound; however, some frequencies will be transmitted through the sample without any absorption occurring. Infrared absorption is related to the vibrational changes that happen inside a molecule when it is exposed to infrared radiation. Therefore, infrared spectroscopy can essentially be described as a vibrational spectroscopy. Different bonds (C–C, C=C, C≡C, C–O, C=O, O–H, and N–H) have diverse vibrational frequencies. If these kinds of bonds are present in an organic molecule, they can be identified by detecting the characteristic frequency absorption band in the infrared spectrum [35]. Fourier Transform Infrared Spectroscopy (FTIR) is a high-resolution analytical tool to identify the chemical constituents and elucidate the structural compounds. FTIR offers a rapid and nondestructive investigation to fingerprint herbal extracts or powders.

2.9. Nuclear Magnetic Resonance Spectroscopy (NMR)

NMR is primarily related to the magnetic properties of certain atomic nuclei; notably the nucleus of the hydrogen atom, the proton, the carbon, and an isotope of carbon. NMR spectroscopy has enabled many researchers to study molecules by recording the differences between the various magnetic nuclei, and thereby giving a clear picture of what the positions of these nuclei are in the molecule. Moreover, it will demonstrate which atoms are present in neighboring groups. Ultimately, it can conclude how many atoms are present in each of these environments [33]. Several attempts have been made in the past by using preparative or semi preparative thin-layer chromatography, liquid chromatography, and column chromatography to isolate individual phenols, the structures of which are determined subsequently by NMR off-line [34].

2.10. Mass Spectrometry for Chemical Compounds Identification

Organic molecules are bombarded with either electrons or lasers in mass spectrometry and thereby converted to charged ions, which are highly energetic. A mass spectrum is a plot of the relative abundance of a fragmented ion against the ratio of mass/charge of these ions. Using mass spectrometry, relative molecular mass (molecular weight) can be determined with high accuracy and an exact molecular formula can be determined with a knowledge of places where the molecule has been fragmented [18]. In previous work, bioactive molecules from pith were isolated and purified by bioactivity-guided solvent extraction, column chromatography, and HPLC [36]. The techniques of UV-visible, IR, NMR, and mass spectroscopy were employed to characterize the structure of the bioactive molecule. Furthermore, molecules may be hydrolyzed and their derivatives characterized. Mass spectrometry provides abundant information for the structural elucidation of the compounds when tandem mass spectrometry (MS) is applied. Therefore, the combination of HPLC and MS facilitates rapid and accurate identification of chemical compounds in medicinal herbs, especially when a pure standard is unavailable [37–40]. Recently, LC/MS has been extensively used for the analysis of phenolic compounds. Electrospray ionization (ESI) is a preferred source due to its high ionization efficiency for phenolic compounds.

3. Lipid Oxidation

Lipid oxidation can occur during the processing, shipping, and storing of many foods. Lipids (such as triglycerides, sterols, and phospholipids) readily become oxidized with exposure to an oxidative environment [41]. Lipid molecules, especially those carrying polyunsaturated double bonds (i.e., linolenic acids), readily undergo oxidation within foods. Oxidatively stable oil with high melting temperature is necessary for solid fat application, and thus, highly saturated seed oil (palmitic acid and stearic acid) would be suitable for this end use [42]. Soybeans provide 56% of the world's oilseed production. However, the percentage of saturated oil is very low in seed plants (about 10%), if compared to unsaturated oil (about 90%) [43]. Palmitic acid improves the oxidative stability of soybean oil, and can also be used to produce *trans*-fat-free shortening, margarine, and cosmetic products. However, this saturated short-chain fatty acid is undesirable for nutrition because its consumption results in an unfavorable lipoprotein profile in blood serum [44]. Stearic acid does not exhibit these cholesterolemic effects on human health [45]. Stearic acid is less likely to be incorporated into cholesterol esters and has a neutral effect on the concentration of blood serum LDL cholesterol [46,47]. Extensive research has been performed in order to increase stearic acid content oil production in the most widely consumed legume crop in the world, soybeans. By employing induced mutagenesis, seed stearic acid content was increased by up to 7 times [48].

Lipid oxidation in food systems can be caused by oxygen free radicals or reactive oxygen species. Free radicals are molecules with one or more unpaired electrons that work independently to cause oxidation [49]. Reactive oxygen species are a perfect example of oxygen free radicals. Reactive oxygen species do not solely contain free radical molecules, but also some non-free radicals that can influence lipid oxidation. Examples of non-free radical reactive oxygen species are hydrogen peroxide (H_2O_2), hydrochloric acid (HCl), ozone (O_3), and molecular oxygen (O_2) [42]. Molecular oxygen can react with linoleic acid about 1450 times faster than triplet oxygen. One of the major causes of oil rancidity is molecular oxygen. Lipid oxidation caused by the chain reaction of free radicals can be illustrated in three stages: initiation, propagation, and termination [42]:

(1) Initiation:

RH + initiator → R

ROOH + initiator → ROO•

(2) Propagation:

R + O_2 → ROO

ROO + RH → ROOH + R•

(3) Termination:

R + R → R-R

ROO• + R → ROOH

The processes above occur in response to several physical or chemical factors including heating, radiation, temperature, metal ion catalysts, reactive oxygen species, and photosensitizers such as chlorophyll. The initiation step, shown in Equation (1), often happens at either an allytic methylene group of an unsaturated fatty acid (RH) or a lipid-hydroperoxide (ROOH). Next, the generated free radical (R•) reacts with oxygen to form a peroxy radical (ROO•). This product can directly react with another lipid molecule to produce a lipid hydroperoxide (ROOH), and thus a lipid free radical (R•). This causes continuously cascading chain reactions to occur until the free radicals are neutralized by other free radicals. This whole stage is shown in Equation (2). In the termination step, there are two radicals that have converted into non-free radical products, and thus will stop the cascade mode of the chain reaction according to Equation (3). Moreover, the reaction chain can also be terminated by some antioxidants or free radical scavengers. Metal ions, especially those of iron and copper, effectively catalyze these reactions [50].

Lipoxygenases (EC 1.13.11) can also act, causing oxidation to produce the peroxides in food materials that contain lipids. Hydrogen peroxide is one of the primary products of the oxidation, and it is very unstable and easily converts into secondary products. The final product of oxidation may include different chemical groups such as aldehydes, ketones, alcohols, acids, or hydrocarbons. These kinds of compounds can have a negative effect on the appearance, quality, and edibility of a food product by changing the texture, color, flavor, and safety of foods, or also by producing unacceptable off odors or off tastes, even negatively affecting the nutritional value [50].

4. Plants as a Source of Antioxidants

Antioxidants can be defined as bioactive compounds that inhibit or delay the oxidation of molecules [42]. Antioxidants are categorized as natural or synthetic antioxidants. Some synthetic antioxidants commonly used are: BHT, BHA, propyl gallate, and tertbutylhydroquinine. Many scientists have concerns about safety because synthetic antioxidants have recently been shown to cause health problems such as liver damage, due to their toxicity and carcinogenicity. Therefore, the development of safer antioxidants from natural sources has increased, and plants have been used as a good source of traditional medicines to treat different diseases. Many of these medicinal plants are indeed good sources of phytochemicals that possess antioxidant activities. Some typical examples of common ingredients that have been used in ethnic foods are tamarind, cardamom, lemon grass, and galangal basil. These spices or herbs have been shown to contain antioxidants [51].

Deterioration of food due to either bacterial or fungal infection has always been a major concern, causing huge losses to food industries and societies throughout the world [51]. Moreover, the spread of food pathogens has become a major public health concern. With an increasing awareness of the negative effects of synthetic preservatives, there has been increased demand for the use of nontoxic, natural preservatives, many of which are likely to have either antioxidant or antimicrobial activities [52,53]. Herbs have always been used for flavor and fragrance in the food industry, and some of them have been found to exhibit antimicrobial properties [54]. Therefore, the call for screening and using plant materials for their antioxidant and antimicrobial properties has increased. Approximately 20% of all plant species have been tested in both pharmacological and biological applications to confirm their safety and advantages [3]. A summary of the types of compounds, plant species, plant parts from which compounds were extracted, etc. can be found in Table S1.

4.1. Presence of Antioxidant in Red Algae

Red algae are aquatic plant species considered one of the oldest groups of eukaryotic algae [55]. The antioxidant activity of a red alga, *Palmaria palmate*, has been studied. The results reported that 9.68 µg of ascorbic acid and 10.3 µg of total polyphenol can equally reduce activity in 1 mg of dulse extracts. The reducing activity was correlated with aqueous/alcohol soluble compounds due to the presence of functional groups such as hydroxyl, carbonyl, etc., which lead to reduced or inhibited oxidation [56].

4.2. Antioxidants from Monocots

Ashawat et al. studied the antioxidant properties of ethanolic extracts of *Areca catechu* and showed that *Areca catechu* had the highest antioxidant activity when compared to other eudicots like *Centella asiatica, Punica granatum*, and *Glycyrrhiza glabra* [57]. Londonkar and kamble studied *Pandanus odoratissimus* L. in order to determine its antioxidant activity [58]. Zahin et al. screened *Acorus calamus* to estimate antioxidant activity and total phenolic contents [59]. The observations confirmed that there was a significant correlation between the phenolic content and antioxidant activity. Another monocot, *O. sanctum*, showed that the inhibition of lipid peroxidation in vivo and in vitro increased proportionally with an increase in the concentration of the extract.

4.3. Antioxidants from Vegetables

Consumption of vegetables has been linked to a reduction in the risk of many diseases, such as cancer and cardiovascular disease, when studied in epidemiological studies [59]. Numerous studies have attempted to screen vegetables for antioxidant activity by using different oxidation systems.

These vegetables include carrots, potatoes, sweet potatoes, red beets, cabbage, Brussels sprouts, broccoli, lettuce, spinach, onions, and tomatoes. In addition to the concise studies, which have used different methodologies to release bioactive compounds, it is becoming increasingly difficult to ignore advanced extraction methods, which have paved the way to extract bioactive compounds rapidly. Despite scientists' successes in showing the activity of vegetables' bioactive compounds, there is little known about the activity of the antioxidant components that have been isolated from these vegetables. Researchers have tended to focus on advanced methods to isolate and measure the activity of antioxidant compounds such as flavonoids, phenolic acids, tocopherols, carotenoids, and ascorbic acid [60].

4.4. Antioxidants from Fruits

Fruit consumption has also been linked to a reduction in the risk of many diseases [61]. Peaches (*Prunus persica* L.) are an economically important fruit in many countries. Studies have shown that phenolic compounds found within various peach genotypes are a major source of potential antioxidants [60]. Interestingly, peaches have shown a great inhibition of low density lipoprotein (LDL) oxidation with a percentage of antioxidant activity of 56–87%. This antioxidant activity can be attributed to its essential compound content including hydroxycinnamic acids, chlorogenic, and neochlorogenic acids, but not to carotenoids such as b-carotene and b-cryptoxanthin. Moreover, low antioxidant activity was found in peach peel. In contrast, Plumb et al. pointed out that hydroxycinnamic acids do not contribute to the inhibition of lipid peroxidation of the liver using plums and peaches because hydroxycinnamic acids had weak ability to scavenge hydroxyl radicals [62].

Grape (*Vitis vinifera* L.) is a fruit crop grown throughout the world. Grapes and its juices have been recently studied by [62]. Phenolic compounds were high in both fresh grapes and commercial grape juices. The percentage of inhibition LDL oxidation was about 22% to 60% for fresh grapes, while it was approximately 68% to 75% for commercial grape juices, when standardized at 10 mg gallic acid equivalents (GAE). According to [63], both grapes and its juices exhibited high oxygen radical absorbance capacity (ORAC), and the anthocyanin pigment malvidin-3,5-diglucoside was a major compound isolated in grapes. Anthocyanins with malvidin nucleus malvidin 3-*O*-(6-*O*-p-coumaroylglucosido)-5-glucoside and phenolics were common compounds isolated from wild grapes (*Vitis coignetiae*). Wangensteen et al. tested the activity of many bioactive compounds by releasing them from grape pomace, and demonstrated that bioactive compounds have the ability to significantly inhibit LDL oxidation in the human body [64]. Grape seeds are an amazing source of polyphenol compounds including monomerics such as catechin, epicatechin, and gallic acid, and polymerics such as procyanidins [65].

Both polyphenols and carotenoids are the major phenolic compounds of apples (*Malus domestica* L.) including caffeic, quinic, and p-coumaric acids. These polyphenols can act as antioxidants. Flavanol monomers and oligomers, as well as quercetin, contribute to the beneficial health aspects of fruits and vegetables [65]. Apple pomace has mainly been used as a major source of polyphenols such as chlorogenic acid [66,67]. In addition phenolics like caffeic, p-coumaroyl quinic, arbutin, p-coumaric acids, and especially flavonol procyanidins have been mentioned as constituents of apple pomace [68]. The ability of procyanidins to work as oxygen radical scavengers, superoxides, and hydroxyl radicals was estimated. Despite the low content in total phenols in apples obtained by using acetone 70%, it has shown strong antioxidant activities towards oxidation of linoleic acid. In this case, the major bioactive compounds obtained were chlorogenic acid and phloretin glycosides; however, Vitamin C was a minor fraction in apple juice [69].

Antioxidant and antibacterial activities of various solvent (ethyl acetate, acetone, methanol, and water) extracts of *Punica granatum* peel were examined by applying the 2,2-diphenyl-1-picrylhydrazyl (DPPH) radical scavenging method. The results obtained showed a significantly higher decreasing power in the methanol extracts and a significantly higher antibacterial activity in the acetone extracts.

Soong and Barlow investigated the antioxidant activity and phenolic content of various fruit seeds [70]. Petroleum ether was used to get rid of the excess fat from the seeds and extraction has been carried out with methanol. The 2,2-azino-bis-3-ethylbenzthiazoline-6-sulfonic acid (ABTS), DPPH, and the ferric reducing ability of plasma (FRAP) methods were used to investigate the antioxidant activity. Abdille et al. examined the antioxidant activity of *Dillenia indica* fruit using different kinds of solvents using DPPH, phospho-molybdenum, and β carotene bleaching methods [71]. The methanol extracts showed the highest antioxidant activity, followed by the ethyl acetate and water extracts. Antioxidant activity of *Syzygium cumini* fruit in vitro has been investigated [71]. Antioxidant activity was measured by DPPH, superoxide, lipid peroxidation, and hydroxyl radical scavenging activity methods. The results brought to light a significant correlation between the concentration of the extract and the percentage of inhibition of free radicals. The antioxidant property of the fruit might be from the presence of antioxidant vitamins, anthocyanins, phenolics, and tannins. It has been reported that blackberry (*Rubus fruticosus* L.) fruit extracts produced in varying climatic regions showed that antioxidant activity depended on the genotype, rather than the climate or season [10]. Juntachote and Berghofer measured the stability of the antioxidant activity of ethanolic extracts for Holy basil and galangal using DPPH, superoxide, β carotene bleaching, reducing power, and iron chelation methods [72]. They found higher antioxidant activity at neutral pH compared to an acidic pH. Holy basil and galangal extracts provided strong iron chelation activity, superoxide anion scavenging activity, and reducing power proportional to the concentration of the extracts. Liyana-Pathirana et al. investigated the antioxidant activity of cherry laurel fruit (*Laurocerasus officinalis* Roem) and its concentrated juice (Pekmez) using in vitro methods such as superoxide, DPPH scavenging activity, and inhibition of LDL oxidation [73]. The results confirmed the presence of a significantly higher antioxidant activity in pekmez compared to the cherry laurel fruit. Employing in vitro methods such as DPPH and superoxide scavenging activity, Orhan et al. measured the antioxidant activity of *Arnebia densiflora* Ledeb and observed that polar extracts had a higher antioxidant activity compared to non-polar extracts [74]. Rathee et al. studied the antioxidant activity of *Mammea longifolia* buds extracted in both methanol and aqueous ethanol. The results found a significant antioxidant activity, and the activity of aqueous ethanol was higher than methanol. The antioxidant activity of leaf extracts of *Annona* species in vitro reveals that *Annona muricata* possessed a higher antioxidant activity compared to *Annona squomosa* [75].

4.5. Cooking Herbs as an Important Source of Antioxidants

The antioxidant activity of 32 herbs belonging to 21 different families has been screened [76]. The finding confirmed that there was a positive correlation between the total antioxidant activity and total phenolic content. Lu and Yeap Foo studied *Salvia officinalis* (L.) for its antioxidant activity and polyphenol content and reported that rosmarinic acid and various catechols were responsible for the radical scavenging activity and caffeic acid was responsible for the xanthine oxidase (EC 1.17.3.2) inhibition [77]. Zhao et al. investigated the antioxidant activity of *Salvia miltiorrhiza* and *Panax notogensing* [78]. The results showed that *Salvia miltiorrhiza* had a higher reducing power and scavenging activities against free radicals, including superoxide and hydroxyl radicals, although it showed weak hydrogen peroxide scavenging.

Furthermore, Javanmardi et al. tested the Iranian *Ocimum* sp. accessions to determine the antioxidant activities and total phenolic contents and demonstrated that the antioxidant activity increased in parallel with the total phenolic content [51].

Evaluation of the pomegranate peel extracts to discover its antioxidant and antimutagenic activities using different solvents such as ethyl acetate, acetone, methanol and water has been carried out [51]. Dried extracts were examined by using the Ames test and the phosphorus-molybdenum method to test both anti-mutagenic and antioxidant activities. The results showed the highest anti-mutagenic and the lowest antioxidant activity in the water extract.

Moreover, the phenolic content and antioxidant activity of parsley (*Petroselinum crispum*) and cilantro (*Coriandrum sativum*) have been tested [79]. The total phenolic content was observed to be different between parsley and cilantro leaves and stems, as well as methanol and water extracts. The methanol leaf extracts exhibited significant antioxidant activity towards both lipid- and water-soluble radicals. The works also investigated the antioxidant activity of aqueous plant extracts using in vitro methods such as DPPH scavenging activity and FRAP. The results revealed a strong correlation between total antioxidant activity and phenolic content and a weak correlation between cupric ion chelators and polyphenols. The antioxidant activity and lipid peroxidation inhibition of *Satureja montana* L. subsp. *Kitaibelii* extracts were tested using hydroxyl radical scavenging. The results obtained showed that there was a significant correlation with total phenolic content [9].

4.6. Antioxidant from Legumes

Antioxidant property of methanol extracts of *Mucuna pruriens* L. (*Fabaceae*) seed extracts has been investigated in vitro using the DPPH radical scavenging method. The results obtained showed a positive correlation between the antioxidant activity and the total phenolic compounds [80]. Siddhuraju and Manian studied horsegram (*Macrotyloma uniflorum* Lam.) seeds to measure the antioxidant and free radical scavenging activity [81]. Acetone extracts had a higher activity of about 70% [81]. Samak et al. studied *Wagatea* sp. to measure its scavenging activities of superoxide and hydroxyl radicals and showed a high oxidation inhibition because it was rich in both phenolic and flavonoid contents. The authors also reported that bark and leaf extracts of *Wagatea* sp. exhibited high scavenging action against super radicals [82].

4.7. Antioxidants from Trees

Antioxidants from trees have been also measured. Phenolics from almond hulls (*Prunus amygdalus* L.) and pine sawdust (*Pinus pinaster* L.) have been extracted employing various methods in order to determine the gram fresh yield of polyphenol compounds and antioxidant activity [83]. The antioxidant activity was measured by the DPPH radical scavenging method. The results showed that ethanol was most appropriate either for phenolics or any bioactive compounds, while methanol was more selective for extracting polyphenolics. The antioxidant activity of juniper (*Juniperus communes*) fruit extracts has been investigated in vitro [84]. The results confirmed that both water and ethanol extract showed strong antioxidant activity. The concentration of 60 µg/mL of water and ethanol extracts exhibited 84% and 92% inhibition, respectively, on the peroxidation of linoleic acid. Ibrahim et al. studied the antioxidant activity of *Cupressus sempervirens* L., and set up goals to isolate quercetin, rutin, cupress flavone, caffeic acid, and para-coumaric acid. The results showed higher antioxidant activity related to quench DPPH and identified these active compounds successively [85].

Higher values of antioxidant activity have been obtained by using a methanolic solvent to extract the bioactive compounds from *Anacardium occidentale*, while other solvents like ethyl acetate gave lower values of antioxidant activity [85]. Kaur et al. studied the Chickrassy *Chukrasia tabularis* A. Juss leaves to confirm its ability to inhibit lipid peroxidation and showed that there was a large inhibition considering its high content of phenolic compounds [86]. Finally, *Acacia nilotica* L. antioxidant activity has been measured using ethyl acetate as a solvent to extract phenolic compounds [86]. The results exhibited the highest antioxidant activity when the concentration of extracts was relatively high.

4.8. Antioxidant from Shrubs

Many shrubs have been shown to contain antioxidant activity. Singh et al. tested several plants to measure the antioxidant activity from different extracts. The antioxidant activity was determined by using peroxide value, thiobarbituric acid, DPPH radical scavenging activity, and reducing power. The results showed that the antioxidant activity of *Coriandrum sativum* L. and *Sarcolobus globosus* L. exhibited high activity by using acetone solvent, and its activity was similar to synthetic antioxidants [87].

Eleven Algerian medicinal plants have been measured for phenolic compound content and antioxidant activity using the ABTS method. The tested plants showed antioxidant activity. *Artemisia campestris* L. had better antioxidant activity than caffeic acid and tocopherol. Moreover,

HPLC analyses exhibited a good correlation between the antioxidant activity and hydroxycinnamic derivative content.

Evaluation of *Vitex negundo* Linn seed antioxidant activity using different methods such as superoxide, hydroxyl, and DPPH scavenging activity has been carried out [87]. The highest antioxidant activity was in both raw and dry heated seed extracts, while lower antioxidant activity was observed in the hydrothermally processed samples.

4.9. Characterization of Antioxidants from Other Eudicots

The nitric oxide and superoxide scavenging activity of green tea have been studied by Nakagawa and Yokozawa [88], who concluded that certain tannins had the ability to exhibit excellent antioxidant activity. Zin et al. estimated the antioxidant activity of the extracts from various parts of Mengkudu (*Morinda citrifolia* L.), including the leaves, fruits, and roots, using different solvents such as methanol and ethyl acetate [89]. Ferric thiocyanate and thiobarbituric acid were used as models to observe and evaluate the antioxidant activity. The results exhibited a higher antioxidant activity in the methanol extract of Mengkudu root, although it was not significantly different from tocopherol and BHT extracts. The methanol extracts of the fruits and leaves showed unassuming activity. According to these scientists, the antioxidant activity in the roots resulted from polar and non-polar compounds, but the antioxidant activity in leaves and fruits was only due to non-polar compounds.

Increase of the antioxidant activity of fennel (*Foeniculum vulgare*) seed extracts in vitro has been shown to be proportional to the increase in the concentration of extract [89]. Nine other extracts of Bolivian plants have been measured for radical scavenging and antioxidant activity using the DPPH and β carotene bleaching methods [90]. It was found that the ethyl acetate fractions had higher radical scavenging and antioxidant activity compared to the other extracts. It has been reported that the bioactive compounds of *Rhodiola rosea* extracted in methanol showed a significant yield of phenolics, about (153 ± 2 mg/g) [91]. Wangensteen et al. investigated the antioxidant activity of *Ss globosus* using DPPH scavenging and inhibition of lipoxygenase [64]. Coriander had a high capacity to inhibit oxidation. There was also a positive correlation between total phenolics and antioxidant activity. Moreover, it was observed that the leaves of the coriander had higher antioxidant activity than the seeds [91].

Antioxidant activity of *Phyllanthus niruri* was estimated using methanol and water as a solvent. The extracts of leaves and fruits exhibited high antioxidant activity by using the inhibition of lipid peroxidation and DPPH scavenging [64]. The results also noticed a higher superoxide scavenging activity in the aqueous extract compared to the methanol extract. Moreover, the antioxidant and free radical scavenging activity of *Phyllantus* species from India in an aqueous extract has been also evaluated [92]. The antioxidant activity was estimated using DPPH, β carotene, superoxide, nitric oxide scavenging, and reducing power methods. The extract of *Coleus aromaticus* exhibited a moderate inhibition on DPPH and nitric oxide scavenging activity.

Panax exhibited strong iron chelating and weak superoxide scavenging. Ajila et al. carried out bioactive compounds and antioxidant activity of mango peel extract [93]. The results showed a higher concentration of anthocyanins and carotenoids in the ripe peel compared to the raw peel, while the raw peel exhibited higher polyphenol content. The range of IC_{50} values of lipid peroxidation and DPPH were 1.39–5.24 μg of gallic acid equivalent. Chen and Yen investigated the antioxidant activity and free radical scavenging capacity of dried guava leaves and fruit [94]. The results confirmed that guava leaf and guava tea extracts had the ability to inhibit oxidation by 94–96% at a concentration of 100 μg/mL. Fruit extracts exhibited less activity compared to leaf extracts, while the scavenging effect increased with an increase in the concentration. Also, there was a correlation between antioxidant activity and phenolic compounds. Dastmalchi et al. investigated the chemical composition and antioxidant activity of water-soluble Moldavian balm (*Dracocephalum moldavica*) in vitro by using DPPH, ABTS, and superoxide activity [95]. The finding confirmed that polar compounds such as caffeic acid and rosmaric acid were responsible for the antioxidant activity observed.

Mulberry leaves were investigated to determine the antioxidant activity using different solvents [95]. The procedure used DPPH and inhibition of lipid peroxidation methods to evaluate its activity. The results showed that the methanolic extract exhibited the highest yield of total phenolics, and it was the most essential antioxidant in all the methods used. The antioxidant activity of kale (*Brassica obraceae* L.) has been screened after removing a fat fraction from the samples [96]. The extraction process used methanol to investigate its antioxidant activity while using DPPH scavenging activity as tested method. The works successfully isolated nine phenolic acids using HPLC and MS, and confirmed that the total phenolic content was correlated with DPPH scavenging activity.

In another study, ethanol has been used to estimate the antioxidant activity of sun-dried cashew nuts (*Anacardium occidentale* L.) skin [97]. First, bioactive compounds were extracted with a protocol including lipid peroxidation, ABTS, and DPPH methods to measure the capability to inhibit oxidation. The results found that epicatechin was the major polyphenol in the extract, which was responsible for antioxidant activity.

Kaviarasan et al. measured the antioxidant and antiradical activity of fenugreek (*Trigonella foenum* ssp. *graecum*) seeds in vitro; the results showed that there was a positive relationship between the antiradical activity and phenolic compound content in the extract [98]. Hexane and methanol were used to extract the bioactive compounds and measured the antioxidant activity of *Pueraria tuberosa* by using ABTS, lipid peroxidation, and superoxide and hydroxyl scavenging activity. An independent study has shown an inhibition of the lipid peroxidation [99].

The rhizome of the lotus (*Nelumbo nucifera* Gaertn.) has been measured for its antioxidant activity in various solvent extracts using β Carotene bleaching and DPPH methods [99]. Methanol extraction had a higher DPPH scavenging activity than acetone. *Helichrysum pedunculatum* has been tested to determine the antioxidant activity, and total phenolic and flavonoid content [100]. The results demonstrated that whenever the amount of phenolic content and flavonoid content was increased, higher antioxidant activity was obtained. Meot-Duros and Magn screened the leaves of *Crithmum maritmum* to show if there was any correlation between the antioxidant activity and phenolic content and found a significant correlation between antioxidant activity and phenolic content when methanol was used as the solvent [101].

Another dicot, *Tricholepis glaberrima* L. (Asteraceae), has been investigated for antioxidant activity using different kinds of extracts [101]. Higher antioxidant activity was found by methanol, and a lower antioxidant activity in both chloroform and aqueous extracts. Sakat et al. investigated *Oxalis corniculata* L. in order to measure the antioxidant and anti-inflammatory activity employing methanol as a solvent. The IC 50 values of DPPH and nitric acid were about 93 and 73.07 µg/mL, respectively [102].

Jain et al. studied *Tabernaemontana divaricata* L. to determine the phytochemical and free radical scavenging activities in vitro. The results indicated that the antioxidant activity was the same in both ethanol and water extracts, but less in petroleum ether [103].

It has been reported that *Ascleipiadaceae* and *Periplocoideae* presented high antioxidant activity, with the presence of a strong correlation between antioxidant activity and phenolic content [103]. Laitonjam and Kongbrailatpam studied the chemical composition and antioxidant activities of *Smilax lanceafolia* by isolating the flavonol glycoside and steroidal saponin, which showed high antioxidant activity [104]. Spinach (*Spinacea olerace* L.) is among the most popular vegetables in the world. It was domesticated and first cultivated in West Asia. According to analytical chemistry, spinach is a source of violaxanthin and neoxanthin antioxidants that cannot be commercially produced [105]. Although they may be present, pigments such as carotenoids can be masked by chlorophyll in greenish vegetables such as spinach [106]. Β-carotene, lutein, violaxanthin, and neoxanthin are the major carotenoids in raw spinach [107]. Pumpkins belong to the family Cucurbitaceae. This family is classified depending on the texture and shape of stems, such as in *Cucurbita pepo*, *Cucurbita moschata*, *Cucurbita maxima*, and *Cucurbita mixta*. Nowadays, the market offers a wide variety of vegetables, with pumpkin being one of them because of its many applications for nutrition or decoration [108].

5. Plants Vitamins and Phenolic Compounds as Antioxidants

5.1. Phenolic Compounds

5.1.1. Phenols and Phenolic Acid

Phenolic acids contain carboxylic acid in the chemical composition. Hydroxycinnamic and hydroxybenzoic acids are both main pillars of phenolic acids, according to Figure 1A. Moreover, scientists have noted that p-coumaric, caffeic, ferulic, and sinapic acids are main components of the hydroxycinnamic acids (Figure 1A).

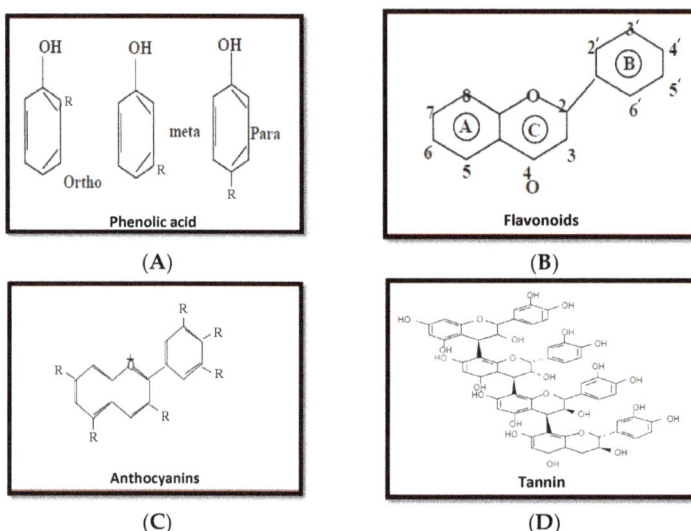

Figure 1. Chemical structure of phenolic acid (**A**), flavonoids (**B**), anthocyanins (**C**), and tannins (**D**).

5.1.2. Flavonoids

Flavonoids have a low molecular weight (Figure 1B). Flavane is an example of a flavonoid. Flavane contains two benzene rings (Figure 1A,B) within its chemical composition. These two rings connect to each other through a pyrane ring (Figure 1C). Flavones, isoflavones, flavonoids, flavonols, flavanones, anthocyanins, and proanthocyanidins are part of flavonoids according to the flavonoid classification (Figure 1B).

5.1.3. Anthocyanins

Anthocyanidins are a simple example of anthocyanins. Anthocyanidins consist of an aromatic ring that is linked to a heterocyclic ring (Figure 1C). Moreover, the heterocyclic ring is connected to the third aromatic ring through a carbon bond [109]. Scientists have noted that anthocyanins are often found in a glycoside form. Moreover, many kinds of anthocyanins are found in nature, making these kinds of phenolic compounds very complex. Scientists have noted that anthocyanins in different kinds of fruit are considered an essential compound that can enrich and increase antioxidant activity (Figure 1C).

5.1.4. Tannins

Tannins are natural products present in several plant families, and have large amounts of phenolic rings in the structure. Tannins are classified into two groups: hydrolyzable and condensed. Condensed tannins contain flavonoids units with several degrees of condensation. Hydrolyzable tannins are considered a mixture of simple phenols with ester linkages in its structure. There are many factors such as alkaline compounds, mineral acids, and enzymes that have the ability to hydrolyze tannins (Figure 1D) [110].

5.2. Vitamins Role in Cancer Prevention

Cancer has been increasing throughout the world. It is the main cause of mortality from year to year. There were 10.4 million new cancer cases registered in 2015, and scientists predict that the number of cancer cases per year will double by 2030 [111]. Recently, many studies have shown rigorous evidence that hydroxyl radicals (OH•) and the superoxide anion ($O_2^- •$) are involved in the development of cancer because they are biological reactive oxygen species. Compounds with high reactive oxygen species reduction activity are likely able to prevent cancer's occurrence [112]. As shown previously, fruits and vegetables are the primary source of natural antioxidants, consisting of different kinds of antioxidant compounds such as Vitamin C, Vitamin E, carotenoids, lutein, and lycopene. Some researchers have confirmed that phenolic compounds and polyphenols are secondary plant metabolites, which are considered the best scavengers to prevent the production of free radicals. The United States has an amazing diversity of plant species. Some of them have been used for traditional medicines for a long period of time because of their various desirable activities. Kiwi and pomegranate plants extracts were screened to show the cytotoxic effects on two tumor cell lines (*L20B* and *RD*). The results have shown that the means of both L20B and RD cultures were significantly different ($p < 0.05$), and kiwi and pomegranate plant extracts exhibited a strong ability to inhibit the growth rate of L20B and RD cell lines. At concentrations of 1000 μg/mL, both extracts showed a high ability to decrease the number of L20B and RD cells when compared with the control [113].

The mixtures of the plant natural products have been screened in order to study their effect on human leukemia cells [114]. The finding confirmed that mixtures of natural products were a good source for human leukemia cell inhibition. Nassr-allah et al. investigated the chemical diversity of natural products from plants in order to test their ability to work as anticancer and antioxidant agents [115]. DPPH assay was used to measure the antioxidant activity for plant extraction while using in vivo and vitro methods in order to measure the anticancer activity. The results confirmed that some natural products from Egyptian flora have the potential for use as therapeutics for diseases such as cancer [116].

The effectiveness of an aqueous extract from willow leaves (*Salix safsaf*, Salicaceae) against human carcinoma cells has been tested in vivo and in vitro [115]. The findings mentioned that the metabolites for the willow extract could inhibit tumors, thereby enhancing apoptosis and causing DNA damage. The anticancer activity of different extracts from the leaves of the drumstick tree (*Moringa oleifera*) was screened in order to test against leukemia and hepatocarcinoma cells in vitro. Primary cells harvested from 10 patients with acute lymphoblastic leukemia (ALL) and 15 with acute myeloid leukemia (AML) were significantly killed by hot water and ethanol extracts. Thus, *Moringa oleifera* may have the potential for use as a natural treatment for diseases such as cancer [117]. Altemimi reported that the phenolic extracts from the olive leaf extract could be used as a source of potential antioxidant and antimicrobial agents [118].

6. Plants as an Antimicrobial Source

The antibacterial activity of *Punica granatum* extracts has been investigated by using various solvents [119]. The water extract had the ability to inhibit *Bacillus subtilis* and *Staphylococcus aureus*, but the organic solvents have the ability to inhibit the growth of all the organisms tested. Shariff et al. estimated the antibacterial activity of *Rauvolfia tetraphylla* and *Physalis minima* leaves. The chloroform extract was a more powerful inhibitor of pathogenic bacteria [120].

Indian medicinal plants have been shown to have antimicrobial activity [120]. About 77 extracts belonging to these plants have been tested for their antimicrobial ability against eight species of bacteria and four species of pathogenic fungi. The findings showed that water extracts of *Lantana camara* L., *Saraca asoca* L., *Acacia nilotica* L., and *Justicia zeylanica* L. caused the highest growth inhibition of all tested bacteria. The antimicrobial activity was the highest, ranging between 9.375 and 37.5 µg/mL and 75.0 to 300.0 µg/mL against both bacterial and fungal pathogens.

Devi et al. investigated *Achyranthes bidentata* Blume to determine its phytochemical content and antibacterial activity [121]. The antibacterial ability of the ethanol extract effectively inhibited *Bacillus subtilis*, *Salmonella typhi*, and *Klebsiella pneumoniae*, but was less effective against *Pseudomonas* species and *Staphylococcus aureus* [122]. Ethanolic extracts of *Gymnema montanum* L. have been studied to measure its antimicrobial properties against *Salmonella typhi*, *Pseudomonas aeruginosa*, and *Candida albicans* [121]. The results indicated the highest presence of antimicrobial properties in the leaf extract of *G. montanum*, correlating to its phenolic compound content. The antimicrobial activity of *Piper ribesoides* L. from methanolic root extract against *Staphylococcus aureus* has been reported [123]. Interestingly, a small amount of 3.125 mg/mL was enough to inhibit harmful bacteria. Leaf extracts of *Caesalpinnia pulcherrimma* (L.) showed higher antioxidant activity in water and ethanol extracts and lower antioxidant activity in petroleum ether extracts [124]. *Torilis japonica* L. fruit has been observed to reduce the amount of spores, and the concentration of the vegetative cell was lower than the detection level. Ghosh et al. studied *Stevia rebaudiana* Bertoni to measure its antimicrobial properties against 10 pathogens [125]. The findings confirmed that *Staphylococcus aureus* was more susceptible than others [24]. Mahesh and Satish screened some important medicinal plants to show the antibacterial activity on human pathogenic bacteria [126]. Water and methanol were used as solvents to extract the phenolic compounds. The finding confirmed that the methanol extract had a higher antimicrobial activity than the aqueous extract [125].

Moreover, leaf extracts of *Acacia nilotica* L., *Sida cordifolia* L., *Tinospora cordifolia* L., *Withania somnifera* L., and *Ziziphus mauritiana* L. have been studied to determine the antibacterial activity against *Bacillus subtlis*, *Escherichia coli*, *Staphylococus aureus*, and *Pseudomonas fluorescens*, as well as studying the antifungal activity against *Aspergillus flavus*, *Dreschlera turcica*, and *Fusarium verticilloides* [126]. The highest antibacterial activity was noticed in *Acacia nilotica* and *Sida cordifoliain* leaves, and the highest antifungal activity was noticed in *Acacia nilotica* bark. Water and methanol extracts of *Samanea saman* (Jacq.) exhibited a significant effect against *Xanthomonas* spp. and human pathogenic bacteria. *Pseudarthria viscida* root has been studied to measure its antimicrobial activity using ethanol as a solvent. The results showed high antimicrobial activity when compared to standard drugs like ciprofloxacin and griseofulvin.

Ehsan et al. reported a high antimicrobial activity against *Staphylococcus aureus* using methanol and ethanol extracts for *Hopea pariviflora* Beddome [127]. Ethanolic extracts of *Bryonopsis laciniosa* have been investigated for their antimicrobial activity against different Gram-positive and Gram-negative bacteria. The growth of *Staphylococcus aureus*, *Micrococus luteus*, and *Bacillus cereus* was inhibited, as shown by a decrease in the growth zone.

Plumbago zeylanica L. has been screened to measure the antibacterial activity in chloroform extracts to show antimicrobial activity against *Escherichia coli*, *S. typhi*, and *Staphylococcus aureus* [127]. However, *Bacillus subtilis* and *Klebsiella* were resistant. Khond et al. studied 55 medicinal plants to measure the antimicrobial activity [128]. The higher antibacterial activities were in the extracts of *Madhuca longifolia* L., *Parkia biglandulosa* L., and *Pterospermum acerifolium* L. compared to the other plants screened. Pavithra et al. screened *Evolvulus nummularius* L. for its antibacterial activity, finding that *Escherichia coli* and *B. subtilis* were the most inhibited by an ethanolic extract [129].

Hygrophila spinosa Andres leaves showed significant antibacterial activity when collected between September to October, with less activity seen during other months [129]. *Artemisia pallens* L. has been studied for its antimicrobial activity against seven species of bacteria [130]. The results found that *Bacillus cereus* was more sensitive to *A. pallens* extracts. Also, a methanolic extract exhibited higher antibacterial activity than the other solvents used. Akroum indicated the antimicrobial activity of some Algerian plants [131]. The results expressed higher antibacterial activity in methanolic extracts of

Linum capitatum, Camellia sinensis, Allium schoenoprasum, Vicia faba, Citrus paradise, Lippia citriodora, Vaccinium macrocarpon, and *Punica granatum.* Bajpai et al. screened the antibacterial activity of *Pongamia pinnata* leaves by using methanol and ethyl acetate extracts to confirm its ability against certain pathogenic bacteria [132]. The results exhibited significant inhibition compared to streptomycin. It has been demonstrated that *Memecylon edule* has higher antibacterial activity in chloroform extracts compared to other extracts [132]. Gram-negative bacteria were more susceptible to the crude extracts compared to Gram-positive bacteria. Bansal et al. studied plants found in arid zones in order to determine the antibacterial efficiency [133]. An ethanolic extract of *Tinospora cordifolia* L. inhibited *Bacillus cereus* and *Staphylococcus aureus.* Kumar et al. reported *Andrographis serpyllifolia* L. to have significant antimicrobial activity against tested organisms in methanol extracts of both aerial parts and root [134].

Memecylon malabaricum, Cochlospermum religiosum, and *Andrographis serpyllifolia* have been rested for their possible antimicrobial activity [135,136]. Moderate activity against both Gram-positive and Gram-negative bacteria was observed. The antimicrobial activity of an ethanolic extract of *Anethum graveolens* was better than the aqueous extract. Khanahmadi et al. [137] found a higher antibacterial activity against Gram-positive bacteria compared to Gram-negative bacteria when an ethanolic extract of *Smyrnium cordifolium* Boiss was used [136,137]. Koperuncholan et al. studied some medicinal plants of the south eastern slopes of the Western Ghats [138]. Gram-positive bacteria were more sensitive than Gram-negative bacteria to the plant extracts. Niranjan et al. screened *Schrebera swietenioides* Roxb to measure the effectiveness against human pathogenic bacteria [139]. Water and methanol extracts were most effective to prohibit growth of all the harmful bacteria tested.

Different studies have isolated tannins and saponins from some Indian medicinal plants, testing the antibacterial activity against *Klebsiella pneumoniae* [139,140]. Ethanol extracts of *Tinospora cordifolia* strongly inhibited *Bacillus cereus* and *Staphylococcus aureus.* Also, significant antibacterial activity from ethanolic extracts of *Coleus aromaticus* L. has been found. The most effective range of inhibition was at concentrations of 25–39 µg/mL. Vinothkumar et al. evaluated a *Andrographis paniculata* L. leaf extract's ability to inhibit the growth of Gram-positive and Gram-negative bacteria. The results found that aqueous extracts inhibited harmful microbes [134].

A positive effect of pumpkin has been observed by investigating its antimicrobial activity against *Staphylococcus aureus, Bacillus subtillus, Escherichia coli,* and *Pseudomonas. aeruginosa.* Three different solvents were used to prepare the extracts: water, chloroform, and alcohol. The results showed that the alcohol extract was more powerful than both water and chloroform extracts. *Staphylococcus aureus* was sensitive to all extracts. Recently, the novel antimicrobial activity of ultrasonicated spinach leaf extracts using random amplification of polymorphic DNA (RAPD) markers and electron microscopy against both Gram-positive and Gram-negative bacteria has been revealed [134]. RAPD is an emerging technique used for diagnostic mutation detection within a genome. The range of the minimum inhibitory concentrations (MICs) of the extracted leaf spinach antimicrobial substances against *Escherichia coli* and *Staphylococcus aureus* was observed between 60 and 100 mg/mL. The optimal extraction conditions were at 45 °C, ultrasound power of 44%, and an extraction time of 23 minutes. The study showed that the treated bacterial cells appeared to be damaged by a reduction in cell number. In fact, it was inferred that spinach leaf extracts exert bactericidal activity by inducing mutations in DNA and causing cell wall disruptions.

7. Conclusions

In summary, plant extracts showed strong antioxidant capacity both in vitro and in vivo, and the extracts can be considered a good source of natural antioxidants and antimicrobials. Polyphenol extraction from plants using fast and appropriate techniques is a low-cost method due to the reduction in the amount of solvent used, in addition to avoiding the need for longer extraction times compared to the conventional extraction method. Moreover, natural bioactive compounds have been found to interfere with and prevent all kinds of cancer. Flavonoids have been shown to work as anti-tumor (benign, melanoma) agents involving a free radicals quenching mechanism (i.e., OH, ROO). In fact, many studies have shown that flavonoids play significant multiple roles including mutagenic, cell damage, and carcinogenic, due to their acceleration of different aging factors. In addition to antioxidant activity, the inhibition of cancer development by phenolic compounds relies on a number of basic cellular mechanisms. More comprehensive studies related to these compounds will enhance pharmaceutical exploration in the field of carcinogenic disease prevention.

Supplementary Materials: The following are available online at www.mdpi.com/2223-7747/6/4/42/s1.

Acknowledgments: The authors would like to thank the Higher Committee for Education Development in Iraq (HCED) for the financial support to achieve this work and Stuart Alan Walters of the Department of Plant, Soil and Agricultural Systems, College of Agricultural Sciences, Southern Illinois University.

Author Contributions: Ammar Altemimi drafted the manuscript; Naoufal Lakhssassi and Azam Baharlouei edited the manuscript and made the corresponding figures and tables; Dennis G. Watson and David A. Lightfoot edited and commented on the manuscript. All authors have read and approved the final manuscript.

Conflicts of Interest: The authors declare no conflicts of interest.

References

1. Jakubowski, W.; Bartosz, G. Estimation of oxidative stress in saccharomyces cerevisae with fluorescent probes. *Int. J. Biochem. Cell Biol.* **1997**, *29*, 1297–1301. [CrossRef]
2. Naczk, M.; Shahidi, F. Phenolics in cereals, fruits and vegetables: Occurrence, extraction and analysis. *J. Pharm. Biomed. Anal.* **2006**, *41*, 1523–1542. [CrossRef] [PubMed]
3. Suffredini, I.B.; Sader, H.S.; Gonçalves, A.G.; Reis, A.O.; Gales, A.C.; Varella, A.D.; Younes, R.N. Screening of antibacterial extracts from plants native to the brazilian amazon rain forest and atlantic forest. *Braz. J. Med. Biol. Res.* **2004**, *37*, 379–384. [CrossRef] [PubMed]
4. Boots, A.W.; Haenen, G.R.; Bast, A. Health effects of quercetin: From antioxidant to nutraceutical. *Eur. J. Pharmacol.* **2008**, *585*, 325–337. [CrossRef] [PubMed]
5. Valko, M.; Rhodes, C.J.; Moncol, J.; Izakovic, M.; Mazur, M. Free radicals, metals and antioxidants in oxidative stress-induced cancer. *Chem. Biol. Interact.* **2006**, *160*, 1–40. [CrossRef] [PubMed]
6. Ames, B.N.; Shigenaga, M.K.; Hagen, T.M. Oxidants, antioxidants, and the degenerative diseases of aging. *Proc. Natl. Acad. Sci. USA* **1993**, *90*, 7915–7922. [CrossRef] [PubMed]
7. Duthie, S.J.; Ma, A.; Ross, M.A.; Collins, A.R. Antioxidant supplementation decreases oxidative DNA damage in human lymphocytes. *Cancer Res.* **1996**, *56*, 1291–1295. [PubMed]
8. Vivekananthan, D.P.; Penn, M.S.; Sapp, S.K.; Hsu, A.; Topol, E.J. Use of antioxidant vitamins for the prevention of cardiovascular disease: Meta-analysis of randomised trials. *Lancet* **2003**, *361*, 2017–2023. [CrossRef]
9. Wong, P.Y.Y.; Kitts, D.D. Studies on the dual antioxidant and antibacterial properties of parsley (*Petroselinum crispum*) and cilantro (*Coriandrum sativum*) extracts. *Food Chem.* **2006**, *97*, 505–515. [CrossRef]
10. Ruan, Z.P.; Zhang, L.L.; Lin, Y.M. Evaluation of the antioxidant activity of syzygium cumini leaves. *Molecules* **2008**, *13*, 2545–2556. [CrossRef] [PubMed]
11. Koffi, E.; Sea, T.; Dodehe, Y.; Soro, S. Effect of solvent type on extraction of polyphenols from twenty three ivorian plants. *J. Anim. Plant Sci.* **2010**, *5*, 550–558.
12. Anokwuru, C.P.; Anyasor, G.N.; Ajibaye, O.; Fakoya, O.; Okebugwu, P. Effect of extraction solvents on phenolic, flavonoid and antioxidant activities of three nigerian medicinal plants. *Nat. Sci.* **2011**, *9*, 53–61.
13. Ballard, T.S.; Mallikarjunan, P.; Zhou, K.; O'Keefe, S. Microwave-assisted extraction of phenolic antioxidant compounds from peanut skins. *Food Chem.* **2010**, *120*, 1185–1192. [CrossRef]
14. Kingston, H.M.; Jessie, L.B. *Introduction to Microwave Sample Preparation*; American Chemical Society: Washington, DC, USA, 1998.
15. Suzara, S.; Costa, D.A.; Gariepyb, Y.; Rochaa, S.C.S.; Raghavanb, V. Spilanthol extraction using microwave: Calibration curve for gas chromatography. *Chem. Eng. Trans.* **2013**, *32*, 1783–1788.
16. Li, H.; Deng, Z.; Wu, T.; Liu, R.; Loewen, S.; Tsao, R. Microwave-assisted extraction of phenolics with maximal antioxidant activities in tomatoes. *Food Chem.* **2012**, *130*, 928–936. [CrossRef]

17. Tsubaki, S.; Sakamoto, M.; Azuma, J. Microwave-assisted extraction of phenolic compounds from tea residues under autohydrolytic conditions. *Food Chem.* **2000**, *123*, 1255–1258. [CrossRef]
18. Christophoridou, S.; Dais, P.; Tseng, L.H.; Spraul, M. Separation and identification of phenolic compounds in olive oil by coupling high-performance liquid chromatography with postcolumn solid-phase extraction to nuclear magnetic resonance spectroscopy (lc-spe-nmr). *J. Agric. Food Chem.* **2005**, *53*, 4667–4679. [CrossRef] [PubMed]
19. Williams, O.J.; Raghavan, G.S.V.; Orsat, V.; Dai, J. Microwave-assisted extraction of capsaicinoids from capsicum fruit. *J. Food Biochem.* **2004**, *28*, 113–122. [CrossRef]
20. Garcia-Salas, P.; Morales-Soto, A.; Segura-Carretero, A.; Fernandez-Gutierrez, A. Phenolic-compound-extraction systems for fruit and vegetable samples. *Molecules* **2010**, *15*, 8813–8826. [CrossRef] [PubMed]
21. Corrales, M.; Toepfl, S.; Butz, P.; Knorr, D.; Tauscher, B. Extraction of anthocyanins from grape by-products assisted by ultrasonics, high hydrostatic pressure or pulsed electric fields: A comparison. *Innov. Food Sci. Emerg. Technol.* **2008**, *9*, 85–91. [CrossRef]
22. Tabaraki, R.; Nateghi, A. Optimization of ultrasonic-assisted extraction of natural antioxidants from rice bran using response surface methodology. *Ultrason. Sonochem.* **2011**, *18*, 1279–1286. [CrossRef] [PubMed]
23. Albu, S.; Joyce, E.; Paniwnyk, L.; Lorimer, J.P.; Mason, T.J. Potential for the use of ultrasound in the extraction of antioxidants from *Rosmarinus officinalis* for the food and pharmaceutical industry. *Ultrason. Sonochem.* **2004**, *11*, 261–265. [CrossRef] [PubMed]
24. Cho, W.I.; Choi, J.B.; Lee, K.; Chung, M.S.; Pyun, Y.R. Antimicrobial activity of torilin isolated from *Torilis japonica* fruit against *Bacillus subtilis*. *J. Food Sci.* **2008**, *73*, 37–46. [CrossRef] [PubMed]
25. Barbero, G.F.; Liazid, A.; Palma, M.; Barroso, C.G. Ultrasound-assisted extraction of capsaicinoids from peppers. *Talanta* **2008**, *75*, 1332–1337. [CrossRef] [PubMed]
26. Luque-García, J.L.; Luque de Castro, M.D. Ultrasound: A powerful tool for leaching. *TrAC Trends Anal. Chem.* **2003**, *22*, 41–47. [CrossRef]
27. Mulinacci, N.; Prucher, D.; Peruzzi, M.; Romani, A.; Pinelli, P.; Giaccherini, C.; Vincieri, F.F. Commercial and laboratory extracts from artichoke leaves: Estimation of caffeoyl esters and flavonoidic compounds content. *J. Pharm. Biomed. Anal.* **2004**, *34*, 349–357. [CrossRef]
28. Altemimi, A.W.; Watson, D.G.; Kinsel, M.; Lightfoot, D.A. Simultaneous extraction, optimization, and analysis of flavonoids and polyphenols from peach and pumpkin extracts using a tlc-densitometric method. *Chem. Cent. J.* **2015**, *9*, 1–15. [CrossRef] [PubMed]
29. Altemimi, A.; Lightfoot, D.A.; Kinsel, M.; Watson, D.G. Employing response surface methodology for the optimization of ultrasound assisted extraction of lutein and β-carotene from spinach. *Molecules* **2015**, *20*, 6611–6625. [CrossRef] [PubMed]
30. Sarajlija, H.; Čkelj, N.; Novotni, D.; Mršić, G.; Ćurić, M.; Brncic, M.; Curic, D. Preparation of flaxseed for lignan determination by gas chromatography-mass spectrometry method. *Czech J. Food Sci.* **2012**, *30*, 45–52.
31. Popova, I.E.; Hall, C.; Kubátová, A. Determination of lignans in flaxseed using liquid chromatography with time-of-flight mass spectrometry. *J. Chromatogr. A* **2009**, *1216*, 217–229. [CrossRef] [PubMed]
32. Zhang, Z.; Pang, X.; Xuewu, D.; Ji, Z.; Jiang, Y. Role of peroxidase in anthocyanin degradation in litchi fruit pericarp. *Food Chem.* **2005**, *90*, 47–52. [CrossRef]
33. Kemp, W. Energy and electromagnetic spectrum. In *Organic Spectroscopy*; Kemp, W., Ed.; Macmillan Press: London, UK, 1991; pp. 1–7.
34. Kemp, W. Infrared spectroscopy. In *Organic Spectroscopy*; Macmillan Press Ltd.: London, UK, 1991; pp. 19–56.
35. Urbano, M.; Luque de Castro, M.D.; Pérez, P.M.; García-Olmo, J.; Gómez-Nieto, M.A. Ultraviolet–visible spectroscopy and pattern recognition methods for differentiation and classification of wines. *Food Chem.* **2006**, *97*, 166–175. [CrossRef]
36. Cherkaoui, A.; Hibbs, J.; Emonet, S.; Tangomo, M.; Girard, M.; Francois, P.; Schrenzel, J. Comparison of two matrix-assisted laser desorption ionization-time of flight mass spectrometry methods with conventional phenotypic identification for routine identification of bacteria to the species level. *J. Clin. Microbiol.* **2010**, *48*, 1169–1175. [CrossRef] [PubMed]
37. Beckey, H.D. *1–Theory of Field Ionization (FI) and Field Emission (FE)*; Pergamon: Bergama, Turkey, 1971.
38. Beckey, H.D. *2–Field Ionization Sources*; Pergamon: Bergama, Turkey, 1971.
39. Beckey, H.D. *3–Application of the FI Mass Spectrometer to Physico-Chemical Problems*; Pergamon: Bergama, Turkey, 1971.

40. Beckey, H.D. *4–Qualitative Analyses with the Fi Mass Spectrometer*; Pergamon: Bergama, Turkey, 1971.
41. Kemp, W. Nuclear magnetic resonance spectroscopy. In *Organic Spectroscopy*; Kemp, W., Ed.; Macmillan Press: London, UK, 1991; pp. 101–240.
42. Halliwell, B.; Murcia, M.A.; Chirico, S.; Aruoma, O.I. Free radicals and antioxidants in food and in vivo: What they do and how they work. *Crit. Rev. Food Sci. Nutr.* **1995**, *35*, 7–20. [CrossRef] [PubMed]
43. Clemente, T.E.; Cahoon, E.B. Soybean oil: Genetic approaches for modification of functionality and total content. *Plant Physiol.* **2009**, *151*, 1030–1040. [CrossRef] [PubMed]
44. Lakhssassi, N.; Zhou, Z.; Liu, S.; Colantonio, V.; AbuGhazaleh, A.; Meksem, K. Characterization of the fad2 gene family in soybean reveals the limitations of gel-based tilling in genes with high copy number. *Front. Plant Sci.* **2017**, *8*, 324. [CrossRef] [PubMed]
45. Mensink, R.P.; Katan, M.B. Effect of dietary trans fatty acids on high-density and low-density lipoprotein cholesterol levels in healthy subjects. *N. Engl. J. Med.* **1990**, *323*, 439–445. [CrossRef] [PubMed]
46. Kris-Etherton, P.M.; Yu, S.; Etherton, T.D.; Morgan, R.; Moriarty, K.; Shaffer, D. Fatty acids and progression of coronary artery disease. *Am. J. Clin. Nutr.* **1997**, *65*, 1088–1090. [PubMed]
47. Kris-Etherton, P.M.; Yu, S. Individual fatty acid effects on plasma lipids and lipoproteins: Human studies. *Am. J. Clin. Nutr.* **1997**, *65*, 1628–1644.
48. Byfield, G.E.; Xue, H.; Upchurch, R.G. Two genes from soybean encoding soluble δ9 stearoyl-acp desaturases. *Crop Sci.* **2006**, *46*, 840–846. [CrossRef]
49. Lakhssassi, N.; Colantonio, V.; Flowers, N.D.; Zhou, Z.; Henry, J.S.; Liu, S.; Meksem, K. Stearoyl-acyl carrier protein desaturase mutations uncover an impact of stearic acid in leaf and nodule structure. *Plant Physiol.* **2017**, *174*, 1531–1543. [CrossRef] [PubMed]
50. St Angelo, A.J. Lipid oxidation on foods. *Crit. Rev. Food Sci. Nutr.* **1996**, *36*, 175–224. [CrossRef] [PubMed]
51. Javanmardi, J.; Stushnoff, C.; Locke, E.; Vivanco, J.M. Antioxidant activity and total phenolic content of iranian ocimum accessions. *Food Chem.* **2003**, *83*, 547–550. [CrossRef]
52. Negi, P.S.; Chauhan, A.S.; Sadia, G.A.; Rohinishree, Y.S.; Ramteke, R.S. Antioxidant and antibacterial activities of various seabuckthorn (hippophae rhamnoides l.) seed extracts. *Food Chem.* **2005**, *92*, 119–124. [CrossRef]
53. Baharlouei, A.; Sharifi-Sirchi, G.R.; Bonjar, G.H.S. Identification of an antifungal chitinase from a potential biocontrol agent, streptomyces plicatus strain 101, and its new antagonistic spectrum of activity. *Philipp. Agric. Sci.* **2010**, *93*, 439–445.
54. Baharlouei, A.; Sharifi-Sirchi, G.R.; Bonjar, G.H.S. Biological control of sclerotinia sclerotiorum (oilseed rape isolate) by an effective antagonist streptomyces. *Afr. J. Biotechnol.* **2011**, *10*, 5785–5794.
55. Ruberto, G.; Baratta, M.T.; Deans, S.G.; Dorman, H.J.D. Antioxidant and antimicrobial activity of *Foeniculum vulgare* and *Crithmum maritimum* essential oils. *Planta Medica* **2000**, *66*, 687–693. [CrossRef] [PubMed]
56. Butterfield, N.J. Implications for the evolution of sex, multicellularity, and the mesoproterozoic/neoproterozoic radiation of eukaryotes. *Paleobiology* **2000**, *26*, 386–404. [CrossRef]
57. Ashawat, M.S.; Shailendra, S.; Swarnlata, S. In vitro antioxidant activity of ethanolic extracts of centella asiatica, punica granatum, glycyrrhiza glabra and areca catechu. *Res. J. Med. Plants* **2007**, *1*, 13–16.
58. Londonkar, R.; Kamble, A. Evaluation of free radical scavenging activity of pandanus odoratissimus. *Int. J. Pharmacol.* **2009**, *5*, 377–380. [CrossRef]
59. Zahin, M.; Aqil, F.; Ahmad, I. The in vitro antioxidant activity and total phenolic content of four indian medicinal plants. *J. Appl. Biol. Sci.* **2007**, *1*, 87–90.
60. Block, G.; Patterson, B.; Subar, A. Fruit, vegetables, and cancer prevention: A review of the epidemiological evidence. *Nutr. Cancer* **1992**, *18*, 1–29. [CrossRef] [PubMed]
61. Pratt, D.E.; Watts, B.M. The antioxidant activity of vegetable extracts i. Flavone aglyconesa. *J. Food Sci.* **1964**, *29*, 27–33. [CrossRef]
62. Plumb, G.W.; García-Conesa, M.T.; Kroon, P.A.; Rhodes, M.; Ridley, S.; Williamson, G. Metabolism of chlorogenic acid by human plasma, liver, intestine and gut microflora. *J. Sci. Food Agric.* **1999**, *79*, 390–392. [CrossRef]
63. Frankel, E.N.M.; Meyer, A. Antioxidants in grapes and grape juices and their potential health effects. *Pharm. Biol.* **1998**, *36*, 1–7. [CrossRef]
64. Wangensteen, H.; Miron, A.; Alamgir, M.; Rajia, S.; Samuelsen, A.B.; Malterud, K.E. Antioxidant and 15-lipoxygenase inhibitory activity of rotenoids, isoflavones and phenolic glycosides from sarcolobus globosus. *Fitoterapia* **2006**, *77*, 290–295. [CrossRef] [PubMed]

65. Monagas, M.; Gómez-Cordovés, C.; Bartolomeä, B.; Laureano, O.; Ricardo da Silva, J.M. Monomeric, oligomeric, and polymeric flavan-3-ol composition of wines and grapes from vitis vinifera l. Cv. Graciano, tempranillo, and cabernet sauvignon. *J. Agric. Food Chem.* **2003**, *51*, 6475–6481. [CrossRef] [PubMed]
66. Khan, S.A.; Beekwilder, J.; Schaart, J.G.; Mumm, R.; Soriano, J.M.; Jacobsen, E.; Schouten, H.J. Differences in acidity of apples are probably mainly caused by a malic acid transporter gene on lg16. *Tree Genet. Genomes* **2013**, *9*, 475–487. [CrossRef]
67. Jugde, H.; Nguy, D.; Moller, I.; Cooney, J.M.; Atkinson, R.G. Isolation and characterization of a novel glycosyltransferase that converts phloretin to phlorizin, a potent antioxidant in apple. *FEBS J.* **2008**, *275*, 3804–3814. [CrossRef] [PubMed]
68. Huang, Y.F.; Doliez, A.; Fournier-Level, A.; Le Cunff, L.; Bertrand, Y.; Canaguier, A.; Morel, C.; Miralles, V.; Veran, F.; Souquet, J.M.; et al. Dissecting genetic architecture of grape proanthocyanidin composition through quantitative trait locus mapping. *BMC Plant Biol.* **2012**, *12*, 30. [CrossRef] [PubMed]
69. Takos, A.M.; Ubi, B.E.; Robinson, S.P.; Walker, A.R. Condensed tannin biosynthesis genes are regulated separately from other flavonoid biosynthesis genes in apple fruit skin. *Plant Sci.* **2006**, *170*, 487–499. [CrossRef]
70. Soong, Y.Y.; Barlow, P.J. Antioxidant activity and phenolic content of selected fruit seeds. *Food Chem.* **2004**, *88*, 411–417. [CrossRef]
71. Abdille, M.H.; Singh, R.P.; Jayaprakasha, G.K.; Jena, B.S. Antioxidant activity of the extracts from *Dillenia indica* fruits. *Food Chem.* **2005**, *90*, 891–896. [CrossRef]
72. Juntachote, T.; Berghofer, E. Antioxidative properties and stability of ethanolic extracts of holy basil and galangal. *Food Chem.* **2005**, *92*, 193–202. [CrossRef]
73. Liyana-Pathirana, C.M.; Shahidi, F.; Alasalvar, C. Antioxidant activity of cherry laurel fruit (laurocerasus officinalis roem.) and its concentrated juice. *Food Chem.* **2006**, *99*, 121–128. [CrossRef]
74. Orhan, I.; Kartal, M.; Naz, Q.; Ejaz, A.; Yilmaz, G.; Kan, Y.; Konuklugil, B.; Şener, B.; Iqbal Choudhary, M. Antioxidant and anticholinesterase evaluation of selected turkish *Salvia* species. *Food Chem.* **2007**, *103*, 1247–1254. [CrossRef]
75. Rathee, J.S.; Hassarajani, S.A.; Chattopadhyay, S. Antioxidant activity of mammea longifolia bud extracts. *Food Chem.* **2006**, *99*, 436–443. [CrossRef]
76. Baskar, R.; Rajeswari, V.; Kumar, T.S. In vitro antioxidant studies in leaves of annona species. *Indian J. Exp. Biol.* **2007**, *45*, 480–485. [PubMed]
77. Lu, Y.; Yeap, F.L. Antioxidant activities of polyphenols from sage (*Salvia officinalis*). *Food Chem.* **2001**, *75*, 197–202. [CrossRef]
78. Zhao, G.R.; Xiang, Z.J.; Ye, T.X.; Yuan, Y.J.; Guo, Z.X. Antioxidant activities of *Salvia miltiorrhiza* and *Panax notoginseng*. *Food Chem.* **2006**, *99*, 767–774. [CrossRef]
79. Singh, R.P.; Chidambara Murthy, K.N.; Jayaprakasha, G.K. Studies on the antioxidant activity of pomegranate (punica granatum) peel and seed extracts using in vitro models. *J. Agric. Food Chem.* **2002**, *50*, 81–86. [CrossRef] [PubMed]
80. Ćetković, G.S.; Djilas, S.M.; Čanadanović-Brunet, J.M.; Tumbas, V.T. Antioxidant properties of marigold extracts. *Food Res. Int.* **2004**, *37*, 643–650. [CrossRef]
81. Siddhuraju, P.; Manian, S. The antioxidant activity and free radical-scavenging capacity of dietary phenolic extracts from horse gram (*Macrotyloma uniflorum* (lam.) verdc.) seeds. *Food Chem.* **2007**, *105*, 950–958. [CrossRef]
82. Samak, G.; Shenoy, R.P.; Manjunatha, S.M.; Vinayak, K.S. Superoxide and hydroxyl radical scavenging actions of botanical extracts of wagatea spicata. *Food Chem.* **2009**, *115*, 631–634. [CrossRef]
83. Barla, A.; Öztürk, M.; Kültür, Ş.; Öksüz, S. Screening of antioxidant activity of three *Euphorbia* species from turkey. *Fitoterapia* **2007**, *78*, 423–425. [CrossRef] [PubMed]
84. Pinelo, M.; Rubilar, M.; Sineiro, J.; Núñez, M.J. Extraction of antioxidant phenolics from almond hulls (*Prunus amygdalus*) and pine sawdust (*Pinus pinaster*). *Food Chem.* **2004**, *85*, 267–273. [CrossRef]
85. Ibrahim, N.A.; El-Seedi, H.R.; Mohammed, M.M. Phytochemical investigation and hepatoprotective activity of *Cupressus sempervirens* L. Leaves growing in egypt. *Nat. Prod. Res.* **2007**, *21*, 857–866. [CrossRef] [PubMed]
86. Kaur, R.; Thind, T.S.; Singh, B.; Arora, S. Inhibition of lipid peroxidation by extracts/subfractions of chickrassy (chukrasia tabularis a. Juss.). *Die Naturwissenschaften* **2009**, *96*, 129–133. [CrossRef] [PubMed]

87. Singh, G.; Marimuthu, P.; de Heluani, C.S.; Catalan, C. Chemical constituents and antimicrobial and antioxidant potentials of essential oil and acetone extract of *Nigella sativa* seeds. *J. Sci. Food Agric.* **2005**, *85*, 2297–2306. [CrossRef]
88. Tiwari, O.P.; Tripathi, Y.B. Antioxidant properties of different fractions of *Vitex negundo* Linn. *Food Chem.* **2007**, *100*, 1170–1176. [CrossRef]
89. Zin, Z.M.; Abdul-Hamid, A.; Osman, A. Antioxidative activity of extracts from mengkudu (*Morinda citrifolia* L.) root, fruit and leaf. *Food Chem.* **2002**, *78*, 227–231. [CrossRef]
90. Oktay, M.; Gülçin, İ.; Küfrevioğlu, Ö.İ. Determination of in vitro antioxidant activity of fennel (*Foeniculum vulgare*) seed extracts. *LWT Food Sci. Technol.* **2003**, *36*, 263–271. [CrossRef]
91. Parejo, I.; Viladomat, F.; Bastida, J.; Rosas-Romero, A.; Saavedra, G.; Murcia, M.A.; Jimenez, A.M.; Codina, C. Investigation of bolivian plant extracts for their radical scavenging activity and antioxidant activity. *Life Sci.* **2003**, *73*, 1667–1681. [CrossRef]
92. Harish, R.; Shivanandappa, T. Antioxidant activity and hepatoprotective potential of phyllanthus niruri. *Food Chem.* **2006**, *95*, 180–185. [CrossRef]
93. Ajila, C.M.; Naidu, K.A.; Bhat, S.G.; Rao, U.J.S.P. Bioactive compounds and antioxidant potential of mango peel extract. *Food Chem.* **2007**, *105*, 982–988. [CrossRef]
94. Chen, H.Y.; Yen, G.C. Antioxidant activity and free radical-scavenging capacity of extracts from guava (*Psidium guajava* L.) leaves. *Food Chem.* **2007**, *101*, 686–694. [CrossRef]
95. Dastmalchi, K.; Damien Dorman, H.J.; Koşar, M.; Hiltunen, R. Chemical composition and in vitro antioxidant evaluation of a water-soluble moldavian balm (*Dracocephalum moldavica* L.) extract. *LWT Food Sci. Technol.* **2007**, *40*, 239–248. [CrossRef]
96. Arabshahi-Delouee, S.; Urooj, A. Antioxidant properties of various solvent extracts of mulberry (*Morus indica* L.) leaves. *Food Chem.* **2007**, *102*, 1233–1240. [CrossRef]
97. Ayaz, F.A.; Ayaz, S.; Alpay-Karaoglu, S.; Gruz, J.; Valentová, K.; Ulrichová, J.; Strnad, M. Phenolic acid contents of kale (*Brassica oleraceae* L. Var. *acephala* DC.) extracts and their antioxidant and antibacterial activities. *Food Chem.* **1976**, *7*, 3. [CrossRef]
98. Kaviarasan, S.; Naik, G.H.; Gangabhagirathi, R.; Anuradha, C.V.; Priyadarsini, K.I. In vitro studies on antiradical and antioxidant activities of fenugreek (trigonella foenum graecum) seeds. *Food Chem.* **2007**, *103*, 31–37. [CrossRef]
99. Pandey, N.; Tripathi, Y.B. Antioxidant activity of tuberosin isolated from *Pueraria tuberose* Linn. *J. Inflamm.* **2010**, *7*, 47. [CrossRef] [PubMed]
100. Yang, D.; Wang, Q.; Ke, L.; Jiang, J.; Ying, T. Antioxidant activities of various extracts of lotus (*Nelumbo nuficera gaertn*) rhizome. *Asia Pac. J. Clin. Nutr.* **2007**, *16*, 158–163. [PubMed]
101. Meot-Duros, L.; Magne, C. Antioxidant activity and phenol content of *Crithmum maritimum* L. Leaves. *Plant Physiol. Biochem.* **2009**, *47*, 37–41. [CrossRef] [PubMed]
102. Sakat, S.; Juvekar, R.A.; Gambhire, M.N.; Juvekar, M.; Wankhede, S. In vitro antioxidant and anti-inflammatory activity of methanol extract of oxalis corniculata linn. *Int. J. Pharm. Pharm. Sci.* **2010**, *2*, 146–155.
103. Jain, S.; Jain, A.; Jain, N.; Jain, D.K.; Balekar, N. Phytochemical investigation and evaluation of in vitro free radical scavenging activity of *Tabernaemontana divaricata* Linn. *Nat. Prod. Res.* **2010**, *24*, 300–304. [CrossRef] [PubMed]
104. Laitonjam, W.S.; Kongbrailatpam, B.D. Studies on the chemical constituents and antioxidant activities of extracts from the roots of smilax lanceaefolia roxb. *Nat. Prod. Res.* **2010**, *24*, 1168–1176. [CrossRef] [PubMed]
105. Inbaraj, B.S.; Chien, J.T.; Chen, B.H. Improved high performance liquid chromatographic method for determination of carotenoids in the microalga *Chlorella pyrenoidosa*. *J. Chromatogr. A* **2006**, *1102*, 193–199. [CrossRef] [PubMed]
106. Gandul-Rojas, B.; Cepero, M.R.; Minguez-Mosquera, M.I. Chlorophyll and carotenoid patterns in olive fruits, *Olea europaea* Cv. Arbequina. *J. Agric. Food Chem.* **1999**, *47*, 2207–2212. [CrossRef] [PubMed]
107. Ren, D.; Zhang, S. Separation and identification of the yellow carotenoids in *Potamogeton crispus* L. *Food Chem.* **2008**, *106*, 410–414. [CrossRef]
108. Alves-Rodrigues, A.; Shao, A. The science behind lutein. *Toxicol. Lett.* **2004**, *150*, 57–83. [CrossRef] [PubMed]
109. Altemimi, A.; Lakhssassi, N.; Abu-Ghazaleh, A.; Lightfoot, D.A. Evaluation of the antimicrobial activities of ultrasonicated spinach leaf extracts using rapd markers and electron microscopy. *Arch. Microbiol.* **2017**, 1–13. [CrossRef] [PubMed]

110. Konczak, I.; Zhang, W. Anthocyanins—More than nature's colours. *J. Biomed. Biotechnol.* **2004**, *2004*, 239–240. [CrossRef] [PubMed]
111. Vermerris, W.; Nicholson, R. *Phenolic Compound Biochemistry Book*; Springer: Dordrecht, The Netherland, 2006.
112. Hursting, S.D.; Slaga, T.J.; Fischer, S.M.; DiGiovanni, J.; Phang, J.M. Mechanism-based cancer prevention approaches: Targets, examples, and the use of transgenic mice. *J. Natl. Cancer Inst.* **1999**, *91*, 215–225. [CrossRef] [PubMed]
113. Cai, Y.; Luo, Q.; Sun, M.; Corke, H. Antioxidant activity and phenolic compounds of 112 traditional chinese medicinal plants associated with anticancer. *Life Sci.* **2004**, *74*, 2157–2184. [CrossRef] [PubMed]
114. Palanisamy, U.; Cheng, H.M.; Masilamani, T.; Subramaniam, T.; Ling, L.T.; Radhakrishnan, A.K. Rind of the rambutan, *Nephelium lappaceum*, a potential source of natural antioxidants. *Food Chem.* **2008**, *109*, 54–63. [CrossRef] [PubMed]
115. Nassr-allah, A.A.; Aboul-enein, K.M.; Lightfoot, D.A.; Cocchetto, A.; El-shemy, H.A. Anti-cancer and anti-oxidant activity of some egyptian medicinal plants. *J. Med. Plants Res.* **2009**, *3*, 799–808.
116. El-Shemy, H.A.; Aboul-Enein, K.M.; Lightfoot, D.A. Predicting in silico which mixtures of the natural products of plants might most effectively kill human leukemia cells? *Evid.-Based Complement. Altern. Med.* **2013**, *2013*, 801501. [CrossRef] [PubMed]
117. El-Shemy, H.A.; Aboul-Enein, A.M.; Aboul-Enein, K.M.; Fujita, K. Willow leaves' extracts contain anti-tumor agents effective against three cell types. *PLoS ONE* **2007**, *2*, e178. [CrossRef] [PubMed]
118. Altemimi, A.B. A study of the protective properties of iraqi olive leaves against oxidation and pathogenic bacteria in food applications. *Antioxidants* **2017**, *6*, 34. [CrossRef] [PubMed]
119. Gil, M.I.; Tomas-Barberan, F.A.; Hess-Pierce, B.; Kader, A.A. Antioxidant capacities, phenolic compounds, carotenoids, and vitamin c contents of nectarine, peach, and plum cultivars from california. *J. Agric. Food Chem.* **2002**, *50*, 4976–4982. [CrossRef] [PubMed]
120. Shariff, N.; Sudarshana, M.S.; Umesha, S.; Hariprasad, P. Antimicrobial activity of *Rauvolfia tetraphylla* and *Physalis minima* leaf and callus extracts. *Afr. J. Biotechnol.* **2006**, *5*, 946–950.
121. Devi, P.U.; Murugan, S.; Suja, S.; Selvi, S.; Chinnaswamy, P.; Vijayanand, E. Antibacterial, in vitro lipid per oxidation and phytochemical observation on achyranthes bidentata blume. *Pak. J. Nutr.* **2007**, *6*, 447–451. [CrossRef]
122. Dabur, R.; Gupta, A.; Mandal, T.K.; Singh, D.D.; Bajpai, V.; Gurav, A.M.; Lavekar, G.S. Antimicrobial activity of some indian medicinal plants. *Afr. J. Tradit. Complement. Altern. Med.* **2007**, *4*, 313–318. [CrossRef] [PubMed]
123. Ramkumar, K.M.; Rajaguru, P.; Latha, M.; Ananthan, R. Ethanol extract of gymnema montanum leaves reduces glycoprotein components in experimental diabetes. *Nutr. Res.* **2007**, *27*, 97–103. [CrossRef]
124. Zakaria, Z.; Sreenivasan, S.; Mohamad, M. Antimicrobial activity of piper ribesoides root extract against staphylococcus aureus. *J. Appl. Biol. Sci.* **2007**, *1*, 87–90.
125. Ghosh, S.; Subudhi, E.; Nayak, S. Antimicrobial assay of *Stevia rebaudiana* bertoni leaf extracts against 10 pathogens. *Int. J. Integr. Biol.* **2008**, *2*, 1–5.
126. Mahesh, B.; Satish, S. Antimicrobial activity of some important medicinal plant against plant and human pathogens. *World J. Agric. Sci.* **2008**, *4*, 839–843.
127. Ehsan, B.R. Antimicrobial activity of the ethanolic extract of bryonopsis laciniosa leaf, stem, fruit and seed. *Afr. J. Biotechnol.* **2009**, *8*, 565–3567.
128. Khond, M.; Bhosale, J.D.; Tasleem, A.; Mandal, T.; Padhi, M.M.; Dabur, R. Screening of some selected medicinal plants extracts for in-vitro antimicrobial activity. *Middle-East J. Sci. Res.* **2009**, *4*, 271–278.
129. Pavithra, P.S.; Sreevidya, N.; Verma, R.S. Antibacterial and antioxidant activity of methanol extract of evolvulus nummularius. *Indian J. Pharmacol.* **2009**, *41*, 233–236. [PubMed]
130. Patra, A.; Jha, S.; Murthy, P.N.; Vaibhav, A.D.; Chattopadhyay, P.; Panigrahi, G.; Roy, D. Anti-inflammatory and antipyretic activities of *Hygrophila spinosa* t. Anders leaves (acanthaceae). *Trop. J. Pharm. Res.* **2009**, *8*, 133–137. [CrossRef]
131. Akroum, S. Antimicrobial activity of some alimentary and medicinal plants. *Afr. J. Microbiol. Res.* **2012**, *6*, 1860–1864.
132. Bajpai, V.K.; Rahman, A.; Shukla, S.; Mehta, A.; Shukla, S.; Arafat, S.M.Y.; Rahman, M.M.; Ferdousi, Z. Antibacterial activity of leaf extracts of pongamia pinnata from India. *Pharm. Biol.* **2009**, *47*, 1162–1167. [CrossRef]

133. Bansal, S. Anti-bacterial efficacy of some plants used in folkaric medicines in arid zone. *J. Pharm. Res.* **2010**, *3*, 2640–2642.
134. Vinothkumar, P.S.K.; Ahmed, P.; Sivamani, K.; Senthilkumar, B. Evaluation of antibacterial activities of andrographis paniculata leaf extract against gram positive and gram negative species by in vitro methods. *J. Pharm. Res.* **2010**, *3*, 1513–1515.
135. Kumar, K.H.; Hullatti, K.K.; Sharanappa, P.; Sharma, P. Computative antimicrobial activity and tlcbioautographic analysis of root and aerial parts of *Andrographis serpyllifolia*. *Int. J. Pharm. Pharm. Sci.* **2010**, *2*, 52–54.
136. Jamuna, B.A.; Rai, V.R.; Samaga, P.V. Evaluation of the antimicrobial activity of three medicinal plants of south India. *Malays. J. Microbiol.* **2011**, *7*, 14–18. [CrossRef]
137. Khanahmadi, M.; Rezazadeh, S.; Taran, M. In vitro antimicrobial and antioxidant properties of *Smyrnum cordifolium* Boiss. (Umbelliferae) extract. *Asian J. Plant Sci.* **2010**, *9*, 99–103.
138. Koperuncholan, M.; Kumar, P.S.; Sathiyanarayanan, G.; Vivek, G. Phytochemical screening and antimicrobial studies of some ethnomedicinal plants in south-eastern slope of western ghats. *Int. J. Med. Res.* **2010**, *1*, 48–58.
139. Niranjan, M.H.; Kavitha, H.U.; Sreedharamurthy, S.; Sudarshana, M.S. Antibacterial activity of *Schrebera swietenioides* roxb. Against some human pathogenic bacteria. *J. Pharm. Res.* **2010**, *3*, 1779–1781.
140. Naveen, S.M.T. Evaluation of antibacterial activity of flower extracts of *Cassia auriculata*. *Ethnobot. Leafl.* **2010**, *14*, 8–20.

© 2017 by the authors. Licensee MDPI, Basel, Switzerland. This article is an open access article distributed under the terms and conditions of the Creative Commons Attribution (CC BY) license (http://creativecommons.org/licenses/by/4.0/).

Review

Antimicrobial Resistance and the Alternative Resources with Special Emphasis on Plant-Based Antimicrobials—A Review

Harish Chandra [1], Parul Bishnoi [2], Archana Yadav [3], Babita Patni [1], Abhay Prakash Mishra [4,*] and Anant Ram Nautiyal [1,*]

1. High Altitude Plant Physiology Research Centre, Hemvati Nandan Bahuguna (H.N.B.) Garhwal University, Srinagar, Garhwal 246174, Uttarakhand, India; hreesh5@gmail.com (H.C.); babita28paatni@gmail.com (B.P.)
2. Department of Microbiology, Bangalore City College, Bangalore 560043, India; bishnoiparul@gmail.com
3. Department of Microbiology, Institute of Biosciences and Biotechnology, Chhatrapati Shahu Ji Maharaj (C.S.J.M.) University, Kanpur 208024, India; archana25578@gmail.com
4. Department of Pharmaceutical Chemistry, Hemvati Nandan Bahuguna (H.N.B.) Garhwal University, Srinagar, Garhwal 246174, Uttarakhand, India
* Correspondence: abhaycimap@gmail.com (A.P.M.); arnautiyal@gmail.com (A.R.N.); Tel.: +91-94-5200-2557 (A.P.M.); +91-94-1292-1400 (A.R.N.)

Academic Editor: Milan S. Stankovic
Received: 8 February 2017; Accepted: 23 March 2017; Published: 10 April 2017

Abstract: Indiscriminate and irrational use of antibiotics has created an unprecedented challenge for human civilization due to microbe's development of antimicrobial resistance. It is difficult to treat bacterial infection due to bacteria's ability to develop resistance against antimicrobial agents. Antimicrobial agents are categorized according to their mechanism of action, i.e., interference with cell wall synthesis, DNA and RNA synthesis, lysis of the bacterial membrane, inhibition of protein synthesis, inhibition of metabolic pathways, etc. Bacteria may become resistant by antibiotic inactivation, target modification, efflux pump and plasmidic efflux. Currently, the clinically available treatment is not effective against the antibiotic resistance developed by some bacterial species. However, plant-based antimicrobials have immense potential to combat bacterial, fungal, protozoal and viral diseases without any known side effects. Such plant metabolites include quinines, alkaloids, lectins, polypeptides, flavones, flavonoids, flavonols, coumarin, terpenoids, essential oils and tannins. The present review focuses on antibiotic resistance, the resistance mechanism in bacteria against antibiotics and the role of plant-active secondary metabolites against microorganisms, which might be useful as an alternative and effective strategy to break the resistance among microbes.

Keywords: antibiotic resistance; antimicrobial; mechanism of action; plant metabolite

1. Introduction

The problem of antibiotic resistance is not limited to the Indian subcontinent only, but is a global problem. To date, no known method is available to reverse antibiotic resistance in bacteria. The discovery and development of the antibiotic penicillin during the 1900s gave a certain hope to medical science, but this antibiotic soon became ineffective against most of the susceptible bacteria. The antibiotic resistance in bacteria is generally a natural phenomenon for adaptation to antimicrobial agents. Once bacteria become resistant to some antibiotic, they pass on this characteristic to their progeny through horizontal or vertical transfer. The indiscriminate and irrational use of antibiotics these days has led to the evolution of new resistant strains of bacteria that are somewhat more lethal compared to the parent strain. Cases of widespread occurrence of resistant bacteria are now very common which leads to many health-related problems [1]. Change in the genetic constitution of these resistant bacteria is so rapid that the effectiveness of common antibiotics may be lost within a period of 5 years [2]. As per the report of the WHO (World Health Organization) resistance was more prevalent in cases of bacterial infections which cause most of the deadly infectious bacterial infections worldwide such as respiratory tract infections, diarrhea, meningitis, syphilis, gonorrhea and tuberculosis [3]. *Staphylococcus aureus* isolated from clinical samples are now showing resistance to more than three drugs and are considered as multiple-drug resistant bacteria [4]. In the case of *Streptococcus pyogenes* isolates, the resistance rates for penicillin are 50% and for erythromycin are 80% globally. The 1539 sputum samples of patients from KwaZulu Natal, South Africa during the period 2005–2006 were screened to study the antibiotic resistance pattern of *M. tuberculosis*; 475 samples were found positive for Tuberculosis (TB) and out of 475 samples, 53 samples were found positive for XDR-TB (Extremely Drug-Resistant Tuberculosis), i.e., they are not susceptible to routine antibiotics used to treat TB [5]. The severity of drug resistance is an alarming situation for scientific and medical professionals, who must search for alternate treatments or develop new drugs to combat drug-resistant bacteria. There is an urgent need of medical assistance for treating these resistant bacteria. The empirical use of antibiotics is one of the factors that contribute to the increase of drug resistance. At times, people use antibiotics without knowing the importance of taking a full course of antibiotics. After taking 2–3 doses of a particular antibiotic, they feel better and then discontinue the course. In comparison to the first line of a drug, the second line or third generation drug are quite costlier and due to the lack of financial assistance these diseases cannot be treated efficiently.

Nowadays, researchers are in search of some novel antimicrobial molecules which have a broad spectrum of activity against both Gram-negative and Gram-positive bacteria without having many or any side effects. They are exploring the variety of medicinal plants which are described in Ayurveda, Charak Samhita, Sushrut Samhita and other literatures available in their respective countries. In the present scenario, some diseases are emerging which are difficult to control by available antimicrobial agents such as XDR strains of *Mycobacterium tuberculosis*, HIV, Hepatitis B, Hepatitis C, Swine flu, Dengue and Japanese encephalitis. There are some antimicrobial drugs which can be used for the treatment of these diseases; however, there are some irreversible side effects associated with them such as liver damage, kidney failure, strokes etc. In recent years, the research on plant-based drugs has increased tremendously and there is some hope seen in certain medicinal plants which can be used for the treatment of these incurable diseases. The aqueous leaf extract of Bhumiamla (*Phyllanthus niruri*) has been reported to have anti-hepatitis B activity. It binds to Hb_SAg (surface antigen) and inhibits DNA polymerase required for the multiplication of the Hepatitis virus [6]. *Allium sativum*, *Acalypha indica*, *Adhatoda vasica*, *Aloe vera* and *Allium cepa* are reported to have antituberculosis activity [7].

2. Mechanism of Resistance to Antibacterial Agents

Bacteria become resistant to antimicrobial drugs through different mechanisms (Figure 1; [8]) which are discussed in the following sections.

2.1. Antibiotic Inactivation

Some pathogenic microorganisms became resistant to β-lactam antibiotic by modifying the antibiotic or releasing some enzymes such as transferases which inhibit or break down the chemical structure of antibiotics [9]. Plasmids are extrachromosomal material present in the bacteria and carry genes which encode for resistance against certain types of antibiotics. Most of the β-lactam ring-containing antibiotics such as penicillin, ampicillin, amoxicillin, imipenem, piperacillin, ceftazidime etc., become ineffective due to the production of the β-lactamase enzyme which causes hydrolysis of the amide bond in the β-lactam ring [10]. In the case of Gram-negative bacteria, the aminoglycoside group of antibiotics becomes ineffective due to the modification of the antibiotic molecule through phosphorylation, adenylation and acetylation. Over 1000 naturally occurring β-lactamase enzymes have been identified so far [11].

2.2. Target Modification

Antimicrobial agents act on a particular site where they bind and alter the normal function; this is called the target site. The bacterial cells become resistant to some antibiotics due to the modification of these target sites. The alteration or modification of the target site may be the result of constitutive and inducible enzymes produced by the bacteria. The pathogenic *Streptococcus* spp. evade the action of MLS$_B$ (macrolides, lincosamide and streptogramin B) antibiotics by preventing them from binding to the 50S ribosomal subunit and block protein synthesis by the post-transcriptional modification of the 23S rRNA component of the 50S ribosomal subunit. This is due to the methylation of the N$_6$ amino group of an adenine residue in 23S rRNA [12]. A mobile transposable genetic element called SCC*mec*, which contains the *mec*A-resistant gene, is responsible for the resistance in methicillin-resistant *S. aureus* (MRSA) [13].

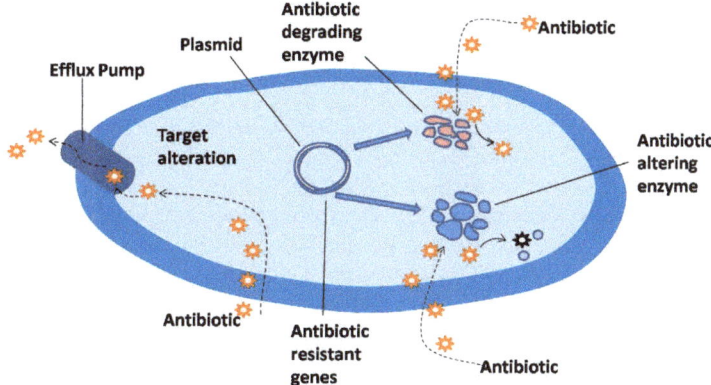

Figure 1. Various ways of resisting the action of antibiotics [8] (Reproduced with permission from IOS Press and D.I. Andersson).

Vancomycin is a glycopeptide antibiotic that inhibits the cell wall synthesis of bacteria by binding to D-Ala-D-Ala, forming a cap which results in the loss of cross linking in the polypeptide chain. Bacterial species become resistant to vancomycin by changing the usual binding site from D-Ala-D-Ala to D-alanyl-D-serine or D-alanyl-D-lactate at the C-terminus [14]. Aminoglycosides, tetracycline, oxazolidinones, chloramphenicol, fusidic acid, streptogramin, and macrolides are examples of antibiotics that inhibit the growth of bacteria by inhibiting protein synthesis or transcription. The resistance of the bacterial species to these antibiotics is due to the development of mechanisms in which a specific target is modified [15]. The resistance in *Enterococcus* species towards oxazolidinones (linezolid) is due to the effect of mutation in the 23S rRNA, which leads to reduced or feeble affinity for binding [16] and mutation in 16S rRNA causes resistance to aminoglycoside [17]. Resistance developed in *M. tuberculosis* against Streptomycin is a result of mutation in the *rpsL* gene which is responsible for encoding the ribosomal protein S12. Similarly, Fluoroquinolones interfere with the action of DNA gyrase and topoisomerase IV. In the case of *E. coli*, DNA gyrase was the primary target of quinolones while action against topoisomerase IV appeared to be limited, i.e., topoisomerase IV was shown to be a secondary target whereas in the case of *Staphylococcus*, topoisomerase IV was the primary target. In both *E. coli* and *S. aureus*, the resistance against quinolones occurred due to alteration made in either the primary target or secondary target or both.

2.3. Efflux Mechanism of Resistance

The intrinsic antibiotic resistance in bacterial genomes is caused by efflux pump proteins encoded by genes that are involved in the maintenance of cellular functions [18]. The information is available in the literature which indicates that active efflux is a mechanism of resistance for almost all antibiotics [19]. Most of the efflux mechanism systems in bacteria are non-drug-specific proteins that identify and expel chemicals, antibacterial agents and structurally unrelated compounds without any changes and degradation of the drug [20]. Expulsion of these antimicrobials or chemicals from the cell results in a low antibiotic concentration which has no or little effect on the growth of bacteria.

The ultrastructure of Gram-negative bacteria revealed that it has a two-layered outer membrane that contains a phospholipid in the inner layer and lipid A moiety of lipopolysaccharide in the outer layer. The penetration and transport of the drug across the outer membrane of Gram-negative bacteria is slightly difficult due to the composition of the outer membrane and transport of the drug is facilitated by porin proteins that form water-filled channels. The entry of antimicrobial molecules through the outer membrane of the bacterial cell may occur either by diffusion through porin or diffusion through the lipid bilayer. The chemical composition of the drug molecule is the most important determinant of the entry mode. For example, chloramphenicol and fluroquinolones penetrated the Gram-negative bacteria with the help of porin [21].

2.4. Plasmidic Efflux

The acquisition of new genetic material from other resistant organisms is responsible for the resistance in some bacteria. The transfer of genetic material between the same bacterial species or different species, other than the transfer through the parent to its progeny, is termed horizontal transfer (HGT). Bacterial species may exchange the genetic material through the processes of transformation, conjugation and transduction. These processes of genetic transmission are facilitated by a mobile genetic element, i.e., transposons. Plasmids may carry resistant genes and transmit these to other bacteria (particularly Gram-negative bacteria) through conjugation. During conjugation, pilus form between two bacterial cells, through which the genetic material or plasmid carrying resistant genes are transferred. Enterococcal pheromone-responsive plasmids constitute the mobile genetic element (MGE). The donor cell in the presence of pheromone produces a proteinaceous structure on the cell surface called aggregation substance (AS) which binds to the enterococcal binding substance (EBS) present on the surface of the recipient. A mating channel is formed between the donor and recipient that enables the transfer of the plasmid DNA. After acquiring the plasmid, the recipient stops the production of pheromone and initiates the production of a specific encoded inhibitor peptide which serves to desensitize the bacterial cell to a low level of endogenous and exogenous pheromone produced by the donor. The pheromone-responsive conjugative plasmid system has been most extensively studied for plasmids pAD1, pCF10, and pAM373 from *E. faecalis*. Pheromone-responsive plasmid contributes to the enterococcal phenotype being an important vehicle of antibiotic resistance in *E. faecalis* [22]. In 1998, the plasmid-mediated *qnr* gene which encodes quinolone resistance in bacteria was discovered, the function of which is to protect DNA gyrase from quinolones [23].

Faced with such difficulties and challenges, there is an urgent requirement to search for new antimicrobial molecules or compounds from plant sources which have a broad spectrum of activity against bacterial species as well as having immunomodulatory action. The Indian subcontinent is well known for its traditional knowledge of medicine since time immemorial; a vast number of medicinal plants are described which have immense potential to treat illnesses caused by bacterial species. In India, the use of onion, garlic and ginger as flavoring agents is well documented and well practiced owing to their medicinal value.

3. Active Compounds of Plants with Antimicrobial Properties

For the alternative antimicrobial drugs, screening of plants as a source is now being conducted all over the world. Antimicrobial properties in plants are attributed to the presence of active compounds, e.g., quinones, phenols, alkaloids, flavonoids, terpenoids, essential oil, tannins, lignans, glucosinolates and some secondary metabolites (see Table A1 for example). Other antimicrobial agents of plants include the peptides forming their defense systems which are similar to human antimicrobial peptides in structure and function. The comprehensive discussions on various plant-active compounds have been reviewed as follows.

Quinones are compounds with a fully conjugated cyclic Dione structure, such as benzoquinones which consist of two isomers of cyclohexadienedione. These compounds have the molecular formula $C_6H_4O_2$. The browning reaction in injured or cut fruits and vegetables is due to the presence of Quinones. The dyeing properties of henna (*Lawsonia inermis*) are due to the presence of quinine and it also possesses antimicrobial activity against *Pseudomonas aeruginosa* [24]. Hypericin, an anthraquinone from *Hypericum perforatum*, had general antimicrobial properties and also showed activity against methicillin-resistant and methicillin-sensitive *Staphylococcus* [25].

Alkaloids are phytochemicals commonly found in Angiosperm and rarely found in Gymnosperm. The importance of the medicinal properties of alkaloids first came into existence when morphine was isolated from *Papaver somniferum* which is generally used as pain killer. Caffeine, quinine, cineline, strychnine, brucine, emetine and narcotine are a few examples of alkaloids that have known medicinal value. Berberine is also an example of an alkaloid found in *Berberis* spp., *Cortex phellodendri* and *Rhizoma coptidis* and has antimicrobial activity against *Streptococcus agalactiae*. The mechanism of action of berberine is due to its ability to intercalate with DNA and disrupt the membrane structure by increasing the membrane permeability of bacteria [26]. Hasubanalactam alkaloid isolated from the tubers of *Stephania glabra* has antimicrobial activity against *Staphylococcus aureus*, *S. mutans*, *Microsporum gypseum*, *M. canis* and *Trichophyton rubrum* [27]. The main disadvantage of alkaloids is their toxicity which gives them a marked therapeutic effect in small quantities. That is why the use of alkaloids-based herbal preparation is not frequently used in folk medicine.

Flavonoids are well known phytochemicals that occur in a wide range of plant parts and products mainly in honey, fruits, seeds, vegetables, wines and tea. These phytochemicals are known to have antimicrobial, antiviral, antiallergic and anti-inflammatory properties [28]. The flavonoids such as kaempferol, rutin and quercetin have antifungal properties [29]. A leguminous plant *Lupinus* spp. contains dihydrofuranoisoflavones which showed antifungal activity against *Botrytis cinerea* and *Aspergillus flavus* [30].

Flavones are hydroxylated phenolics containing one carbonyl group (two in quinones), while the addition of a 3-hydroxyl group yields a flavonol. The antimicrobial activity of six flavonoids isolated from *Galium fissurense*, *Viscum album* ssp. *album* and *Cirsium hypoleucum* was shown against extended-spectrum β-lactamase, producing multidrug-resistant bacteria *K. pneumoniae* [31].

Coumarins are phenolic substances made of fused benzene and an alpha pyrone ring [32]. Coumarin isolated from *Angelica lucida* L. is active against the oral pathogens *Streptococcus mutans* and *S. viridians* [33]. Pyranocoumarins isolated from *Ferulago campestris* showed antibacterial activity against nine bacterial strains and the same clinically isolated Gram-positive and Gram-negative bacterial strains [34].

Essential oils are another example of plant secondary metabolites that have compounds with isoprene structure, also known as terpenes, with the typical formula $C_{10}H_{16}$. Different types of terpenes are known such as diterpenes, triterpenes and tetraterpenes (C20, C30, and C40), as well as hemiterpenes (C5) and sesquiterpenes (C15). When the compounds contain oxygen as an additional element, they are called terpenoids.

The terpenoids, also known as isoprenoids, are basically a different class of naturally-occurring organic chemicals similar to terpenes. These compounds are multicyclic structures and differ from one another in their basic carbon chains as well as in functional groups. These are the largest group of natural products and can be found in all classes of living things. Plant terpenoids are used for their aromatic qualities. They play a role in traditional herbal remedies and are under investigation for antibacterial, antineoplastic, and other pharmaceutical functions. The characteristic smell of *Eucalyptus*—a smell of cinnamon, cloves, and ginger—is due to the presence of terpenoids. Examples of well-known terpenoids include menthol, citral, camphor, salvinorin A in the plant *Salvia divinorum*, and the cannabinoids found in *Cannabis*.

Terpenenes or terpenoids are active against bacteria, fungi, viruses, and protozoa. *Trichodesma amplexicaule* contains a mixture of terpenoids: β-sitosterol, α-amyrin, lupeol, hexacosanoic acid, ceryl alcohol and hexacosane. The terpenoids extracted from the bark of *Acacia nilotica* have antimicrobial activity against *S. viridans*, *S. aureus*, *E. coli*, *B. subtilis* and *Shigella sonnei* [35]. The essential oil of *Cymbopogon citratus* has moderate activity against *C. albicans* and low activity against *P. aeruginosa*, *E. coli*, *S. aureus* and *T. mentagrophy* [36]. Essential oils are more active against Gram-positive bacteria than Gram-negative bacteria; the possible mechanism of action is membrane permeabilisers.

The term tannin (from *tanna*, German word for oak or fir tree) refers to the use of wood tannins from oak galls and these serve as the source of tannic acid. Tannins have an ability to combine with proteins, resulting in the tanning of animal hides into leather. Chemically, tannin is a large polyphenolic compound containing hydroxyls and carboxyl groups. Tannins present in plant impart astringent (clean the skin and constrict the skin pores) properties and cause a puckering feeling in the mouth when taken orally, e.g., red wine and unripened fruits. The presence of tannins in plants has a defensive role against predation by animals.

It has been well documented that consumption of red wine and green tea, which are good sources of tannins, can cure or prevent a variety of illness by enhancing the immune system [37]. The plant extracts containing tannins cause activation of phagocytic cells and anti-infective actions. Tannins have properties that inhibit the growth and protease activity of ruminal bacteria by binding the cell wall of bacteria [38]. The tannin of Sorghum has antimicrobial activity against *S. aureus*, *Salmonella typhimurium*, *A. niger*, *A. flavus* and *Saccharomyces cerevisae* [39].

Antimicrobial activities of peptides were first reported in 1942. Chemically, antimicrobial peptides have disulfide bonds and are positively charged. These inhibit the growth of bacteria by forming ion channels in the bacterial membrane. The positive charge of antimicrobial peptides binds to negatively charged molecules such as phospholipids, teichoic acid and lipopolysacharide and cause change in the membrane resulting in the death of the bacterial cell. Thionins are the first plant antimicrobial peptides known to kill the plant pathogens; their mechanism of action is to alter the membrane permeability of the microbial cell. Other examples of antimicrobial peptides which have shown promising antimicrobial substance include wheat α-thionin, lipid transfer proteins (LTPs), maltose binding protein (MBP)-1 (isolated from maize) [40] and Ib-AMPs (isolated from *Impateins balsamina*). Fabatin isolated from the fava bean contains 47 peptide residues that have shown antimicrobial activity against *Ps. aeruginosa*, *E. coli*, and *E. hirae* but have not shown any activity against *Saccharomyces* or *Candida* spp. [41]. Pseudothionin (Pth-st1) peptide, isolated from *Solanum tuberosum*, has antifungal and antibacterial activity against *Fusarium solani*, *Clavibacter michiganensis* and *Ps. solanacearum* respectively [42].

Lignans are a group of dimericphenylpropanoids first introduced in 1948 by Howarth. These can be formed by the condensation process of two cinnamyl alcohol/cinnamic acids through the β-carbon of the aliphatic chain. The lignans of some plants are known to have antimicrobial activity. Lignans isolated from *Pseudolarix kaempferi* were reported to have antimicrobial activity against *Candida albican* and *S. aureus* [43]. Dibenzocyclooctadiene lignin isolated from *Schissandra chinensis* inhibits the growth of *Chlamydia trachomatis* and *C. pneumonia* [44].

Glucosinolates are secondary metabolites that consist of sulphur and nitrogen, mainly produced by the Brassicaceae family. Sinigrin is one of the important glucosinolates present in broccoli, mustard and Brussels sprouts and is reported to have antifungal, antimicrobial, anticancer, antioxidant and anti-inflammatory activity [45]. Glucosinolates such as glucoiberverine, glucoiberin and glucoerucin were isolated from the seeds and leaves of *Lobularia libyca* and analysed for antimicrobial activity against *C. albicans* and *Ps. aeruginosa*. It was shown that seed hydrolysates have significant activity against both tested microorganisms [46].

The last few decades have seen a notable shift to a natural health care system and more and more people are resorting to the use of plant-based drugs. Scientific validation of the traditional health care system prevalent in tribal societies, ethnobotanical literature and plants described in Ayurveda, using modern analytical tools, is currently an active area of research. There is growing interest in testing the efficacy of medicinal plants for treating various ailments; also, individual plants as well as combinations of medicinal plants against bacterial species which have become resistant to multiple drugs are being tested. In fact, some of the Allopathic practicing doctors recommend plant-based medicine, especially in the case of liver disease (Jaundice) and other ailments such as joint pain and calculus in the kidney. Recently, Teixobactin was discovered which shows excellent activity against *S. aureus* and *M. tuberculosis* and no resistance is shown to this antibiotic [47].

4. Conclusions

Traditional medicines, including plants, have emerged as a boon in medical sciences as they are readily available and have almost no side effects. Plant derivatives have even been proved to cure HIV infection, although only to a certain extent. The identification and isolation of active compounds from the plants is still a challenge for most of the countries rich in plant diversity. *Taxus baccata* is one such example which grows on the upper Himalayas and has active compounds to cure cancer completely by inhibiting the uncontrolled proliferation of the cell. Continuous efforts are being made to explore the plant kingdom in order to find wonder drugs that could save human life from noxious microbial and viral infections such as XDR tuberculosis and HIV infection respectively. Diseases such as Hepatitis B, Hepatitis C, HIV, Swine flu, Dengue, and XDR tuberculosis are a few examples which still pose challenges in allopathic medicines. However, there is evidence which suggests that some medicinal plants are very effective in the treatment of these diseases, as mentioned in this review. The chances of urinary tract infection (UTI) in women are increased during pregnancy. In case a pregnant lady contracts UTI, then the administration of antibiotics is necessary but most of the antibiotics used for the treatment of the infection are contradictory as they may cause severe problems to the fetus. So, it is very difficult for any gynecologist or physician to suggest antibiotics. The most common causal microorganism of UTI is *E. coli* and it can be controlled or eliminated from the urinary tract by use of some medicinal plants such as *Cinnamomum* spp., decoction of Coriander leaf, *Berberis aristata*, *Mimosa pudica*, *Solanum nigrum*, *Raphanus sativus*, *Syzygium aromaticum*, etc. (Table A1). In the last 10–15 years, people have been paying more attention to herbal formulation due to its properties of little or no side effects and action on the root cause of disease. Owing to this, an upsurge in the demand for herbal-based medicines, cosmetics and nutraceuticals has been noticed in India in the last few years.

Acknowledgments: Authors are thankful to Pratti Prasad, High Altitude Plant Physiology Research Centre (HAPPRC), H.N.B. Garhwal University (A Central University), Srinagar, Garhwal-246174 (Uttarakhand), India.

Author Contributions: Harish Chandra and Abhay Prakash Mishra did the literature review and wrote the first draft of the manuscript; Anant Ram Nautiyal gave guidance and evaluated critically the literature review; Parul Bishnoi and Archana Yadav gave scientific involvement on the text; Babita Patni proposed the theme and was associated with the revision of the paper.

Conflicts of Interest: The authors declare no conflict of interest.

Appendix A

Table A1. Examples of certain plants and their antimicrobial components.

S. No.	Name of Plants	Antimicrobials	Microbes treated	References
1.	Acacia nilotica	Terpenoids, flavanoid, Saponins, Tannins	S. viridans, S. aureus, E. coli, B. subtilis, Shigella sonnei, MDR E. coli, C. albican, K. pneumoniae	[35,48]
2.	Allium cepa	Flavanoid, Polyphenol	MDR Pseudomonas aeruginosa, S. Typhi, E. coli	[49]
3.	Allium sativum	Organosulphur compounds (Phenolic compounds), Allicin	Campylobacter jejuni, MDR E. coli, C. albican, Entamoeba histolytica, Giardia lamblia	[50,51]
4.	Angelica lucida L.	Coumarins	S. viridans, S. mutans	[33]
5.	Chelidonium majus	Glycoprotein	B. cereus, Staphylococcus spp.	[52]
6.	Cinnamomum spp.	Cinnamaldehyde (essential oil)	Legionella pneumophila, MDR E. coli, C. albican, K. pneumoniae	[53]
7.	Cirsium hypoleucum	Flavones	MDR K. pneumoniae	[31]
8.	Curcuma longa	Curcuminoid (A phenolic compound), turmerone, curlone, Essential oil, curcumins, turmeric oil	S. typhi, E. coli, S. aureus, B. cereus, B. subtilis, Ps. aeruginosa, B. coagulans, A. niger, P. digitatum, Antifungal and antiviral activity	[54]
9.	Cymbopogon citratus	Essential oil	C. albicals, Aspergillus flavus, A. parasiticus	[36]
10.	Galium fissurense	Flavones	MDR K. pneumoniae	[31]
11.	Hypericum perforatum	Hypericin (anthraquinone)	Methicillin Resistant Staphylococcus aureus and Methicillin sensitive Staphylococcus	[25]
12.	Lawsonia inermis	Quinones	MDR Pseudomonas aeruginosa	[24]
13.	Medicago sativa	Saponins, Canavanine	Enterococcus faecium, S. aureus, Antifungal	[35]
14.	Mentha longifoilia	Essential oil	MDR Staphylococcus aureus	[36]
15.	Ocimum basilicum	Essential oil	MDR Staphylococcus aureus, S. Typhi, Aeromonas hydrophila, Pseudomonas spp.	[57]
16.	Onobrychis sativa	AMP's (antimicrobial peptides)	E. faecium, S. aureus	[55]
17.	Origanum vulgare	Essential oil	B. subtilis, B. cereus, MDR Staphylococcus aureus	[58]
18.	Piper longum	Piperine, Saponin, alkaloid	MDR B. subtilis, Shigella sonnei	[59]
19.	Raphanus sativum	Rs-AFP2 (Antifungal peptide)	C. albicans	[60]
20.	Rhazya stricta	Alkaloids and Non alkaloids	MDR E. coli, K. pneumoniae (ESBL), E. faecium(VRE)	[61]
21.	Rosmarinus officinalis	Essential oil	Streptococcus mutans	[58]
22.	Sanguisorba officianalis	Alkaloids, antimicrobial peptides	Ps. aeruginosa, E. coli	[52]
23.	Sorghum spp.	Tannins	S.aureus, S. typhimurium, A. niger, A. flavus, S. cerevisiae	[39]
24.	Stephania glabra	Alkaloids	S. aureus, S. mutans, Microsporum gypseum, M. canis and Trichophyton rubrum	[27]
25.	Syzygium aromaticum	Essential oil, Eugenol	Streptococcus mutans, Staphylococcus aureus, Lactobacillus acidophilus, Candida albicans and Saccharomyces cerevisiae, MDR E. coli, K. pneumoniae	[39]
26.	Vetiveria zizniodies	Vetivone (vetiver oil)	Enterobacter spp.	[62]
27.	Viscum album	Flavones	MDR K. pneumoniae	[31]
28.	Zingiber officinale	Gingerol	E. coli, Enterobacter spp., P. aeruginosa, Proteus spp., Klebsiella spp., S. aureus and Bacillus spp.	[49,63]

MDR-Multidrug Resistance, ESBL-Extended Spectrum Beta Lactamase, VRE-Vancomycin Resistant Enterococci.

1. Iwu, M.W.; Duncan, A.R.; Okunji, C.O. New antimicrobials of plant origin. In *Perspectives on New Crops and New Uses*; Janick, J., Ed.; ASHS Press: Alexandria, VA, USA, 1999; pp. 457–462.
2. Bush, K. Antibacterial drug discovery in the 21st century. *Clin. Microbiol. Inf.* **2004**, *10*, 10–17. [CrossRef] [PubMed]
3. World Health Organization (WHO). *Antimicrobial Resistance*; Fact Sheet No. 194; WHO: Geneva, Switzerland, 2002.
4. Styers, D.; Sheehan, D.J.; Hogan, P.; Sahm, D.F. Laboratory-based surveillance of current antimicrobial resistance patterns and trends among *Staphylococcus aureus*: 2005 status in the United States. *Ann. Clin. Microb. Antimicrob.* **2006**, *5*. [CrossRef] [PubMed]
5. Gandhi, N.R.; Moll, A.; Sturm, A.W.; Pawinski, R.; Govender, T.; Lalloo, U.; Zeller, K.; Andrews, J.; Friedland, G. Extensively drug-resistant tuberculosis as a cause of death in patients co-infected with tuberculosis and HIV in a rural area of South Africa. *Lancet* **2006**, *368*, 1575–1580. [CrossRef]
6. Venkateswaran, P.S.; Millman, I.; Blumberg, B.S. Effect of an extract from *Phyllanthus niruri* on hepatitis B and woodchuck hepatitis viruses: In vitro and in vivo studies. *Proc. Natl. Acad. Sci. USA* **1987**, *84*, 274–278. [CrossRef] [PubMed]
7. Gupta, R.; Thakur, B.; Singh, P.; Singh, H.B.; Sharma, V.D.; Katoch, V.M.; Chauhan, S.V.S. Anti-tuberculosis activity of selected medicinal plants against multi drug resistant *Mycobacterium tuberculosis* isolates. *Ind. J. Med. Res.* **2010**, *131*, 809–813.
8. Anderson, D.I. The ways in which bacteria resist antibiotics. *Int. J. Risk Saf. Med.* **2005**, *17*, 111–116.
9. Wright, G.D. Bacterial resistance to antibiotics: Enzymatic degradation and modification. *Adv. Drug Deliv. Rev.* **2005**, *57*, 1451–1470. [CrossRef] [PubMed]
10. Wilke, M.S.; Lovering, A.L.; Strynadka, N.C.J. β-Lactam antibiotic resistance: A current structural perspective. *Curr. Opin. Microbiol.* **2005**, *8*, 525–533. [CrossRef] [PubMed]
11. Bush, K.; Fisher, J.F. Epidemiological expansion, structural studies, and clinical challenges of new β-lactamase from Gram-negative bacteria. *Ann. Rev. Microbiol.* **2011**, *65*, 455–478. [CrossRef] [PubMed]
12. Kataja, J.; Seppala, H.; Skurnik, M.; Sarkkinen, H.; Huovinen, P. Different erythromycin resistance mechanisms in group C and group G *Streptococci*. *Antimicrob. Agents Chemother.* **1998**, *42*, 1493–1494. [PubMed]
13. Okuma, K.; Iwakawa, K.; Turnidge, J.D.; Grubb, W.B.; Bell, J.M.; O'Brien, F.G.; Coombs, G.W.; Pearman, J.W.; Tenover, F.C.; Kapi, M.; et al. Dissemination of New Methicillin-Resistant *Staphylococcus aureus* Clones in the Community. *J. Clin. Microbiol.* **2002**, *40*, 4289–4294. [CrossRef] [PubMed]
14. Hiramatsu, K. Vancomycin-resistant *Staphylococcus aureus*: A new model of antibiotic resistance. *Lancet Infect. Dis.* **2001**, *1*, 147–155. [CrossRef]
15. Happi, C.T.; Gbotosho, G.O.; Folarin, O.A.; Akinboye, D.O.; Yusuf, B.O.; Ebong, O.O.; Sowunmi, A.; Kyle, D.E.; Milhous, W.; Wirth, D.T.; et al. Polymorphisms in *Plasmodium falciparum dhfr* and *dhps* genes and age related in vivo sulfaxine-pyrimethamine resistance in malaria-infected patients from Nigeria. *Acta Trop.* **2005**, *95*, 183–193. [CrossRef] [PubMed]
16. Wang, G.; Taylor, D.E. Site-specific mutations in the 23S rRNA gene of *Helicobacter pylori* confer two types of resistance to macrolide-lincosamide-streptogramin B antibiotics. *Antimicrob. Agents Chemother.* **1998**, *42*, 1952–1958. [PubMed]
17. Suzuki, Y.; Katsukawa, C.; Tamaru, A.; Abe, C.; Makino, M.; Mizuguch, Y.; Taniguchi, I.H. Detection of kanamycin-resistant *Mycobacterium tuberculosis* by identifying mutations in the 16S rRNA gene. *J. Clin. Microbiol.* **1998**, *36*, 1220–1225. [PubMed]
18. Lomovskaya, O.; Bostian, K.A. Practical applications and feasibility of efflux pump inhibitors in the clinic—A vision for applied use. *Biochem. Pharmacol.* **2006**, *7*, 910–918. [CrossRef] [PubMed]
19. Lin, J.; Michel, L.O.; Zhang, Q. Cme ABC functions as a multidrug efflux system in *Campylobacter jejuni*. *Antimicrob. Agents Chemother.* **2002**, *46*, 2124–2131. [CrossRef] [PubMed]
20. Kumar, A.; Schweizer, H.P. Bacterial resistance to antibiotics: Active efflux and reduced uptake. *Adv. Drug Deliv. Rev.* **2005**, *57*, 1486–1513. [CrossRef] [PubMed]
21. Nikaido, H. Molecular basis of bacterial outer membrane permeability revisited. *Microbiol. Mol. Biol. Rev.* **2003**, *67*, 593–656. [CrossRef] [PubMed]

22. Wardal, E.; Sadowy, E.; Hryniewicz, W. Complex nature of *Enterococcal* pheromone-responsive plasmids. *Pol. J. Microbiol.* **2010**, *59*, 79–87. [PubMed]
23. Martinez-Martinez, L.; Pascual, A.; Jacoby, G.A. Quinolone resistance from a transferable plasmid. *Lancet* **1998**, *351*, 797–799. [CrossRef]
24. Habbal, O.; Hasson, S.S.; El-Hag, A.H.; Al-Mahrooqui, Z.; Al-Hashmi, N.; Al-Balushi, M.S.; Al-Jabri, A.A. Antibacterial activity of *Lawsonia inermis* Linn (Henna) against *Pseudomonas aeruginosa*. *Asian Pac. J. Trop. Biomed.* **2011**, *1*, 173–176. [CrossRef]
25. Dadgar, T.; Asmar, M.; Saifi, S.; Mazandarani, M.; Bayat, H.; Moradi, A.; Bazueri, M.; Ghaemi, F. Antibacterial activity of certain Iranian medicinal plants against methicillin resistant and methicillin sensitive *Staphylococcus aureus*. *Asian J. Plant Sci.* **2006**, *5*, 861–866.
26. Peng, L.; Kang, S.; Yin, Z.; Jia, R.; Song, X.; Li, L.; Li, Z.; Zou, Y.; Liang, X.; Li, L.; et al. Antibacterial activity and mechanism of berberine against *Streptococcus agalactiae*. *Int. J. Clin. Exp. Pathol.* **2015**, *8*, 5217–5223. [PubMed]
27. Semwal, D.K.; Rawat, U. Antimicrobial hasubanalactam alkaloid from *Stephania glabra*. *Planta Med.* **2009**, *75*, 378–380. [CrossRef] [PubMed]
28. Gabor, M. Anti-inflammatory and anti-allergic properties of flavonoids. In *Plant Flavonoids in Biology and Medicine: Biochemical, Pharmacological and Structure-Activity Relationships*; Cody, V., Middleton, V.E., Harbourne, J.B., Eds.; Alan R. Liss: New York, NY, USA, 1986; pp. 471–480.
29. Beschia, M.; Leonte, A.; Oancea, I. Phenolic compounds with biological activity. *Bull. Univ. Galati Faso* **1984**, *6*, 23–27.
30. Thara, S.; Ingham, J.; Nakahara, S.; Mizutani, J.; Harborne, J.B. Fungitoxic dihydrofuranoisoflavones and related compounds in white lupin, *Lupinus albus*. *Phytochemistry* **1984**, *23*, 1889–1900. [CrossRef]
31. Ozcelik, B.; Deliorman, O.D.; Ozgen, S.; Ergun, F. Antimicrobial Activity of Flavonoids against Extended-Spectrum Beta Lactamase (ESBL)-Producing *Klebsiella pneumoniae*. *Trop J. Pharm. Res.* **2008**, *7*, 1151–1157. [CrossRef]
32. O'Kennedy, R.; Thornes, R.D. *Coumarins: Biology, Applications and Mode of Action*; John Wiley & Sons, Inc.: New York, NY, USA, 1997.
33. Widelski, J.; Popova, M.; Graikou, K.; Glowniak, K.; Chinou, I. Coumarins from *Angelica lucida* L.—Antibacterial Activities. *Molecules* **2009**, *14*, 2729–2734. [CrossRef] [PubMed]
34. Basile, A.; Sorbo, S.; Spadaro, V.; Bruno, M.; Maggio, A.; Faraone, N.; Rosselli, S. Antimicrobial and antioxidant activities of Coumarins from the roots of *Ferulago campestris* (Apiaceae). *Molecules* **2009**, *14*, 939–952. [CrossRef] [PubMed]
35. Banso, A. Phytochemical and antibacterial investigation of bark extracts of *Acacia nilotica*. *J. Med. Plants Res.* **2009**, *3*, 82–85.
36. Ragaso, C.Y. Antimicrobial and Cytotoxic terpenoid from *Cymbopogon citratus* Stapf. *Philipp. Sci.* **2008**, *45*, 111–122. [CrossRef]
37. Serafini, M.; Ghiselli, A.; Ferro-Luzzi, A. Red wine, tea and anti-oxidants. *Lancet* **1994**, *344*, 626. [CrossRef]
38. Jones, S.B., Jr.; Luchsinger, A.E. *Plant Systematics*, 2nd ed.; McGraw-Hill Book Co.: New York, NY, USA, 1986.
39. Moneim, A.; Suleman, E.; Issa, F.M.; Elkhalifa, E.A. Quantitative determination of tannin content in some sorghum cultivars and evaluation of its antimicrobial activity. *Res. J. Microbiol.* **2007**, *2*, 284–288.
40. Garcia-Olmedo, F.; Molina, A.; Sefura, A.; Moreno, M. The defensive role of nonspecific lipid transfer proteins in plants. *Trends Microbiol.* **1995**, *3*, 72–74. [CrossRef]
41. Zhang, Y.; Lewis, K. Fabatins: New antimicrobial plant peptides. *FEMS Microbiol. Lett.* **1997**, *149*, 59–64. [CrossRef] [PubMed]
42. Moreno, M.; Segura, A.; Garcia-Olmedo, F. Pseudothionin-St1, a potato peptide active against potato pathogens. *Eur. J. Biochem.* **1994**, *223*, 135–139. [CrossRef] [PubMed]
43. He, W.J.; Chu, H.B.; Zhang, Y.M.; Han, H.J.; Yan, H.; Zeng, G.Z.; Fu, Z.H.; Olubanke, O.; Tan, N.H. Antimicrobial, cytotoxic lignans and terpenoids from the twigs of *Pseudolarix kaempferi*. *Planta Med.* **2011**, *77*, 1924–1931. [CrossRef] [PubMed]
44. Hakala, E.; Hanski, L.; Uvell, H.; Yrjonen, T.; Vuorela, H.; Elofsson, M.; Vuorela, P.M. Dibenzocyclooctadiene lignans from *Schisandra* spp. selectively inhibit the growth of the intracellular bacteria *Chlamydia pneumoniae* and *Chlamydia trachomatis*. *J. Antibiot.* **2015**, *68*, 609–614. [CrossRef] [PubMed]

45. Mazumdaer, A.; Dwivedi, A.; du Plessis, J. Sinigrin and its therapeutic bebefits. *Molecules* **2016**, *21*, 416. [CrossRef] [PubMed]
46. Al-Gendy, A.A.; Nematallah, K.A.; Zaghloul, S.S.; Ayoub, N.A. Glucosinolates profile, volatile constituents, antimicrobial and cytotoxic activities of *Lobularia libyca*. *Pharm. Biol.* **2016**, *54*, 3257–3263. [CrossRef] [PubMed]
47. Losee, L.L.; Tanja, S.; Aaron, J.P.; Amy, L.S.; Ina, E.; Brian, P.C.; Anna, M.; Till, F.S.; Dallas, E.H.; Slava-Michael, J.; et al. A new antibiotic kills pathogen without detectable resistance. *Nature* **2015**, *517*, 455–459.
48. Riaz, S.; Faisal, M.; Hasnain, S.; Khan, N.A. Antibacterial and cytotoxic activities of *Acacia nilotica* L (Mimosaceae) methanol extract against extended spectrum beta lactamase producing *E. coli* and *Klebsiella* species. *Trop. J. Pharm. Res.* **2011**, *10*, 785–791. [CrossRef]
49. Adesino, G.O.; Jibo, S.; Aggu, V.E.; Ehinmidu, J.O. Antibacterial activity of fresh juices of *Allium cepa* and *Zingiber ofiicinale* against multidrug resistant bacteria. *Int. J. Pharm. Biol. Sci.* **2011**, *2*, 289–295.
50. Lu, X.; Rasco, B.A.; Jabal, J.M.; Aston, D.E.; Lin, M.; Konkel, M.E. Investigating antibacterial effects of garlic (*Allium sativum*) concentrate and garlic-derived organosulfur compounds on *Campylobacter jejuni* by using Fourier transform infrared spectroscopy, Raman spectroscopy, and electron microscopy. *Appl. Environ. Microbiol.* **2011**, *77*, 5257–5269. [CrossRef] [PubMed]
51. Ankri, S.; Mirelman, D. Antimicrobial properties of allicin from garlic. *Microb. Infect.* **1999**, *1*, 125–129. [CrossRef]
52. Janovska, D.; Kubikova, K.; Kokoska, L. Screening for antimicrobial activity of some medicinal plants species of traditional Chinese medicine. *Czech J. Food Sci.* **2003**, *21*, 107–110.
53. Chang, C.; Chang, W.; Chang, S.; Cheng, S. Antibacterial activities of plant essential oils against *Legionella pneumophila*. *Water Res.* **2008**, *42*, 278–286. [CrossRef] [PubMed]
54. Gul, P.; Bakht, J. Antimicrobial activity of turmeric extract and its potential use in food industry. *J. Food Sci. Technol.* **2015**, *52*, 2272–2279. [CrossRef] [PubMed]
55. Aliahmadi, A.; Roghanian, R.; Emtiazi, G.; Mirzajani, F.; Ghassempour, A. Identification and primary characterization of a plant antimicrobial peptide with remarkable inhibitory effects against antibiotic resistant bacteria. *Afr. J. Biotechnol.* **2012**, *11*, 9672–9676.
56. Al-Ali, K.; Abdelrazik, M.; Hemeg, H.; Ozbak, H. Antibacterial activity of four herbal extract against methicillin resistant bacterial strains isolated from patient in Almadinah hospitals, Saudi Arabia. *Int. J. Acad.Sci. Res.* **2015**, *3*, 34–40.
57. Wan, J.; Wilcock, A.; Coventry, M.J. The effect of essential oil of basil on the growth of *Aeromonas hydrophila* and *Pseudomonas fluorescens*. *J. Appl. Microbiol.* **1998**, *84*, 152–158. [CrossRef] [PubMed]
58. Falco, E.D.; Mancini, E.; Roscigno, G.; Mignola, E.; Taglialatela-Scafati, O.; Senatore, F. Chemical composition and biological activity of essential oils of *Origanum vulgare* L. subsp. *vulgare* L. under different growth conditions. *Molecules* **2013**, *18*, 14948–14960. [PubMed]
59. Kumar, V.; Shriram, V.; Mulla, J. Antibiotic resistance reversal of multiple drug resistant bacteria using *Piper longum* fruit extract. *J. Appl. Pharm. Sci.* **2013**, *3*, 112–116.
60. Aerts, A.M.; Carmona-Gutierrez, D.; Lefevre, S.; Govaert, G.; Francois, I.E.; Madeo, F.; Santos, R.; Cammue, B.P.; Thevissen, K. The antifungal plant defensin RsAFP2 from radish induces apoptosis in a metacaspase independent way in *Candida albicans*. *FEBS Lett.* **2009**, *583*, 2513–2516. [CrossRef] [PubMed]
61. Khan, R.; Baeshen, M.N.; Saini, K.S.; Bora, R.S.; Al-Hejin, A.M.; Baeshen, N.A. Antibacterial activities of *Rhazyastricta* leaf extracts against multidrug resistant human pathogens. *Biotechnol. Biotechnol. Equip.* **2016**, *30*, 1016–1025. [CrossRef]
62. Srivastava, J.; Chandra, H.; Singh, N. Allelopathic response of *Vetiveria zizanioides*(L.) Nash on members of Enterobacteriaceae and *Pseudomonas* spp. *Environmentalist* **2007**, *27*, 253–260. [CrossRef]
63. Karuppiah, P.; Rajaram, S. Antibacterial effect of *Allium sativum* cloves and *Zingiber officinale* rhizomes against multiple drug resistant chemical pathogens. *Asian Pac. J. Trop. Biomed.* **2012**, *2*, 597–601. [CrossRef]

© 2017 by the authors. Licensee MDPI, Basel, Switzerland. This article is an open access article distributed under the terms and conditions of the Creative Commons Attribution (CC BY) license (http://creativecommons.org/licenses/by/4.0/).

Article

Influence of Boiling, Steaming and Frying of Selected Leafy Vegetables on the In Vitro Anti-inflammation Associated Biological Activities

K. D. P. P. Gunathilake [1,2], K. K. D. S. Ranaweera [2] and H. P. V. Rupasinghe [3,*]

[1] Department of Food Science & Technology, Faculty of Livestock, Fisheries & Nutrition, Wayamba University of Sri Lanka, Makandura, Gonawila 10250, Sri Lanka; gunathilakep@Dal.Ca
[2] Department of Food Science and Technology, Faculty of Applied Sciences, University of Sri Jayewardenepura, Gangodawila, Nugegoda 60170, Sri Lanka; kkdsran@yahoo.com
[3] Department of Plant, Food, and Environmental Sciences, Faculty of Agriculture, Dalhousie University, Truro, NS B2N 5E3, Canada
* Correspondence: vrupasinghe@dal.ca; Tel.: +1-902-893-6623

Received: 14 February 2018; Accepted: 13 March 2018; Published: 16 March 2018

Abstract: The aim of the present study was to evaluate the effect of cooking (boiling, steaming, and frying) on anti-inflammation associated properties *in vitro* of six popularly consumed green leafy vegetables in Sri Lanka, namely: *Centella asiatica*, *Cassia auriculata*, *Gymnema lactiferum*, *Olax zeylanica*, *Sesbania grnadiflora,* and *Passiflora edulis*. The anti-inflammation associated properties of methanolic extracts of cooked leaves were evaluated using four *in vitro* biological assays, namely, hemolysis inhibition, proteinase inhibition, protein denaturation inhibition, and lipoxygenase inhibition. Results revealed that the frying of all the tested leafy vegetables had reduced the inhibition abilities of protein denaturation, hemolysis, proteinase, and lipoxygenase activities when compared with other food preparation methods. Steaming significantly increased the protein denaturation and hemolysis inhibition in *O. zeylanica* and *P. edulis*. Steaming of leaves increased inhibition activity of protein denaturation in *G. lactiferum* (by 44.8%) and *P. edulis* (by 44%); hemolysis in *C. asiatica*, *C. auriculata,* and *S. grandiflora*; lipoxygenase inhibition ability in *P. edulis* (by 50%), *C. asiatica* (by 400%), and *C. auriculata* leaves (by 250%); proteinase inhibition in *C. auriculata* (100%) when compared with that of raw leaves. In general, steaming and boiling in contrast to frying protect the health-promoting properties of the leafy vegetables.

Keywords: plant-food; processing; nutraceuticals; inflammation; health

1. Introduction

Many degenerative human diseases, such as cancer, inflammation, and cardiovascular diseases have been recognized as a consequence of free radical damage. [1] Inflammation is a part of the complex biological response of vascular tissues to harmful stimuli, which is frequently linked with pain and involves many biological occurrences, such as an increase of vascular permeability, an increase of protein denaturation, and membrane alteration [2]. Numerous recent studies have shown that chronic inflammation is associated with a wide range of progressive diseases such as cancer, neurological disease, metabolic disorder, and cardiovascular disease [3]. Therefore, there have been many studies undertaken on how to delay or prevent the onset of these chronic diseases, as these lead to global health problems. The most likely and practical way to fight against degenerative diseases, such as inflammation, is to improve body antioxidant status, which could be achieved by higher consumption of vegetables and fruits [4]. Foods from plant origin usually contain natural antioxidants that can scavenge free radicals [5]. Green leafy vegetables are rich sources of minerals and antioxidant vitamins [6], as well as other antioxidant compounds, such as polyphenols and carotenoids [7,8] Among the leafy vegetables that are available in Sri Lanka, *Cassia auriculata, Gymnema lactiferum, Olax zeylanica, Sesbania grnadiflora, Passiflora edulis,* and *Centella asiatica* are reported to possess higher antioxidant activities, as described in Gunathilake and Ranaweera [7]. Although green leafy vegetables are considered as potential sources of dietary antioxidants and bioactives, only a few studies may have been reported on evaluating the impact of cooking on the anti-inflammatory properties. Therefore, this study aimed to investigate the influence of cooking of selected leafy vegetables on their anti-inflammation associated biological properties in vitro, which were measured by heat-induced hemolysis assay, protein denaturation assay, proteinase inhibition assay, and lipoxygenase inhibition assay.

2. Results and Discussion

2.1. Protein Denaturation

Denaturation of proteins is a well-documented cause of inflammation [9]. As part of the investigation on the changes in the anti-inflammatory activity, the ability of cooked leaf extracts on protein denaturation was studied. Figure 1 shows the effect of cooking of leafy vegetables on protein denaturation inhibitory activity. Protein denaturation inhibition ability significantly increased ($p < 0.05$) in steamed leaves of *O. zeylanica* (by 44.8%) and *P. edulis* (by 44%) when compared with that of their fresh leaves. Interestingly, frying of all the tested leafy vegetables resulted in the lowest protein denaturation inhibition ability when compared with other processing methods. Boiled leaves of *O. zeylanica, C. auriculata, S. grandiflora,* and *G. lactiferum* have shown significantly lower ($p < 0.05$) inhibition ability when compared with their fresh leaves, whereas boiled leaves of *C. asiatica,* and *P. edulis* have shown a significantly higher ($p < 0.05$) protein denaturation inhibition ability. Significantly lower protein denaturation inhibition ability was observed in all of the cooked leaves of *C. auriculata, S. grandiflora,* and *G. lactiferum* when compared with their fresh leaves. In a study, methanol extract of whole plant of *Oxalis corniculata* Linn (Family: Oxalidaceae) was assessed by Sakat et al. [9] for its anti-inflammatory activity by in vitro methods and was reported that the extract showed anti-inflammatory activity by inhibiting the heat-induced albumin denaturation with the IC_{50} values of 288.04 ± 2.78 µg/mL, respectively. Based on another study, a flavonoid-rich fraction of *M. myristica* have shown an albumen denaturation in a dose-dependent manner and the maximum inhibition of denaturation of albumin was found to around 75.38% ± 0.56% at 350 µg/mL: IC_{50} value of 258 µg/mL, while a standard anti-inflammatory drug (aspirin) showed a maximum inhibition of 98.41% ± 0.13% at the same concentration [10]. However, the precise mechanism of this membrane stabilization was yet to be elucidated. It has been proposed that the plant extracts might inhibit the release of the lysosomal content of neutrophils at the site of inflammation [11]. According to Chou [12], these neutrophils lysosomal constituents include bactericidal enzymes and proteinases which upon extracellular release cause further tissue inflammation and damage.

Table 1 shows the correlation of anti-inflammatory properties with polyphenols, flavonoids, and carotenoids of green leafy vegetables. The changes in protein denaturation ability with the cooking treatments may be related to the changes in polyphenols and flavonoids content. In a study, it was found that heating of green leafy vegetables reduced the vitamin C content, thus reducing properties and free radicals scavenging properties [5].

Table 1. Pearson correlations between major bioactives (total phenolics, total flavonoids, total carotenoids) and % inhibition of protein denaturation, hemolysis denaturation, lipoxygenase activity, and proteinase activity of cooked leafy vegetables.

Parameters	r	P
Total phenolics versus protein denaturation	0.646	0.001
Total phenolics versus hemolysis	0.294	0.024
Total phenolics versus lipoxygenase activity	0.558	0.000

Table 1. Cont.

Parameters	r	P
Total phenolics versus proteinase activity	0.594	0.001
Total flavonoids versus protein denaturation	0.519	0.000
Total flavonoids versus hemolysis	0.444	0.000
Total flavonoids versus lipoxygenase activity	0.592	0.001
Total flavonoids versus proteinase activity	0.666	0.000
Total carotenoids versus protein denaturation	0.106	0.420
Total carotenoids versus hemolysis	0.203	0.120
Total carotenoids versus lipoxygenase activity	0.564	0.000
Total carotenoids versus proteinase activity	0.634	0.000

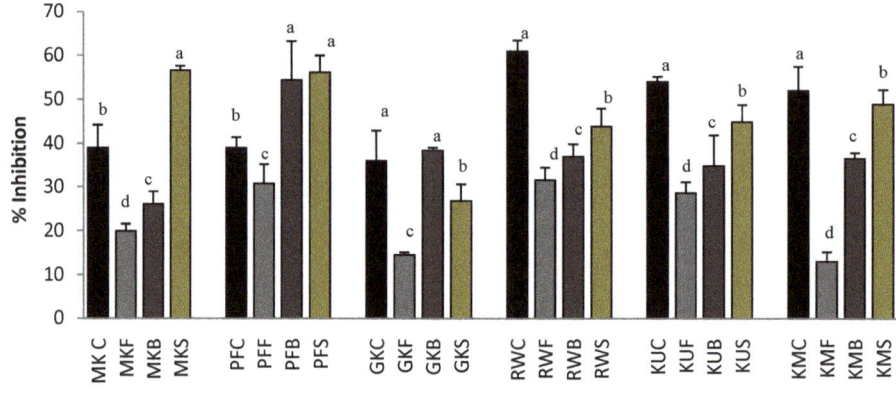

Figure 1. Protein denaturation inhibition ability of raw and cooked extracts of some GLV. MK, *O. zeylanica*; RW, *C. auriculata*; KM, *S. grandiflora*; KU, *G. lactiferum*; PF, *P. edulis*; GK, *C. asiatica*. C-fresh leaves; F-fried; B-boiled; S-steamed. Data are presented as means ± standard deviations of three replicate determinations. Columns with different letters for each vegetable are significantly different ($p < 0.05$).

2.2. Heat-Induced Hemolysis

According to Chippada and co-authors [13], lysosomal enzymes that are released during inflammation leads to the tissue injury by damaging the macromolecules, such as proteins, lipids, DNA, etc. Further, it damages tissues by lipid peroxidation of membranes, which are assumed to be responsible for certain pathological conditions as heart attacks and rheumatoid arthritis, etc. Therefore, the stabilization of lysosomal membrane is vital in controlling the inflammatory response by inhibiting the release of lysosomal constituents of activated neutrophil, such as bactericidal enzymes and proteases, which may lead to further tissue inflammation and damage upon extracellular release or by stabilizing the lysosomal membrane [13]. The membrane of the human's red blood cell is analogous to the lysosomal membrane, and its stabilization implies that the extract may as well stabilize lysosomal membranes. The in vitro bioassay that was used in this study determines the stabilization of human red blood cell membrane by hypo tonically induced membrane lysis, and this can be taken as an in vitro measure of anti-inflammatory activity of the many drugs or various plant extracts [13].

The heat-induced hemolysis inhibition abilities of raw and cooked leaf samples are shown in Figure 2. Similarly, the frying process significantly reduced ($p < 0.05$) the hemolysis inhibition ability of *O. zeylanica*, *C. auriculata*, *S. grandiflora*, *G. lactiferum*, *P edulis*, and *C. asiatica* by 4.0%, 5.5%, 6.0%,

6.2%, 7.0%, and 5.1%, respectively. Boiled leaves of *C. asiatica, C. auriculata,* and *S. grandiflora* have shown a significantly higher ($p < 0.05$) hemolysis inhibition ability hen compared with that of raw and their other cooked leaves. Steamed leaves of *O. zeylanica, P. edulis,* and *G. lactiferum* have shown the similar or higher hemolytic inhibition ability than that of their raw and other cooked leaves. Steamed and boiled leaves of *S. grandiflora* showed significantly higher ($p < 0.05$) hemolytic inhibition ability when compared with that of its raw leaves. In a previous study, the methanolic extract of the whole plant of *Oxalis corniculata* Linn has been shown the red blood cells membrane stabilization with the IC_{50} values of 467.1 ± 9.6 µg/mL [9]. Table 1 shows a poor correlation of heat-induced hemolysis with polyphenols, flavonoids, and carotenoids of green leafy vegetables.

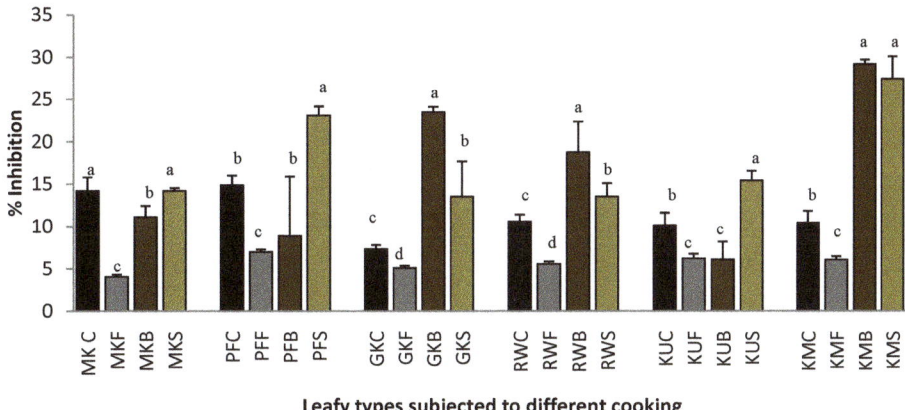

Figure 2. Heat-induced hemolysis inhibition ability of raw and cooked extracts of some GLV. Values represent means of triplicate readings. MK, *O. zeylanica*; RW, *C. auriculata*; KM, *S. grandiflora*; KU, *G. lactiferum*; PF, *P. edulis*; GK, *C. asiatica*. C-fresh leaves; F-fried; B-boiled; S-steamed. Data are presented as means ± standard deviations of three replicate determinations. Columns with different letters for each vegetable are significantly different ($p < 0.05$).

2.3. Lipoxygenase Inhibitory Activity.

The mechanism of anti-inflammation involves a series of events in which metabolism of Arachidonic acid plays an important role [10] Arachidonic acid is cleaved from the membrane phospholipids upon appropriate stimulation of neutrophils, and can be converted to leukotrienes and prostaglandins through the action of lipoxygenase and cyclooxygenase pathways, respectively [10]. Lipoxygenase enzymes catalyze the oxidation of Arachidonic acid (linoleic acid) to produce leukotrienes that are important mediators in a variety of inflammatory events [14]. In a previous study, it was reported that the essential oil of *Cymbopogon giganteus* from Benin has potential to be used as an anti-inflammatory agent towards lipoxygenase inhibition [14]. Therefore, use of in vitro inhibition of lipoxygenase could be a good model for the screening of plants with inflammatory potentials. Figure 3 shows the lipoxygenase inhibition ability of raw and cooked leafy types, and the results clearly showed that the lipoxygenase inhibition ability had reduced during frying in all of the leaf varieties when compared with their fresh, boiled, and steamed leaves. Boiling of leaves increased the lipoxygenase inhibition ability significantly ($p < 0.05$) in *O. zeylanica, P. edulis, C. asiatica,* and *C. auriculata* leaves when compared with that of their raw leaves. However, boiling has increased the lipoxygenase inhibition ability in *S. grandiflora* as compared with its steamed leaves, though it is lower than its fresh leave. Interestingly, cooked leaves of *G. lactiferum* have shown significantly lower ($p < 0.05$) lipoxygenase inhibition ability when compared with that of its raw leaves.

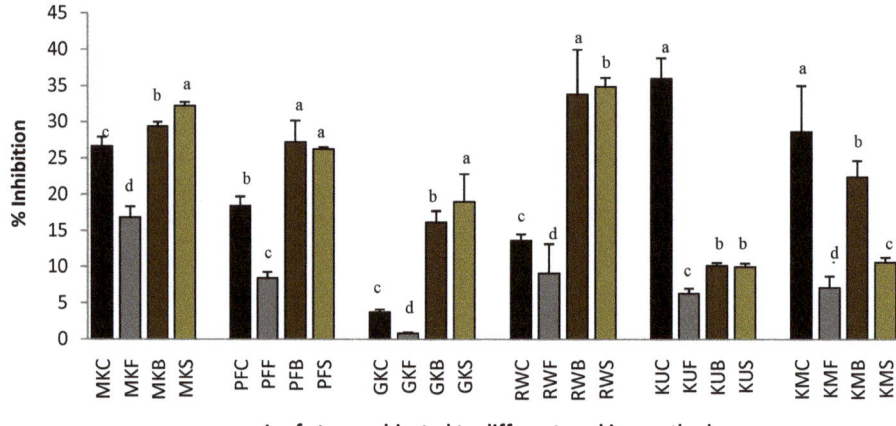

Figure 3. Lipoxygenase inhibition ability of raw and cooked extracts of some GLV. Values represent means of triplicate readings. MK, *O. zeylanica*; RW, *C. auriculata*; KM, *S. grandiflora*; KU, *G. lactiferum*; PF, *P. edulis*; GK, *C. asiatica*. C-fresh leaves; F-fried; B-boiled; S-steamed. Data are presented as means ± standard deviations of three replicate determinations. Columns with different letters for each vegetable are significantly different ($p < 0.05$).

2.4. Proteinase Inhibitory Activity

Plant extracts have been reported to inhibit protein denaturation. Although, the precise mechanism of this membrane stabilization was yet to be elucidated, but it has been proposed that the extract might inhibit the release of the lysosomal content of neutrophils at the site of inflammation [9]. These neutrophils lysosomal constituents include bactericidal enzymes and proteinases, which, upon extracellular release, cause further tissue inflammation and damage [12].

Figure 4 demonstrates the proteinase inhibition ability of raw and cooked leafy types. Fried leaves of *P. edulis*. *C. asiatica,* and *G. lactiferum* showed significantly lower ($p < 0.05$) inhibition % when compared with that of their raw, boiled, and steamed leaves. Raw leaves of *G. lactiferum* and *S. grandiflora* have shown higher proteinase inhibition ability compared with their all cooked forms. However, all of the cooked leaves of *C. auriculata* have shown similar or higher proteinase inhibition ability compared with that of its raw leaves. For examples, boiling and steaming of leaves of *C. auriculata*, increased the proteinase inhibitory activity by 54.2% and 56.7%, respectively. Boiled leaves of *O. zeylanica* exhibited a similar ability of proteinase inhibition to its raw leaves. However, fried and steamed leaves of *O. zeylanica* showed significantly lower ($p < 0.05$) proteinase inhibition ability when compared with that of boiled and raw leaves. About 80% reduction in proteinse inhibitory activity was observed in *Gymnema lactiferum*. According to the Table 1, anti-inflammatory associated activities are correlated (<0.5) with polyphenols, carotenoids, and flavonoids of green leafy vegetables. Therefore, the changes in proteinase inhibition ability with the cooking treatments may be related to the changes in polyphenols, flavonoids, and carotenoids content in these leaves.

In this study, it was found that boiling and steaming process might increase or decrease the anti-inflammatory activities. The frying process reduces the anti-inflammatory activity of all leafy types. The changes in anti-inflammatory activities of these studied leafy types in the cooking process may be due to changes in bioactives, such as polyphenols, flavonoids, and carotenoids. Wachtel-Galor and co-authors [15] have reported that an increase in polyphenols, such as flavonoids, after subsequent boiling or steaming, may be related to an enhanced availability for extraction, to a more efficient release of polyphenols or flavonoids compounds from intracellular proteins and altered cell wall structures. However, according to Palermo et al. [16], the more intense the cooking treatment, such as frying, the greater the flavonoid degradation. Further, High frying temperatures, in fact, could cause the oil to produce hydroperoxide free radicals and accelerate the degradation of carotenoids, as well lead to a reduction in their bioactivity [17]. Accordingly, variation in losses and gains of phenolics like bioactives due to cooking treatments in studied leafy types could be due to the types of cooking, the nature of leaves, and the forms of the bioactives that are present in the plant matrices.

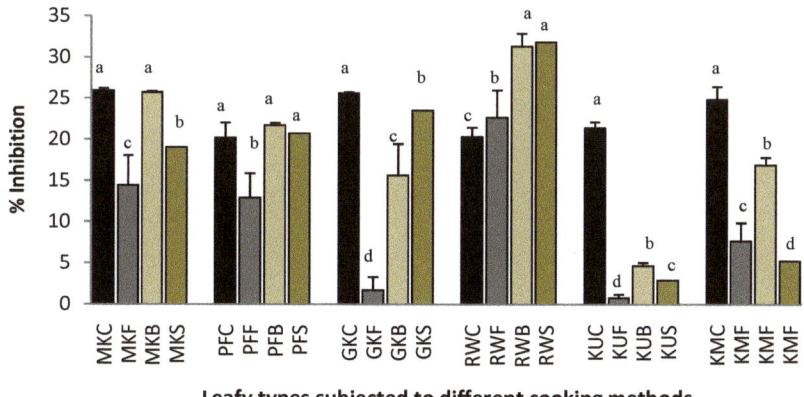

Figure 4. Proteinase inhibition ability of raw and cooked extracts of some GLV. Values represent means of triplicate readings. MK, *O. zeylanica*; RW, *C. auriculata*; KM, *S. grandiflora*; KU, *G. lactiferum*; PF, *P. edulis*; GK, *C. asiatica*. C-fresh leaves; F-fried; B-boiled; S-steamed. Data are presented as means ± standard deviations of three replicate determinations. Columns with different letters for each vegetable are significantly different ($p < 0.05$).

3. Materials and Methods

3.1. Materials

Fresh green leafy vegetable samples; Ranawara (*Cassia auriculata*), mella (*Olax Zeylanica*), Gotukola (*Centella asiatica*), Ceylon cow tree (*Gymnema lactiferum*), Kathurumurunga (*Sesbania grandiflora*), and Passion fruit (*Passiflora edulis*) were collected in Gampaha and Kurunegala districts in Sri Lanka. All of the chemicals were of analytical grade and were purchased from Sigma Aldrich, St. Louis, MO, USA through Analytical Instrument Pvt Ltd., Colombo, Sri Lanka.

3.2. Preparation of Cooked Samples

Cleaned leaf samples were subjected to different cooking treatments separately at atmospheric pressure. Cooking conditions were selected based on preliminary trials. The cleaned and washed leaves were cut into small pieces, and the samples (400 g) were divided into four parts (100 g each), keeping one portion as control (uncooked, stored at 4 °C in the refrigerator until use for within 24 h), and the rest was subjected to different cooking treatments, as indicated below. Briefly, for boiling, leaf samples (100 g) were added to boiling tap water (150 mL) in a covered stainless-steel pot and were cooked on a moderate flame for 5 min, and then samples were drained off and cooled rapidly on plenty of ice. For steaming, leaf samples were placed on a perforated tray in a stainless steel steamer covered over boiling water for 5 min, and then samples were rapidly cooled on ice. For frying, leaves were added to 500 mL of white coconut oil ("N Joy", Adamji and Lukmangi pvt Ltd., Colombo, Sri Lanka) in a stainless steel pan at 170 °C and stirred for three minutes until the sample became crisp-tender. At the end of each trial, the samples were drained off and dabbed with blotting paper to allow for the absorption of exceeding oil. The cooked leaf samples were homogenized and stored at −18 °C. As anti-inflammatory activities were calculated according to dry matter basis, moisture contents of the cooked samples were determined according to the method described by Turkmen et al. [18].

3.3. Preparation of Methanolic Extracts.

Methanolic extracts of leaves were prepared according to the method described by Gunathilake & Ranaweera [8]. Briefly, one gram of cooked leafy vegetable samples was weighed and mixed with 8mL of 70% methanol and vortexed at high speed for thirty minutes, and then centrifuged (Hettich, EBA 20, Hettich GmbH & Co., Tuttlingen, Germany) for 10 min at 792 g. The extracts were subsequently filtered through a filter paper (Whatman No. 42, Whatman Paper Ltd, Maidstone, UK. The solvents that remained in crude extracts were removed using a rotary evaporator (HAHNVAPOR, Model HS-2005 V, HAHNSHIN Scientific, Seoul, Korea) at 40 °C. The prepared concentrated extracts were oven dried at 40 °C for 12 h and were stored at −18 °C in air-tight screw-capped cryogenic vials until they assayed within one week. Extracts were dissolved in methanol to obtain a concentration of 3 mg/mL for each assay.

3.4. Anti-Inflammatory Properties

3.4.1. Heat-Induced Hemolysis

Erythrocytes suspension was prepared by the method described by Shinde et al. [19], with some modifications [20]. Briefly, blood was obtained from a healthy human volunteer and transferred to heparinized centrifuge tubes and centrifuged at 3000 rpm for 5 min and washed three times with equal volume of normal saline (0.9% sodium chloride). The volume of the blood was measured and reconstituted as a 10% (v/v) suspension with isotonic buffer solution (10 mM sodium phosphate buffer pH 7.4, the composition of the buffer solution (g/L) was NaH_2PO_4 (0.2), Na_2HPO_4 (1.15) and NaCl (9.0). Heat-induced hemolysis was carried out, as described by Okoli et al. [21], with some modifications. About 0.05 mL of blood cell suspension and 0.05 mL extracts of cooked leaves were mixed with 2.95 mL phosphate buffer (pH 7.4), and the mixture was mixed gently and incubated at 54 °C for 20 min in a water bath. At the end of the incubation, the reaction mixture was centrifuged at 2500 rpm for 3 min and the absorbance of the supernatant measured at 540 nm using a UV/VIS spectrometer (Optima, SP-3000, Tokyo, Japan). Phosphate buffer solution without sample was used as the control. The level of hemolysis was calculated using the following relation Equation (1):

$$\% \text{ inhibition of hemolysis} = 100 \times (1 - A2/A1) \quad (1)$$

where, A1 = Absorption of the control sample, A2 = Absorption of test sample solution.

3.4.2. Effect on Protein Denaturation

The test was performed following the method described by Gambhire et al. [22], with some modifications [20]. Briefly, 0.2 mL of 1% bovine albumin, 4.780 mL of phosphate buffered saline (PBS, pH 6.4), and 0.02 mL of cooked leaf extract was mixed gently, and was incubated at 37 °C for 15 min in a water bath, and then the reaction mixture was heated at 70 °C for 5 min. After cooling, the absorbance of the solutions was measured at 660 nm using a UV/VIS spectrometer. Phosphate buffer solution without sample was used as the control, and the percentage inhibition of protein denaturation was calculated by using the following formula Equation (2)

$$\% \text{ inhibition of denaturation} = 100 \times (1 - A2/A1) \quad (2)$$

where A1 = Absorption of the control sample and A2 = Absorption of the test sample

3.4.3. Proteinase Inhibitory Activity

The test was performed according to the modified method of Sakat et al. [9], with some modifications, as suggested by [20]. Briefly, 0.06 mg trypsin, 1 mL of 20 mm Tris-HCl buffer (pH 7.4), 0.02 mL cooked leaf extract, and 0.980 mL methanol were mixed, and the reaction mixture was incubated at 37 °C for 5 min, and then 1 mL of 0.8% (w/v) casein was added. The mixture was incubated further for an additional 20 min. About 2 mL of 70% perchloric acid was added to terminate the reaction. Cloudy suspension was centrifuged, and the absorbance of the supernatant was read at 210 nm against buffer as blank. The percentage of inhibition of proteinase activity was calculated.

Phosphate buffer solution without sample was used as the control. The percentage inhibition of protein denaturation was calculated by using the following Equation (3):

$$\% \text{ inhibition of denaturation} = 100 \times (1 - A2/A1) \tag{3}$$

where A1 = Absorption of the control sample and A2 = Absorption of the test sample

3.4.4. Lipoxygenase Inhibition Assay

Lipoxygenase was assayed, according to the method described by Wu [23], with some modifications being mentioned in Gunathilake [20]. Briefly, 1 mL sodium borate buffer (0.1 M, pH 8.8) and 10 µL lipoxygenase (8000 U/mL) was incubated with 10 µL cooked leaf extract in a 1 mL cuvette at room temperature for 5 min. The reaction was started by the addition of linoleic acid substrate (10 µL, 10 mmol). The absorbance of the resulting mixture was measured at 234 nm, and the phosphate buffer solution without sample was used as the control, and the percentage inhibition of lipoxygenase was calculated using the following Equation (4):

$$\% \text{ Inhibition} = 100 \times (\text{absorbance of the control} - \text{absorbance of the sample})/\text{absorbance of the control} \tag{4}$$

3.5. Analysis of Phenolics, Flavonoids and Carotenoids

Analysis of polyphenols was measured, as described in Gunathilake and Rupasinghe [24], flavonoid content using Gunathilake et al. [6], and the carotenoid content by using the method that was described in Gunathilake and Ranaweera [8]. However, data are not shown in this paper, and the data were used for the correlation studies with anti-inflammatory data.

3.6. Statistical Analysis

All data are presented as the mean ± standard deviation for all in vitro assays done. All of the samples were analyzed in triplicate, and one-way analysis of variance (ANOVA) was performed using MINITAB 15 software (Minitab Inc, State College, PA, USA). When there were significant differences ($p < 0.05$), multiple mean comparisons were carried out using LSD method. Pearson's correlation coefficients (r), with the level of significance ($P \leq 0.05$) (2-tailed) for total polyphenols, flavonoids, and carotenoids versus studied anti-inflammatory results were estimated using MINITAB 15 software. Polyphenols, flavonoids, and carotenoids content of the extracts that were used for correlation studies are based on the same study; however, data are not shown.

4. Conclusions

The present study clearly indicates that the in vitro anti-inflammatory associated biological activities of studied green leafy vegetables are modified, increased or decreased, by boiling, steaming, and frying process, depending upon the vegetable species. Among the cooking methods, the frying of all leafy vegetables has reduced the inhibition abilities of protein denaturation, hemolysis, proteinase, and lipoxygenase activities when compared with other cooking methods that were studied. Steaming significantly increased the protein denaturation and hemolysis inhibition in *O. zeylanica* and *P. edulis*.

Boiling of leaves increased the inhibitory activity of protein denaturation in *C. asiatica* and *P. edulis*; hemolysis in *C. asiatica, C. auriculata,* and *S. grandiflora*; lipoxygenase inhibition ability in *O. zeylanica, P. edulis, C. asiatica* and *C. auriculata* leaves; proteinase inhibition in *C. auriculata* when compared with that of raw and their other cooked leaves. The results of the study can be used as a database, providing information on the effects of different cooking methods on the health promotion potential of green leafy vegetables studied.

Acknowledgments: The authors would like to acknowledge National Science Foundation of Sri Lanka for the financial support under the Competitive Research Grant Scheme (Project No.: RG/AG/2014/04).

Author Contributions: K.D.P.P.G. performed all the experiments, analyzed the data, and wrote the manuscript. All the authors contributed to the designing of the experiments and proofreading the manuscript.

Conflicts of Interest: The authors declare no conflict of interest.

References

1. Kaliora, A.C.; Dedoussis, G.V.Z.; Schmidt, H. Dietary antioxidants in preventing atherogenesis—A review. *J. Atheroscler.* **2006**, *187*, 1–17. [CrossRef] [PubMed]
2. Ferrero-Millani, L.; Nelsen, O.H.; Anderson, P.S.; Girardin, S.E. Chronic inflammation: Importance of NOD2 and NALP3 in interleukin-1β generation. *Clin. Exp. Immunol.* **2007**, *147*, 227–235. [CrossRef] [PubMed]
3. Calder, P.C.; Albers, R.; Antoine, J.M.; Blum, S.; Bourdet-Sicard, R.; Ferns, G.A.; Folkerts, G.; Friedmann, P.S.; Frost, G.S.; Guarner, F.; et al. Inflammatory disease processes and interactions with nutrition. *Br. J. Nutr.* **2009**, *101*, 1–45. [CrossRef] [PubMed]
4. Zhang, M.; Hettiarachchy, N.S.; Horax, R.; Kannan, A.; Praisoody, A.; Muhundan, A. Phytochemicals, the antioxidant and antimicrobial activity of *Hibiscus sabdariffa, Centella asiatica, Moringa oleifera* and *Murraya koenigii* leaves. *J. Med. Plants Res.* **2011**, *5*, 6672–6680.
5. Oboh, G. Effect of blanching on the antioxidant properties of some tropical green leafy vegetables. *LWT Food Sci. Technol.* **2005**, *38*, 513–517. [CrossRef]
6. Gunathilake, K.D.P.P.; Ranaweera, K.K.D.S.; Rupasinghe, H.P.V. Change of phenolics, carotenoids, and antioxidant capacity following simulated gastrointestinal digestion and dialysis of selected edible green leaves. *Food Chem.* **2018**, *245*, 371–379. [CrossRef] [PubMed]
7. Gunathilake, K.D.P.P.; Ranaweera, K.K.D.S. Antioxidative properties of 34 green leafy vegetables. *J. Funct. Foods* **2016**, *26*, 176–186. [CrossRef]
8. Andarwulan, N.; Kurniasih, D.; Apriady, R.A.; Rahmat, H.; Roto, A.V.; Bolling, B.W. Polyphenols, carotenoids, and ascorbic acid in underutilized medicinal vegetables. *J. Funct. Foods* **2012**, *4*, 339–347. [CrossRef]
9. Sakat, S.; Juvekar, A.R.; Gambhire, M.N. In vitro antioxidant and anti-inflammatory activity of methanol extract of Oxalis corniculata Linn. *Int. J. Pharm. Pharm. Sci.* **2010**, *2*, 146–155.
10. Akinwunmi, K.F.; Oyedapo, O.O. In vitro Anti-inflammatory Evaluation of African Nutmeg (*Monodora myristica*) Seeds. *Eur. J. Med. Plants* **2015**, *8*, 167–174. [CrossRef]
11. Govindappaaga, S.S.; Poojashri, M.N.; Sadananda, T.S.; Chandrappa, C.P. Antimicrobial, antioxidant and in vitroantiinflammatory activity of ethanol extract and active phytochemical screening of *Wedelia trilobata* (L) Hitchc. *J. Pharm. Phytother.* **2011**, *3*, 43–51.
12. Chou, C.T. The anti inflammatory effect of *Tripterygium wilfordii* Hook F on adjuvant induced paw edema in rats and inflammatory mediator release. *Phytother. Res.* **1997**, *11*, 152–154. [CrossRef]
13. Chippada, S.C.; Vangalapati, M. Antioxidant, an anti-inflammatory and anti-arthritic activity of *Centella asiatica* extracts. *J. Chem. Biol. Phys. Sci.* **2011**, *1*, 260.
14. Alitonou, G.A.; Avlessi, F.; Sohounhloue, D.K.; Agnaniet, H.; Bessiere, J.M.; Menut, C. Investigations on the essential oil of *Cymbopogon giganteus* from Benin for its potential use as an anti-inflammatory agent. *Int. J. Aromather.* **2006**, *16*, 37–41. [CrossRef]
15. Wachtel-Galor, S.; Wong, K.W.; Benzie, I.F. The effect of cooking on *Brassica* vegetables. *Food Chem.* **2008**, *110*, 706–710. [CrossRef]
16. Palermo, M.; Pellegrini, N.; Fogliano, V. The effect of cooking on the phytochemical content of vegetables. *J. Sci. Food Agric.* **2014**, *94*, 1057–1070. [CrossRef] [PubMed]

17. Mayeaux, M.; Xu, Z.; King, J.M.; Prinyawiwatkul, W. Effects of cooking conditions on the lycopene content in tomatoes. *J. Food Sci.* **2006**, *71*, 461–464. [CrossRef]
18. Turkmen, N.; Sari, F.; Velioglu, Y.S. The effect of cooking methods on total phenolics and antioxidant activity of selected green vegetables. *Food Chem.* **2005**, *93*, 713–718. [CrossRef]
19. Shinde, U.A.; Phadke, A.S.; Nari, A.M.; Mungantiwar, A.A.; Dikshit, V.J.; Saraf, M.N. Membrane stabilization activity—A possible mechanism of action for the anti-inflammatory activity of *Cedrus deodora* wood oil. *Fitoterapia* **1999**, *70*, 251–257. [CrossRef]
20. Gunathilake, K.D.P.P. antioxidant and anti-inflammatory of selected green leafy vegetables in sri lanka. Ph.D. Thesis, University of Sri Jayewardenepura, Nugegoda, Sri Lanka, 2017.
21. Okoli, C.O.; Akah, P.A.; Onuoha, N.J.; Okoye, T.C.; Nwoye, A.C.; Nworu, C.S. *Acanthus montanus*: An experimental evaluation of the antimicrobial, anti-inflammatory and immunological properties of a traditional remedy for furuncles. *BMC Complement. Altern. Med.* **2008**, *8*, 27. [CrossRef] [PubMed]
22. Gambhire, M.; Juvekar, A.; Wankhede, S. Evaluation of the anti-inflammatory activity of methanol extract of *Barleria cristata* leaves by in vivo and in vitro methods. *Int. J. Pharm.* **2009**, *7*, 1–6.
23. Wu, H. Affecting the activity of soybean lipoxygenase-1. *J. Mol. Graph.* **1996**, *14*, 331–337. [CrossRef]
24. Gunathilake, K.D.P.P.; Rupasinghe, H.P.V. Optimization of water based-extraction methods for the preparation of bioactive-rich ginger extract using response surface methodology. *Eur. J. Med. Plants* **2014**, *4*, 893. [CrossRef]

© 2018 by the authors. Licensee MDPI, Basel, Switzerland. This article is an open access article distributed under the terms and conditions of the Creative Commons Attribution (CC BY) license (http://creativecommons.org/licenses/by/4.0/).

Article

Antibacterial Properties of Flavonoids from Kino of the Eucalypt Tree, *Corymbia torelliana*

Motahareh Nobakht, Stephen J. Trueman *, Helen M. Wallace, Peter R. Brooks, Klrissa J. Streeter and Mohammad Katouli

Centre for Genetics, Ecology and Physiology, University of the Sunshine Coast, Maroochydore DC, QLD 4558, Australia; mnobakht@research.usc.edu.au (M.N.); hwallace@usc.edu.au (H.M.W.); pbrooks@usc.edu.au (P.R.B.); kstreete@usc.edu.au (K.J.S.); mkatouli@usc.edu.au (M.K.)
* Correspondence: strueman@usc.edu.au; Tel.: +61-7-5456-5033

Received: 29 August 2017; Accepted: 11 September 2017; Published: 14 September 2017

Abstract: Traditional medicine and ecological cues can both help to reveal bioactive natural compounds. Indigenous Australians have long used kino from trunks of the eucalypt tree, *Corymbia citriodora*, in traditional medicine. A closely related eucalypt, *C. torelliana*, produces a fruit resin with antimicrobial properties that is highly attractive to stingless bees. We tested the antimicrobial activity of extracts from kino of *C. citriodora*, *C. torelliana* × *C. citriodora*, and *C. torelliana* against three Gram-negative and two Gram-positive bacteria and the unicellular fungus, *Candida albicans*. All extracts were active against all microbes, with the highest activity observed against *P. aeruginosa*. We tested the activity of seven flavonoids from the kino of *C. torelliana* against *P. aeruginosa* and *S. aureus*. All flavonoids were active against *P. aeruginosa*, and one compound, (+)-(2S)-4′,5,7-trihydroxy-6-methylflavanone, was active against *S. aureus*. Another compound, 4′,5,7-trihydroxy-6,8-dimethylflavanone, greatly increased biofilm formation by both *P. aeruginosa* and *S. aureus*. The presence or absence of methyl groups at positions 6 and 8 in the flavonoid A ring determined their anti-*Staphylococcus* and biofilm-stimulating activity. One of the most abundant and active compounds, 3,4′,5,7-tetrahydroxyflavanone, was tested further against *P. aeruginosa* and was found to be bacteriostatic at its minimum inhibitory concentration of 200 µg/mL. This flavanonol reduced adhesion of *P. aeruginosa* cells while inducing no cytotoxic effects in Vero cells. This study demonstrated the antimicrobial properties of flavonoids in eucalypt kino and highlighted that traditional medicinal knowledge and ecological cues can reveal valuable natural compounds.

Keywords: antibiotic resistance; antimicrobial activity; cytotoxicity; ethnobotany; *Eucalyptus*; natural products; *Pseudomonas aeruginosa*; stingless bees; *Tetragonula*; traditional medicine

1. Introduction

The long-term use of antibiotics has led to widespread bacterial resistance, and so there has been a global drive to develop new antibiotics [1–4]. Traditional medicine based on natural products has led to the discovery of new drugs, and ecological cues can also reveal bioactive natural compounds [5–8]. Eucalypt kino is a trunk exudate produced by eucalypt trees (*Angophora*, *Corymbia* and *Eucalyptus* spp.) that contains high levels of potentially useful polyphenol compounds. Kino is characterised by its deep rich coloration, high tannin content, polyphenol composition and astringency [9,10]. Indigenous Australians have used kino from *Corymbia* and *Eucalyptus* trees to cure ailments such as diarrhoea, scabies, and haemorrhage [11]. Kino exudates from *Corymbia dichromophloia* trees have also been used as a treatment for toothache, cold and flu, and heart, lung, and bronchial diseases [12]. Kino from *C. citriodora* trees has been used traditionally as a treatment for chronic bowel inflammation [11], and kino from *C. intermedia* trees has been used to treat wounds [13]. Leaf or bark extracts from naturalized *C. torelliana* trees in Nigeria have been used to treat gastrointestinal disorders, wounds, and coughs [14,15].

The biological activity of eucalypt kinos used in traditional medicine has, until recent times, received little attention. Two compounds from *C. citriodora* and *C. maculata* kino, aromadendrin 7-methyl ether and ellagic acid, have long been known to possess antimicrobial activity against the Gram-positive bacterium, *Staphylococcus aureus* [16]. Aqueous kino extracts from 15 eucalypt species have recently been tested for their antimicrobial activity against *S. aureus*, *Bacillus subtilis*, *Kocuria rhizophila*, *Pseudomonas aeruginosa*, *Escherichia coli*, and *Saccharomyces cerevisiae*. Extracts from *C. maculata*, *C. ficifolia*, and *C. calophylla* kino exhibited strong activity against the Gram-positive bacteria, although no activity was observed from any of the eucalypt species against the Gram-negative bacteria [17]. Aqueous and ethanolic extracts from *C. intermedia* leaves also exhibit antimicrobial activity against *S. aureus* and the unicellular fungus, *Candida albicans* [13]. Volatile components from *C. citriodora* essential oil have strong activity against *Mycobacterium tuberculosis* [18]. Extracts of *C. torelliana* leaves or bark have antibacterial activity against a wide range of species including *M. tuberculosis* and non-tuberculous *Mycobacteria* spp., *S. aureus*, *E. coli*, *P. aeruginosa*, *Klebsiella* sp. and *Helicobacter pylori* [14,15,19,20]. *C. torelliana* also has an unusual mutualistic relationship with stingless bees that disperse its seeds and use its fruit resin to construct their nests [21–26]. The fruit resin from *C. torelliana* has recently been found to possess antimicrobial properties [27–29]. Bees prefer the fruit resin from *C. torelliana* to fruit resin from other species, and this fruit resin may protect their nest from pathogenic microbes [26–28]. However, extracts from the kino of *C. torelliana* and *C. citriodora* have not previously been tested for their antimicrobial activity.

In this study, we investigated the antimicrobial activity of extracts from kino of *C. citriodora*, *C. torelliana* and their widely planted hybrid, *C. torelliana* × *C. citriodora*. We examined their activity against three Gram-negative bacteria, *P. aeruginosa*, *E. coli*, and *Salmonella typhimurium*, two Gram-positive bacteria, *S. aureus* and *B. cereus*, and one fungus, *Candida albicans*. Furthermore, we investigated the antibacterial activity against *P. aeruginosa* and *S. aureus* of seven individual flavonoids (Figure 1) isolated from the kino of *C. torelliana* [30]. We also assessed whether the kino extracts and one of the most-abundant and active flavonoids in *C. torelliana* kino had cytotoxic effects.

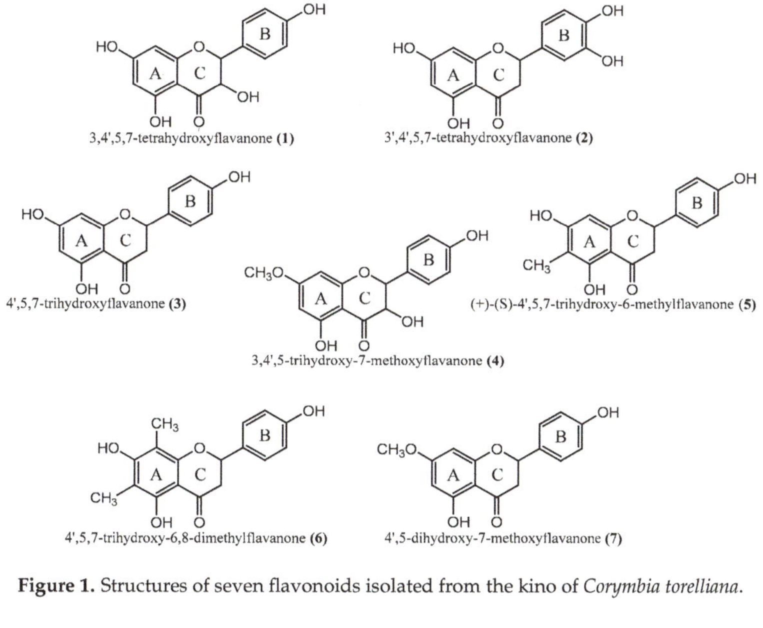

Figure 1. Structures of seven flavonoids isolated from the kino of *Corymbia torelliana*.

2. Results and Discussion

2.1. Antimicrobial Activity and Cytotoxicity of Crude Extracts from Corymbia Trees

Ethanolic extracts from kinos of *Corymbia citriodora*, *C. torelliana* × *C. citriodora*, and *C. torelliana* showed strong antimicrobial activity against all of the tested microorganisms (Figure 2). The extent of the inhibition zone varied among the extracts and microorganisms, but the highest inhibition with all three extracts was obtained against *P. aeruginosa*. This is the first report of antimicrobial activity of kino from *C. citriodora*, *C. torelliana* × *C. citriodora* and *C. torelliana* against Gram-negative bacteria, Gram-positive bacteria and *C. albicans*. These results are similar to findings that aqueous kino extracts from the closely related species, *C. maculata*, and the more-distantly related species, *C. ficifolia* and *C. calophylla*, are active against the Gram-positive bacteria, *S. aureus*, *B. subtilis*, and *K. rhizophila* [17]. Aqueous and ethanolic extracts from the leaves of another distantly-related species, *C. intermedia*, also have strong activity against *S. aureus* and *C. albicans* [13]. However, the extracts from *C. maculata*, *C. ficifolia*, *C. calophylla*, and *C. intermedia* had little or no activity against *P. aeruginosa* [13,17]. The use of aqueous rather than ethanolic extracts [17] and the sampling of leaves rather than kino [13] may explain the lack of activity of other *Corymbia* extracts against Gram-negative bacteria. Alternatively, the differences may be the result of specific chemical compounds that are present in *C. citriodora*, *C. torelliana* × *C. citriodora*, and *C. torelliana* kino.

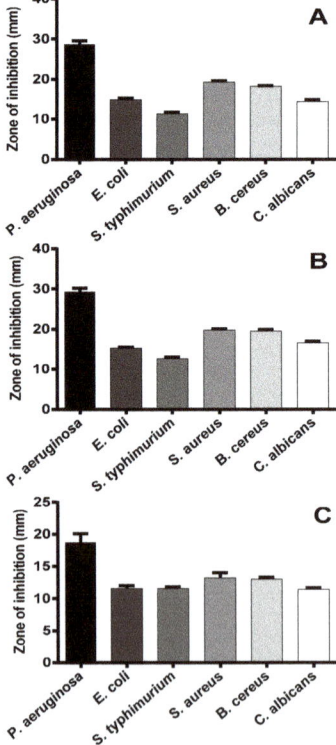

Figure 2. Antimicrobial activity of 400 μg of kino extract from (**A**) *Corymbia citriodora*, (**B**) *C. torelliana* × *C. citriodora*, and (**C**) *C. torelliana* against six microorganisms. Zones of inhibition are presented as mean + S.E. (n = 21 trees for *C. citriodora* and the hybrid; n = 3 trees for *C. torelliana*). Means among the three kino extracts do not differ significantly (ANOVA, $p > 0.05$).

Different strains of a bacterial species may exhibit different levels of susceptibility to an antimicrobial agent. The highest activity of the crude extracts was against *P. aeruginosa* and so we extended our screening by testing the extracts against four strains of *P. aeruginosa* (i.e., strains C1, C8, C11, and C19) that represented four different clonal groups, isolated recently from clinical cases in our laboratory [31]. The kino extracts showed strong activity against all four strains, except that C11 was resistant to the *C. torelliana* extract (Table 1). This strain was also highly resistant to ticarcillin and intermediately resistant to aztreonam and ticarcillin-clavulanic acid. Ticarcillin is a fourth generation of penicillin, a β-lactam antibiotic. This group of antimicrobial agents inhibits bacteria by penetrating the cytoplasmic membrane and attaching to penicillin binding proteins [32]. Resistance of bacteria to this antibiotic normally develops through a mechanism that inhibits the antibiotic from reaching this target. The resistance of the C11 strain to crude extract of *C. torelliana* suggests that the active component of *C. torelliana* kino might operate by a similar mechanism to these antibiotics. The MIC of kino extracts from *C. citriodora*, *C. torelliana* × *C. citriodora* and *C. torelliana* was 200 μg/mL against each of the bacteria in this study (data not presented). This suggests that, irrespective of the *Corymbia* species, the type and concentration of active compounds in the extracts were similar. At this MIC, 200 μg/mL, the extracts had bacteriostatic activity.

Table 1. Antimicrobial activity of a 1.0% (v/v) solution of kino extract from *Corymbia citriodora*, *C. torelliana* × *C. citriodora*, or *C. torelliana* against four clinical strains of *Pseudomonas aeruginosa* (C1, C8, C11, and C19) representing different clonal types. Antibiotic susceptibility profile of isolates is also given.

Corymbia Species or Hybrid	Strain/Zone of Inhibition (mm) *			
	C1	C8	C11	C19
C. citriodora	12 ± 0	18 ± 1	15 ± 1	11 ± 0
C. torelliana × *C. citriodora*	11 ± 0	16 ± 1	19 ± 1	12 ± 0
C. torelliana	11 ± 0	11 ± 1	1 ± 1 **	12 ± 0
Antibiotic Susceptibility Profile †				
Amikacin (30 μg)	S	S	S	S
Aztreonam (30 μg)	S	I	I	I
Ceftazidime (30 μg)	S	S	S	S
Cefepime (30 μg)	S	S	S	S
Piperacillin (100 μg)	S	I	S	R
Piperacillin-tazobactam (100/10 μg)	S	R	S	I
Ticarcillin (75 μg)	I	I	R	I
Gentamicin (10 μg)	S	I	S	S
Ciprofloxacin (5 μg)	S	S	S	S
Norfloxacin (10 μg)	S	S	S	S
Imipenem (10 μg)	R	S	S	S
Ticarcillin-clavulanic acid (75/10 μg)	I	R	I	R

* Zones of inhibition are presented as mean ± S.E. ($n = 9$ trees for *C. citriodora* and the hybrid; $n = 3$ trees for *C. torelliana*). Means within a *P. aeruginosa* strain do not differ significantly (ANOVA, $p > 0.05$); ** Distance (mm) from the rim of the well; † Antibiotic susceptibility profile is classified as susceptible (S), intermediate (I) or resistant (R) according to CLSI guidelines [33].

The kino extracts were also tested for their ability to inhibit biofilm formation by the bacteria. All extracts increased biofilm formation of the bacterial strains (Figure 3). These results were unexpected, as we anticipated that extracts capable of inhibiting microbial growth would have also reduced their biofilm formation. However, similar results have been observed during testing of aqueous extracts from the neem tree, *Azadirachta indica*, against two yeast strains [34]. These researchers concluded that increased biofilm formation could be related to an increased level of hydrophobicity, which is a non-specific mechanism for adhesion of bacteria to surfaces. The use of ethanolic extracts could have partly effected the high level of biofilm formation in the current study, as ethanol has potential to increase hydrophobicity. The preparation of aqueous extracts was not feasible due to low solubility of *C. citriodora*, *C. torelliana* × *C. citriodora*, and *C. torelliana* kino in water. From a clinical perspective, biofilm formation is important for survival of bacteria that colonize the host and it assists in physical resistance to phagocytosis and tolerance to antibiotics [35,36]. Under these conditions, antibiotics are prevented from diffusing through the physical barrier formed by the exopolymeric substances in biofilms [37]. Nonetheless, the high antimicrobial activity of all extracts against *P. aeruginosa* indicates the presence of an active compound (or compounds) with anti-*Pseudomonas* activity in the kino from these *Corymbia* species.

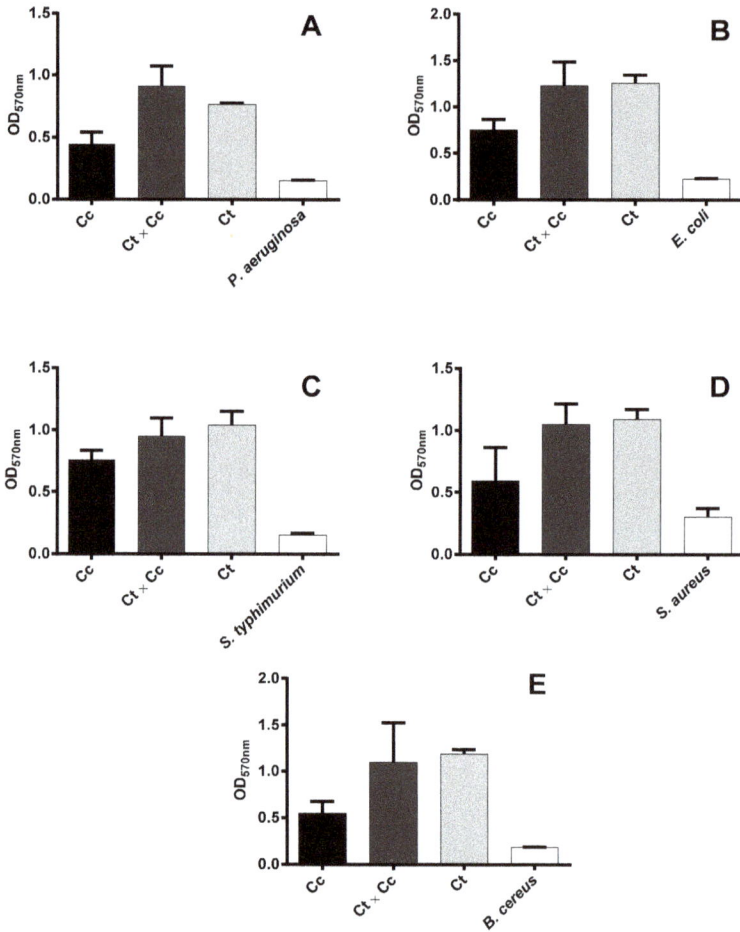

Figure 3. Biofilm formation in the presence of kino extracts from *Corymbia citriodora* (Cc), *C. torelliana* × *C. citriodora* (Ct × Cc), and *C. torelliana* (Ct) by (**A**) *Pseudomonas aeruginosa*, (**B**) *Escherichia coli*, (**C**) *Salmonella typhimurium*, (**D**) *Staphylococcus aureus*, and (**E**) *Bacillus cereus*. Optical densities at 570 nm wavelength (OD_{570nm}) are presented as mean + S.E. (n = 3 trees for Cc, Ct × Cc, and Ct; n = 3 for the kino-free control). Means among the three kino extracts do not differ significantly (ANOVA, $p > 0.05$).

The cytotoxicity of *C. citriodora*, *C. torelliana* × *C. citriodora*, and *C. torelliana* kino was tested using 1000 µg of the extracts against Vero cells. No cytopathic effect (CPE), such as detachment or rounding of the cells, was observed. However, cells showed morphological changes, characterized by shrinking, within 48 h. Studies investigating the CPE of bacterial toxins on eukaryotic cells, including Vero cells, have defined cytotoxicity as 50% or more of the cells showing CPE such as rounding or disruption of the cell monolayer within 4 h [38–40].

2.2. Antibacterial Activity of Flavonoids from Corymbia torelliana

Seven flavonoids isolated from the kino of *C. torelliana* (Figure 1) possessed antibacterial activity against *P. aeruginosa* (Table 2). The activity of these flavonoids against this Gram-negative bacterium is highly significant because the outer membrane of Gram-negative bacteria possesses narrow porin channels that slow the penetration of small hydrophilic solutes and increase their tolerance to antibiotics [41]. Our results suggest that flavonoids from the kino of *C. torelliana* have a mechanism that overcomes this barrier.

Table 2. Antimicrobial activity of seven flavonoids [3,4′,5,7-tetrahydroxyflavanone (**1**), 3′,4′,5,7-tetrahydroxyflavanone (**2**), 4′,5,7-trihydroxyflavanone (**3**), 3,4′,5-trihydroxy-7-methoxyflavanone (**4**), (+)-(2S)-4′,5,7-trihydroxy-6-methylflavanone (**5**), 4′,5,7-trihydroxy-6,8-dimethylflavanone (**6**), and 4′,5-dihydroxy-7-methoxyflavanone (**7**)] from *Corymbia torelliana* kino against *Pseudomonas aeruginosa* and *Staphylococcus aureus*.

Bacterium	Zone of Inhibition (mm)						
	Compound						
	1	2	3	4	5	6	7
P. aeruginosa	20.3 ± 1.8	6.7 ± 6.7	19.7 ± 1.9	12.3 ± 6.3	18.7 ± 1.2	24.7 ± 2.9	20.3 ± 2.8
S. aureus	inactive	inactive	inactive	inactive	12.7 ± 1.8	inactive	inactive

Means ($\pm$ S.E.) among seven flavonoids within *P. aeruginosa* do not differ significantly (ANOVA, $p > 0.05$).

Only one of the seven compounds, (+)-(2S)-4′,5,7-trihydroxy-6-methylflavanone (**5**), was active against *S. aureus* (Table 2). The other six flavonoids might inhibit *S. aureus* growth at concentrations higher than 50 µg/well, although there can also be relationships between the structure of flavonoids and their antibacterial activity [42]. Hydroxyl groups in the chemical structure increase the activity of flavonoids against methicillin-resistant *S. aureus*, while methoxy groups reduce their activity [43]. However, we did not find the presence of hydroxyl groups to be a defining character in the anti-*Staphylococcus* activity of our flavonoids. For example, (+)-(2S)-4′,5,7-trihydroxy-6-methylflavanone (**5**) and 4′,5,7-trihydroxy-6,8-dimethylflavanone (**6**) both have three hydroxyl groups, located at positions 5 and 7 of the A ring and position 4′ of the B ring, while neither compound contains a methoxy group. The unique aspect of compound (**5**) is that it contains a single methyl group, located at position 6 of the A ring. We did not investigate possible relationships between the concentration of flavonoids and their antibacterial effects due to difficulties in obtaining sufficient quantities of flavonoids from *C. torelliana* extracts. However, there can be relationships between the concentration of flavonoids and their antibacterial activity [44].

One of the compounds, 4′,5,7-trihydroxy-6,8-dimethylflavanone (**6**), greatly increased biofilm formation by both *P. aeruginosa* and *S. aureus* (Figure 4). This compound is unique in possessing two methyl groups, located at positions 6 and 8 of the A ring (Figure 1). The stimulation of biofilm formation by this flavonoid was highly unexpected. However, extracts of *Azadirachta indica* increase biofilm formation by *C. albicans* [34] and some phenolics and aminoglycosides, at sub-inhibitory concentrations, increase biofilm formation by *P. aeruginosa* and *E. coli* [45,46]. A relationship between the existence of two methyl groups and enhanced biofilm formation has not been reported previously. However, it could be concluded that this phenomenon of increased biofilm formation is entirely, or at least partly, due to a hydrophobicity effect rather than the methyl groups.

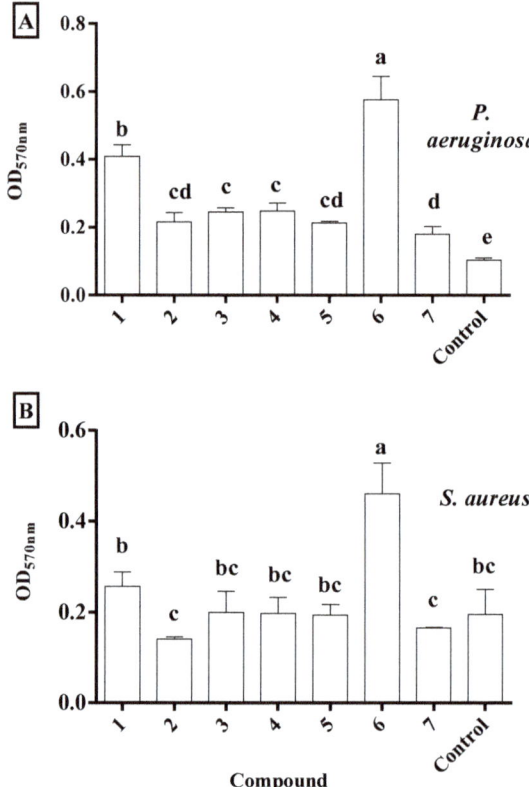

Figure 4. Biofilm formation (OD_{570nm}) in the presence of 100 µg of 3,4′,5,7-tetrahydroxyflavanone (1), 3′,4′,5,7-tetrahydroxyflavanone (2), 4′,5,7-trihydroxyflavanone (3), 3,4′,5-trihydroxy-7-methoxyflavanone (4), (+)-(2S)-4′,5,7-trihydroxy-6-methylflavanone (5), 4′,5,7-trihydroxy-6,8-dimethylflavanone (6), and 4′,5-dihydroxy-7-methoxyflavanone (7).

2.3. Anti-Pseudomonas Activity and Cytotoxicity of 3,4′,5,7-Tetrahydroxyflavanone

3,4′,5,7-tetrahydroxyflavanone (**1**) inhibited the growth of *P. aeruginosa*, with a minimum inhibitory concentration of 200 µg/mL (Table 3). This compound was bacteriostatic against *P. aeruginosa*. It significantly increased biofilm formation by *P. aeruginosa* at 48 h in comparison with the control, both in this experiment at masses of 200, 100, and 50 µg (Figure 5) and in the previous experiment at a mass of 100 µg (Figure 4). Further investigation is warranted to determine the mechanisms behind the biofilm-stimulating effect of this compound with *P. aeruginosa*. 3,4′,5,7-tetrahydroxyflavanone reduced adhesion of *P. aeruginosa* by 19%, 38%, and 35% at 200 µg, 100 µg, and 50 µg final mass, respectively (Table 3). This is an important result because adhesion to host tissue is an important step in bacterial survival and colonization [47]. The flavonoid induced no cytotoxic effects in Vero cell culture assay after 48 h (Table 3). Crude extracts from *C. torelliana* kino also had no cytotoxic effects on human colorectal epithelial adenocarcinoma (Caco-2) cells (Section 2.1, above). These results, therefore, confirm the potential of 3,4′,5,7-tetrahydroxyflavanone as a new antibacterial agent against *P. aeruginosa*. This is a significant finding because kino extracts have previously been found inactive against Gram-negative bacteria [16,17]. It was concluded that high molecular weight compounds, such as tannins, in the eucalypt kino could not penetrate the outer membrane of Gram-negative bacteria [17]. However, results in the current study suggest that compounds such as 3,4′,5,7-tetrahydroxyflavanone from the kino of *C. torelliana* can overcome this outer membrane barrier. This flavanonol has also been identified in kino from the distantly related species, *C. calophylla* and *C. gummifera* [48,49].

Table 3. Minimum inhibitory concentration (MIC) and reduction in adhesion of *Pseudomonas aeruginosa* in the presence of 3,4′,5,7-tetrahydroxy-flavanone from *Corymbia torelliana* kino.

MIC (µg/mL)	Control Adhesion cfu (Mean ± S.E.)	Adhesion Difference (%)		
		Final Mass (µg)		
		200	100	50
200	3.86 ± 0.16	−19	−38	−35

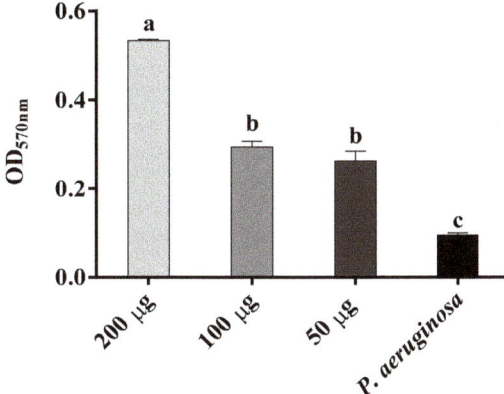

Figure 5. Biofilm formation of *Pseudomonas aeruginosa* in the presence of 200 µg, 100 µg, and 50 µg (final mass) of 3,4′,5,7-tetrahydroxyflavanone from kino exudate of *Corymbia torelliana*. Optical densities at 570 nm wavelength (OD_{570nm}) are presented as mean + S.E. Means with different letters are significantly different (ANOVA and Tukey's HSD test; $p < 0.05$; $n = 3$).

These results confirm that natural products from traditional medicine are promising leads to finding new sources of antibiotic drugs. Natural products have often played a key role in the formulation of new drugs [50] and there is particular interest in polyphenols such as flavonols, flavan-3-ols, and tannins for their potential antimicrobial effects [51]. Indigenous Australians have used kino exudates from eucalypt trees including *C. citriodora* to cure various ailments, and eucalypt kinos are well known for their polyphenol content [11]. The successful discovery of new sources of drugs based on traditional knowledge can also be guided by ecological cues [7]. *C. torelliana* is a geographically-restricted species that is closely related to the traditional medicinal species, *C. citriodora*, but which has a unique mutualism with stingless bees [21]. These bees are strongly attracted to the resin of *C. torelliana* fruits, which they use to construct their nests, and the bees disperse the seeds of this species [22–26]. Resin from the fruit of *C. torelliana* has also been shown to possess antimicrobial properties [27,28]. We were, therefore, guided by both traditional knowledge of the medicinal properties of this eucalypt group and ecological information on the attractiveness to bees of one particular species in this group, to identify potentially valuable antibacterial compounds.

3. Materials and Methods

3.1. Antimicrobial Activity and Cytotoxicity of Crude Extracts from Corymbia Trees

Fresh kino samples were obtained from *C. citriodora* subsp. *variegata* and *C. torelliana* × *C. citriodora* subsp. *variegata* (21 trees each) in a forestry plantation at Binjour (25°30′ S, 151°27′ E), Australia, established by the Queensland Department of Agriculture and Fisheries [52,53]. Samples were also collected from three *C. torelliana* trees on the Sunshine Coast (26°42' S, 153°02' E), Australia. *C. citriodora*, and hybrids between *C. citriodora* and *C. torelliana*, are grown extensively in forestry plantations [54–60] but *C. torelliana* is rarely grown in plantations, partly because of its invasive potential [21,25]. Therefore, the three samples of *C. torelliana* were obtained from isolated trees in amenity plantings. Kino samples were collected from naturally-occurring, freely-flowing trunk exudates into clean vials, transported to the laboratory on ice, and stored in the dark at −20 °C until testing. Collection of samples from older, crystallized exudates was avoided due to the effect of long-term sunlight exposure on the chemical composition of kino [61].

Crude extracts (1.0%, w/v) were prepared from the kino of each tree in ethanol (70%, v/v) at laboratory temperature and filtered through cotton wool to remove coarse debris. The kino extracts were then examined for their antimicrobial activity using well-diffusion methods [62,63]. Microbial suspensions were prepared by inoculating single colonies of the type culture strains, *P. aeruginosa* (ATCC 27853), *E. coli* (ATCC 25922), *S. typhimurium* (ATCC 13311), *S. aureus* (ATCC 25923), *B. cereus* (ATCC 11788), and a laboratory strain of *Candida albicans* into tryptone soya broth (TSB) (Oxoid, Australia). Cultures were incubated overnight at 37 °C on a rotary shaker (150 rpm). The bacterial suspensions were diluted using phosphate buffered saline (PBS, pH = 7.4) to approximately 1.0×10^9 colony forming units (cfu)/mL using a spectrophotometer at 600 nm wavelength. From each bacterial suspension, 2.5 mL was transferred to 247.5 mL of molten Mueller-Hinton agar at approximately 55°C, and thoroughly mixed to obtain a uniform concentration of 1.0×10^7 cfu/mL. Each of the six agar suspensions was poured into one sterilized acrylic plate (30 cm × 30 cm internal diameter and 1 cm depth) and allowed to set. Using a sterile cork borer of 4 mm external diameter, 49 holes were cut in each plate and the holes were inoculated with 40 µL of a 1.0% (w/v) solution of kino from each tree, providing a final mass of 400 µg of kino in each well. Plates were covered and incubated for 18 h at 37 °C, and the zone of inhibition for each kino extract was measured. A zone of inhibition >5 mm was considered as positive [64]. Ethanol (70%, v/v) was used as a negative control, and this provided no growth inhibition in any experiment. We also used *P. aeruginosa* (ATCC 27853) as the positive control for testing four clinical strains of *P. aeruginosa* (see below).

Based on the initial antimicrobial activity results, 40 µL of a 1.0% (w/v) solution of kino (final mass of 400 µg) from crude extract of *C. citriodora*, *C. torelliana* × *C. citriodora* or *C. torelliana* was also tested against four wild clinical strains of *P. aeruginosa*, representing different clonal groups. These strains were obtained from a study investigating the clonality of *P. aeruginosa* strains and their virulence properties [31]. The four clinical strains were also tested for their resistance to twelve antimicrobial agents according to Clinical Laboratory Standard Institute (CLSI) guidelines [33], and the results were compared with the antimicrobial activities of the kino extracts. The antimicrobial impregnated disks (Oxoid) included piperacillin (100 µg), ticarcillin (75 µg), piperacillin–tazobactam (100/10 µg), ticarcillin–clavulanic acid (75/10 µg), ceftazidime (30 µg), cefepime (30 µg), aztreonam (30 µg), imipenem (10 µg), gentamicin (10 µg), amikacin (30 µg), ciprofloxacin (5 µg) or norfloxacin (10 µg). Disks were placed on Mueller-Hinton agar that had been inoculated with bacterial suspension at a concentration of 1.0×10^7 cfu/mL. Plates were then incubated at 37 °C for 16–18 h, after which the diameter of the inhibition zone was measured. The strains were classified as susceptible (S), intermediate (I) or resistant (R) to the antibiotics according to CLSI guidelines [33].

MIC values of the extracts were measured using standard methods [65]. Fresh cultures of the microorganisms were prepared and added to Mueller-Hinton broth to give a final concentration of 1.0×10^7 cfu/mL. Five concentrations of kino samples (1.0%, 0.8%, 0.5%, 0.3%, and 0.1%; w/v) were prepared from three trees of each taxon and tested against each microorganism, giving final masses of 400 µg, 320 µg, 200 µg, 120 µg, and 40 µg of kino extract per tube. The tubes were incubated on a rotary shaker (150 rpm) at 37 °C for 18 h, and the growth of the microorganisms was recorded visually. A loopful of the last tube showing no growth was then subcultured on tryptone soy agar (TSA) to determine whether the kino was bactericidal or bacteriostatic. Plates were incubated at 37 °C for 18 h.

Biofilm formation in the presence of kino was tested in 96-well tissue culture plates. Suspensions of bacteria grown on TSA agar were prepared to a final concentration of 1.0×10^6 cfu/mL. As a control, an aliquot of 200 µL of each suspension was added to a well of the plate and incubated without shaking for 48 h at 37 °C. The highest concentration of prepared kino extract (i.e. 1.0%, w/v) from three trees of each taxon was used for biofilm formation assay. 100 µL of sample was added to 100 µL of TSB in each well, and bacterial growth was measured at 600 nm wavelength before staining. The growth medium was removed and plates were rinsed twice with PBS to rinse away non-adhering bacteria. The plates were then allowed to dry. Bacterial biofilm in the wells was stained with 220 µL of crystal violet (0.3%) for 10 min, excess stain was removed with tap water, and the dye was solubilised using 250 µL of acetone/ethanol (20/80, v/v) by shaking at 200 rpm for 10 min. Dissolved crystal violet was measured at 570 nm wavelength. The kino samples from each tree were tested in triplicate. The average optical density (OD) of all three wells at 570 nm wavelength was calculated.

Cytotoxicity of the crude extracts was tested against Vero cells (ATCC CCL-81) derived from African Green Monkey kidneys. Cells were maintained at 37 °C and 5% CO_2 in Eagle's minimal essential medium (EMEM) (Lonza, Australia) supplemented with 10% foetal bovine serum (FBS; Lonza) and 1% penicillin/streptomycin solution (Lonza). 200 µL of cell suspension was seeded into 96-well tissue culture plates and grown to 100% confluence in EMEM without antibiotics. Three concentrations of compound (200 µg, 100 µg, and 50 µg) were tested and all tests were performed in triplicate. Cells were visually examined for cytotoxic effects after 4, 24, and 48 h using an inverted phase contrast microscope (×400). The kino extracts were deemed cytotoxic if cell rounding and cell death (detachment from the bottom of wells) occurred in more than 50% of cells [38].

3.2. Antibacterial Activity of Flavonoids from Corymbia torelliana

Fresh kino samples from *C. torelliana* trees on the Sunshine Coast (see Section 3.1, above) were extracted in ethyl acetate/water (4/3; v/v). The extracts were stored at −20 °C until fractionation by preparative HPLC. The dry extract (100 mg) was dissolved in acetonitrile/water (1/1; v/v) for fractionation by preparative chromatography. Seven flavonoids were recovered and identified by spectroscopic and spectrometric methods including UV, 1D, and 2D NMR, and UPLC-HR-MS, as described previously [30]: 3,4′,5,7-tetrahydroxyflavanone (**1**), 3′,4′,5,7-tetrahydroxyflavanone (**2**), 4′,5,7- trihydroxyflavanone (**3**), 3,4′,5-trihydroxy-7methoxyflavanone (**4**), (+)-(2S)-4′,5,7-trihydroxy-6-methylflavanone (**5**), 4′,5,7-trihydroxy-6,8-dimethylflavanone (**6**), and 4′,5-dihydroxy-7-methoxyflavanone (**7**) (Figure 1). Two of these compounds, (**1**) and (**4**), are flavanonols, while the other compounds are flavanones.

Each flavonoid was prepared in ethanol (70%; v/v) at a final mass of 50 µg and tested for its antibacterial activity against *P. aeruginosa* (ATCC 27853) and *S. aureus* (ATCC 25923) using a well-diffusion method [63]. Bacterial suspensions were prepared in PBS (1.0×10^9 cfu/mL) after overnight growth in TSB. Molten Mueller-Hinton agar was inoculated with bacterial suspension to give a concentration of 1.0×10^7 cfu/mL, and then allowed to set. Holes cut in the plate were inoculated with 50 µL of a 0.1% (w/v) solution of each flavonoid, providing a final mass of 50 µg in each well. This mass was chosen due to the concentration of pure compounds available after several rounds of extraction. Plates were covered and incubated for 18 h at 37 °C, and the zone of inhibition was measured. Ethanol (70%; v/v) provided no growth inhibition in any experiment. No differences in solubility in ethanol were observed among the seven flavonoids.

Anti-biofilm activity of the seven flavonoids was tested in 96-well tissue culture plates using *P. aeruginosa* (ATCC 27853) and *S. aureus* (ATCC 25923). Fresh bacterial suspensions (1.0×10^6 cfu/mL) were prepared in PBS (pH 7.4), with their concentrations determined by observing the OD_{600} value. This suspension was diluted 1:100 in sterile TSB, and 200 µL of suspension was inoculated into a sterile 96-well plate, avoiding the outermost wells to minimise the possibility of desiccation. 100 µL of each flavonoid solution (100 µg of compound) was added to 100 µL of TSB (diluted 1:50) inoculated with bacteria. The plates were processed and assessed for biofilm formation using the method described in

Section 3.1 (above). All tests were performed in triplicate wells. The mean (±standard error) zone of inhibition or optical density of the three wells was calculated for each sample.

3.3. Anti-Pseudomonas Activity and Cytotoxicity of 3,4',5,7-tetrahydroxyflavanone

One of the most-abundant and active flavonoids, 3,4',5,7-tetrahydroxyflavanone (**1**), was tested further against one of the most susceptible bacteria. *P. aeruginosa* (ATCC 27853) was grown and maintained in TSB. Cultures were incubated overnight at 37 °C on a rotary shaker (150 rpm). MICs were measured using methods described previously [65]. Fresh cultures of each bacterium were prepared and added to Mueller-Hinton broth to give a final concentration of c. 1.0×10^7 cfu/mL. Three concentrations of the flavonoid were prepared: 200 µg, 100 µg, and 50 µg per mL. Bacteria were grown at 100 rpm at 37 °C for 18 h, and the presence or absence of growth was determined visually. The last tube showing no growth was then subcultured on TSA plates to determine whether the MIC was bactericidal or bacteriostatic. These plates were incubated at 37 °C and the growth or lack of growth was observed after 18 h.

Antibiofilm activity of 3,4',5,7-tetrahydroxyflavanone was tested in 96-well tissue culture plates against the same bacterial strains (above). Fresh bacterial suspensions (1.0×10^6 cfu/mL) were prepared in sterile PBS (pH 7.4). This suspension was diluted 1:100 in sterile TSB and 200 µL was inoculated into a sterile tissue culture plate, avoiding the outermost wells. Each plate included positive controls (bacteria without any compounds) and negative controls (only TSB broth). 100 µL of each flavonoid solution (to reach 200 µg, 100 µg, and 50 µg final mass) was added to 100 µL of TSB (diluted 1:50) inoculated with bacterium. The plates were processed and assessed for biofilm formation using the method described in Section 3.1 (above). All tests were performed in triplicate and the average OD was calculated.

The ability of *P. aeruginosa* to adhere to Caco-2 cells (derived from human colon adenocarcinoma) was tested in the presence and the absence of 3,4',5,7-tetrahydroxyflavanone. Cells were grown on glass coverslips (12 mm diameter, 1 mm thick) to 75% confluence in EMEM supplemented with 10% FBS and 1% penicillin/streptomycin in a 24-well culture plate (Nunc, Australia) at 37 °C in 5% CO_2. Cells were rinsed three times with 1 mL of EMEM to remove residual antibiotics, and the medium was replaced with antibiotic-free culture medium. Bacteria were grown in TSB at 110 rpm for 4 h at 37 °C. Bacterial suspensions were centrifuged at 3000 rpm for 10 min and the bacterial pellets were resuspended in sterile PBS (pH 7.4) and adjusted to a concentration of c. 1×10^9 cfu/mL. 100 µL of bacterial suspension was inoculated per well, and the plates were incubated at 37 °C in 5% CO_2 for 90 min. Non-adherent bacteria were removed by washing the cells three times with sterile PBS. Cells were fixed with 95% ethanol for 5 min, air dried, Gram-stained, and examined by light microscopy ($\times 1000$). Bacterial adhesion was assessed [66], with the percentage of adherent bacteria determined by the presence of bacteria on 100 randomly selected cells, and the degree of bacterial adhesion assessed by counting the number of attached bacteria on 25 randomly selected Caco-2 cells. Strains that adhered to <10% of Caco-2 cells were deemed non-adherent. These tests were performed in duplicate. Cytotoxicity of 3,4',5,7-tetrahydroxyflavanone was tested in vitro against Vero cells as described above. 200 µL of cell suspension was seeded into 96-well tissue culture plates and grown to 100% confluence in medium without antibiotics. The flavonoid was tested at 200 µg, 100 µg and 50 µg final mass, with the tests performed in triplicate. Cells were examined for cytotoxic effects after 4, 24 and 48 h using and the cytotoxicity was interpreted as described above. These tests were performed in triplicate.

3.4. Data Analysis

Data were analysed by analysis of variance (ANOVA) followed by post-hoc Tukey's Honestly Significant Difference (HSD) test when the ANOVA detected significant differences ($p < 0.05$) among the means.

Acknowledgments: We thank Christina Neuman, Eva Hatje, Tim Smith, Tom Lewis, David Walton, Katie Roberts, Luke Verstraten, Nicholas Evans, Brooke Dwan, Tracey McMahon, and Bruce Randall for assistance. Motahareh Nobakht was supported by an Australian Postgraduate Award, an International Postgraduate Research Scholarship, and a top-up scholarship from Queensland Education and Training International.

Author Contributions: M.K., H.M.W., P.R.B., S.J.T. and M.N. conceived the study. M.N. and M.K. designed and performed the microbiology and cytotoxicity experiments with K.J.S. providing assistance. M.N., S.J.T. and H.M.W. designed and performed the plant sampling and statistical analyses. M.N. and P.R.B. designed and performed the flavonoid extraction and purification. All authors contributed to writing the paper.

Conflicts of Interest: The authors declare no conflict of interest.

References

1. Rappuoli, R. From Pasteur to genomics: Progress and challenges in infectious diseases. *Nat. Med.* **2004**, *10*, 1177–1185. [CrossRef] [PubMed]
2. Schmidt, F.R. The challenge of multidrug resistance: Actual strategies in the development of novel antibacterials. *Appl. Microbiol. Biotechnol.* **2004**, *63*, 335–343. [CrossRef] [PubMed]
3. Bell, B.G.; Schellevis, F.; Stobberingh, E.; Goossens, H.; Pringle, M. A systematic review and meta-analysis of the effects of antibiotic consumption on antibiotic resistance. *BMC Infect. Dis.* **2014**, *14*, 13. [CrossRef] [PubMed]
4. Moloney, M.G. Natural products as a source for novel antibiotics. *Trends Pharmacol. Sci.* **2016**, *37*, 689–701. [CrossRef] [PubMed]
5. Siller, G.; Rosen, R.; Freeman, M.; Welburn, P.; Katsamas, J.; Ogbourne, S.M. PEP005 (ingenol mebutate) gel for the topical treatment of superficial basal cell carcinoma: Results of a randomized phase IIa trial. *Australas. J. Dermatol.* **2010**, *51*, 99–105. [CrossRef] [PubMed]
6. Russo, P.; Frustaci, A.; Fini, M.; Cesario, A. From traditional European medicine to discovery of new drug candidates for the treatment of dementia and Alzheimer's disease: Acetylcholinesterase inhibitors. *Curr. Med. Chem.* **2013**, *20*, 976–983. [CrossRef] [PubMed]
7. Ogbourne, S.M.; Parsons, P.G. The value of nature's natural product library for the discovery of New Chemical Entities: The discovery of ingenol mebutate. *Fitoterapia* **2014**, *98*, 36–44. [CrossRef] [PubMed]
8. Hamilton, K.D.; Brooks, P.R.; Ogbourne, S.M.; Russell, F.D. Natural products isolated from *Tetragonula carbonaria* cerumen modulate free radical-scavenging and 5-lipoxygenase activities in vitro. *BMC Complement. Altern. Med.* **2017**, *17*, 232. [CrossRef] [PubMed]
9. Maiden, J. The gums, resins and other vegetable exudations of Australia. *J. R. Soc. NSW* **1901**, *1*, 161–212.
10. Penfold, A. *The Eucalypts*; Interscience Publishers: New York, NY, USA, 1961.
11. Locher, C.; Currie, L. Revisiting kinos—An Australian perspective. *J. Ethnopharmacol.* **2010**, *128*, 259–267. [CrossRef] [PubMed]
12. Reid, E.J.; Betts, T.J. *The Records of Western Australian Plants Used by Aboriginals as Medicinal Agents*; Western Australian Institute of Technology: Perth, Australia, 1977.
13. Packer, J.; Naz, T.; Yaegl Community Elders; Harrington, D.; Jamie, J.F.; Vemulpad, S.R. Antimicrobial activity of customary medicinal plants of the Yaegl Aboriginal community of northern New South Wales, Australia: A preliminary study. *BMC Res. Notes* **2015**, *8*, 276. [CrossRef] [PubMed]
14. Adeniyi, C.B.A.; Lawal, T.O.; Mahady, G.B. In vitro susceptibility of *Helicobacter pylori* to extracts of *Eucalyptus camaldulensis* and *Eucalyptus torelliana*. *Pharm. Biol.* **2009**, *47*, 99–102. [CrossRef] [PubMed]
15. Lawal, T.O.; Adeniyi, B.A.; Adegoke, A.O.; Franzblau, S.G.; Mahady, G.B. In vitro susceptibility of *Mycobacterium tuberculosis* to extracts of *Eucalyptus camaldulensis* and *Eucalyptus torelliana* and isolated compounds. *Pharm. Biol.* **2012**, *50*, 92–98. [CrossRef] [PubMed]
16. Satwalekar, S.S.; Gupta, T.R.; Narasimha, P.L. Chemical and antibacterial properties of kinos from *Eucalyptus* spp. Citriodorol—The antibiotic principle from the kino of *E. citriodora*. *J. Ind. Inst. Sci.* **1956**, *39*, 195–212.
17. Von Martius, S.; Hammer, K.A.; Locher, C. Chemical characteristics and antimicrobial effects of some *Eucalyptus* kinos. *J. Ethnopharmacol.* **2012**, *144*, 293–299. [CrossRef] [PubMed]
18. Ramos Alvarenga, R.F.; Wan, W.; Inui, T.; Franzblau, S.G.; Pauli, G.F.; Jaki, B.U. Airborne antituberculosis activity of *Eucalyptus citriodora* essential oil. *J. Nat. Prod.* **2014**, *77*, 603–610. [CrossRef] [PubMed]
19. Adeniyi, B.A.; Odufowoke, R.O.; Olaleye, S.B. Antibacterial and gastroprotective properties of *Eucalyptus torelliana* [Myrtaceae] crude extracts. *Int. J. Pharmacol.* **2006**, *2*, 362–365.

20. Lawal, T.O.; Adeniyi, B.A.; Idowu, O.S.; Moody, J.O. In vitro activities of *Eucalyptus camaldulensis* Dehnh. and *Eucalyptus torelliana* F. Muell. against non-tuberculous mycobacteria species. *Afr. J. Microbiol. Res.* **2011**, *5*, 3652–3657.
21. Wallace, H.M.; Trueman, S.J. Dispersal of *Eucalyptus torelliana* seeds by the resin-collecting stingless bee, *Trigona carbonaria*. *Oecologia* **1995**, *104*, 12–16. [CrossRef] [PubMed]
22. Wallace, H.M.; Howell, M.G.; Lee, D.J. Standard yet unusual mechanisms of long-distance dispersal: Seed dispersal of *Corymbia torelliana* by bees. *Divers. Distrib.* **2008**, *14*, 87–94. [CrossRef]
23. Wallace, H.; Lee, D.J. Resin-foraging by colonies of *Trigona sapiens* and *T. hockingsi* (Hymenoptera: Apidae, Meliponini) and consequent seed dispersal of *Corymbia torelliana* (Myrtaceae). *Apidologie* **2010**, *41*, 428–435. [CrossRef]
24. Leonhardt, S.D.; Wallace, H.M.; Schmitt, T. The cuticular profiles of Australian stingless bees are shaped by resin of the eucalypt tree *Corymbia torelliana*. *Austral Ecol.* **2011**, *36*, 537–543. [CrossRef]
25. Wallace, H.M.; Leonhardt, S.D. Do hybrid trees inherit invasive characteristics? Fruits of *Corymbia torelliana* × *C. citriodora* hybrids and potential for seed dispersal by bees. *PLoS ONE* **2015**, *10*, e0138868. [CrossRef] [PubMed]
26. Leonhardt, S.D.; Baumann, A.-M.; Wallace, H.M.; Brooks, P.; Schmitt, T. The chemistry of an unusual seed dispersal mutualism: Bees use a complex set of olfactory cues to find their partner. *Anim. Behav.* **2014**, *98*, 41–51. [CrossRef]
27. Drescher, N.; Wallace, H.M.; Katouli, M.; Massaro, C.F.; Leonhardt, S.D. Diversity matters: How bees benefit from different resin sources. *Oecologia* **2014**, *176*, 943–953. [CrossRef] [PubMed]
28. Massaro, C.F.; Smyth, T.J.; Smyth, W.F.; Heard, T.; Leonhardt, S.D.; Katouli, M.; Wallace, H.M.; Brooks, P. Phloroglucinols from anti-microbial deposit-resins of Australian stingless bees (*Tetragonula carbonaria*). *Phytother. Res.* **2015**, *29*, 48–58. [CrossRef] [PubMed]
29. Massaro, C.F.; Katouli, M.; Grkovic, T.; Vu, H.; Quinn, R.J.; Heard, T.A.; Carvalho, C.; Manley-Harris, M.; Wallace, H.M.; Brooks, P. Anti-staphylococcal activity of C-methyl flavanones from propolis of Australian stingless bees (*Tetragonula carbonaria*) and fruit resins of *Corymbia torelliana* (Myrtaceae). *Fitoterapia* **2014**, *95*, 247–257. [CrossRef] [PubMed]
30. Nobakht, M.; Grkovic, T.; Trueman, S.J.; Wallace, H.M.; Katouli, M.; Quinn, R.J.; Brooks, P.R. Chemical constituents of kino extract from *Corymbia torelliana*. *Molecules* **2014**, *19*, 17862–17871. [CrossRef] [PubMed]
31. Streeter, K.; Neuman, C.; Thompson, J.; Hatje, E.; Katouli, M. The characteristics of genetically related *Pseudomonas aeruginosa* from diverse sources and their interaction with human cell lines. *Can. J. Microbiol.* **2016**, *62*, 233–240. [CrossRef] [PubMed]
32. Tomasz, A. Mode of action of β-lactam antibiotics—A microbiologist's view. In *Antibiotics—Handbook of Experimental Pharmacology*; Demain, A.L., Solomon, A.N., Eds.; Springer: Berlin, Germany, 1983; pp. 15–96.
33. Clinical and Laboratory Standards Institute. *Performance Standards for Antimicrobial Susceptibility Testing; Twenty-Second Informational Supplement*; CLSI document M100-S22; Clinical and Laboratory Standards Institute: Wayne, PA, USA, 2012.
34. Polaquini, S.R.B.; Svidzinski, T.I.E.; Kemmelmeier, C.; Gasparetto, A. Effect of aqueous extract from Neem (*Azadirachta indica* A. Juss) on hydrophobicity, biofilm formation and adhesion in composite resin by *Candida albicans*. *Arch. Oral Biol.* **2006**, *51*, 482–490. [CrossRef] [PubMed]
35. Mittal, R.; Aggarwal, S.; Sharma, S.; Chhibber, S.; Harjai, K. Urinary tract infections caused by *Pseudomonas aeruginosa*: A minireview. *J. Infect. Public Health* **2009**, *2*, 101–111. [CrossRef] [PubMed]
36. Høiby, N.; Ciofu, N.; Bjarnshol, T. *Pseudomonas aeruginosa* biofilms in cystic fibrosis. *Future Microbiol.* **2010**, *5*, 1663–1674. [CrossRef] [PubMed]
37. Lynch, A.S.; Robertson, G.T. Bacterial and fungal biofilm infections. *Annu. Rev. Med.* **2008**, *59*, 415–428. [CrossRef] [PubMed]
38. Fiorentini, C.; Barbieri, E.; Falzano, L.; Mattaresse, P.; Baffone, W.; Pianetti, A.; Katouli, M.; Kühn, I.; Möllby, R.; Bruscolini, F.; et al. Occurrence, diversity and pathogenicity of mesophilic *Aeromonas* in estuarine waters of the Italian coast of the Adriatic Sea. *J. Appl. Microbiol.* **1998**, *85*, 501–511. [CrossRef] [PubMed]
39. Snowden, L.A.; Wernbacher, L.; Stenzel, D.; Tucker, J.; McKay, D.; O'Brien, M.; Katouli, M. Prevalence of environmental *Aeromonas* in South-East Queensland, Australia: A study of their interactions with human monolayer Caco-2 cells. *J. Appl. Microbiol.* **2006**, *101*, 964–975. [CrossRef] [PubMed]

40. Hatje, E.; Neuman, C.; Katouli, M. Interaction of *Aeromonas* strains with lactic acid bacteria using Caco-2 cells. *Appl. Environ. Microbiol.* **2014**, *80*, 681–686. [CrossRef] [PubMed]
41. Plésiat, P.; Nikaido, H. Outer membranes of Gram-negative bacteria are permeable to steroid probes. *Mol. Microbiol.* **1992**, *6*, 1323–1333. [CrossRef] [PubMed]
42. Tsuchiya, H.; Sato, M.; Miyazaki, T.; Fujiwara, S.; Tanigaki, S.; Ohyama, M.; Tanaka, T.; Iinuma, M. Comparative study on the antibacterial activity of the phytochemical flavanones against methicillin-resistant *Staphylococcus aureus*. *J. Ethnopharmacol.* **1996**, *50*, 27–34. [CrossRef]
43. Alcaraz, L.E.; Blanco, S.E.; Puig, O.N.; Tomas, F.; Ferretti, F.H. Antibacterial activity of flavonoids against methicillin-resistant *Staphylococcus aureus* strains. *J. Theor. Biol.* **2000**, *205*, 231–240. [CrossRef] [PubMed]
44. Silva, J.F.M.; Souza, M.C.; Matta, S.R.; Andrade, M.R.; Vidal, F.V.N. Correlation analysis between phenolic levels of Brazilian propolis extracts and their antimicrobial and antioxidant activities. *Food Chem.* **2006**, *99*, 431–435. [CrossRef]
45. Hoffman, L.R.; D'Argenio, D.A.; MacCoss, M.J.; Zhang, Z.; Jones, R.A.; Miller, S.I. Aminoglycoside antibiotics induce bacterial biofilm formation. *Nature* **2005**, *436*, 1171–1175. [CrossRef] [PubMed]
46. Plyuta, V.; Zaitseva, J.; Lobakova, E.; Zagoskina, N.; Kuznetsov, A.; Khmel, I. Effect of plant phenolic compounds on biofilm formation by *Pseudomonas aeruginosa*. *APMIS* **2013**, *121*, 1073–1081. [CrossRef] [PubMed]
47. Sousa, L.P.; Silva, A.F.; Calil, N.O.; Oliveira, M.G.; Silva, S.S.; Raposo, N.R.B. In vitro inhibition of *Pseudomonas aeruginosa* adhesion by xylitol. *Braz. Arch. Biol. Technol.* **2011**, *54*, 877–884. [CrossRef]
48. Hillis, W. The chemistry of the Eucalypt kinos. Part I. chromatographic resolution. *Aust. J. Basic Appl. Sci.* **1951**, *3*, 385–397.
49. Hillis, W. The chemistry of the Eucalypt kinos. Part II. Aromadendrin, kaempferol and ellagic acid. *Aust. J. Sci. Res.* **1952**, *2*, 379–386.
50. Butler, M.S. The role of natural product chemistry in drug discovery. *J. Nat. Prod.* **2004**, *67*, 2141–2153. [CrossRef] [PubMed]
51. Daglia, M. Polyphenols as antimicrobial agents. *Curr. Opin. Chem. Biol.* **2012**, *23*, 174–181. [CrossRef] [PubMed]
52. Hayes, R.A.; Piggott, A.M.; Smith, T.E.; Nahrung, H.F. *Corymbia* phloem phenolics, tannins and terpenoids: Interactions with a cerambycid borer. *Chemoecology* **2014**, *24*, 95–103. [CrossRef]
53. Nahrung, H.F.; Smith, T.E.; Wiegand, A.N.; Lawson, S.A.; Debuse, V.J. Host tree influences on longicorn beetle (Coleoptera: Cerambycidae) attack in subtropical *Corymbia* (Myrtales: Myrtaceae). *Environ. Entomol.* **2014**, *43*, 37–46. [CrossRef] [PubMed]
54. Loumouamou, A.N.; Silou, T.; Mapola, G.; Chalcat, J.C.; Figuéredo, G. Yield and composition of essential oils from *Eucalyptus citriodora* × *Eucalyptus torelliana*, a hybrid species growing in Congo-Brazzaville. *J. Essent. Oils Res.* **2009**, *21*, 295–299. [CrossRef]
55. Dickinson, G.R.; Wallace, H.M.; Lee, D.J. Reciprocal and advanced generation hybrids between *Corymbia citriodora* and *C. torelliana*: Forestry breeding and the risk of gene flow. *Ann. For. Sci.* **2013**, *70*, 1–10. [CrossRef]
56. Trueman, S.J.; McMahon, T.V.; Bristow, M. Production of cuttings in response to stock plant temperature in the subtropical eucalypts, *Corymbia citriodora* and *Eucalyptus dunnii*. *New For.* **2013**, *44*, 265–279. [CrossRef]
57. Trueman, S.J.; McMahon, T.V.; Bristow, M. Nutrient partitioning among the roots, hedge and cuttings of *Corymbia citriodora* stock plants. *J. Plant Nutr. Soil Sci.* **2013**, *13*, 977–989. [CrossRef]
58. Trueman, S.J.; McMahon, T.V.; Bristow, M. Biomass partitioning in *Corymbia citriodora*, *Eucalyptus cloeziana* and *E. dunnii* stock plants in response to temperature. *J. Trop. For. Sci.* **2013**, *25*, 504–509.
59. Lopes, E.D.; Laia, M.L.; Santos, A.S.; Soares, G.M.; Leite, R.W.P.; Martins, N.S. Influência do espaçamento de plantio na produção energética de clones de *Corymbia* e *Eucalyptus*. *Floresta* **2017**, *47*, 95–104. [CrossRef]
60. Wendling, I.; Brooks, P.R.; Trueman, S.J. Topophysis in *Corymbia torelliana* × *C. citriodora* seedlings: Adventitious rooting capacity, stem anatomy, and auxin and abscisic acid concentrations. *New For.* **2015**, *46*, 107–120. [CrossRef]
61. Maiden, J.H. Botany Bay of Eucalyptus kino. *Pharm. J. Trans.* **1889**, *3*, 221–321.
62. Kudi, A.; Umoh, J.; Eduvie, L.; Gefu, J. Screening of some Nigerian medicinal plants for antibacterial activity. *J. Ethnopharmacol.* **1999**, *67*, 225–228. [CrossRef]

63. Boyanova, L.; Gergova, G.; Nikolov, R.; Derejian, S.; Lazarova, E.; Katsarova, N.; Mitov, I.; Krastev, Z. Activity of Bulgarian propolis against 94 *Helicobacter pylori* strains in vitro by agar-well diffusion, agar dilution and disc diffusion methods. *J. Med. Microbiol.* **2005**, *54*, 481–483. [CrossRef] [PubMed]
64. Palombo, E.A.; Semple, S. Antibacterial activity of traditional Australian medicinal plants. *J. Ethnopharmacol.* **2001**, *77*, 151–157. [CrossRef]
65. Wiegand, I.; Hilpert, K.; Hancock, R.E. Agar and broth dilution methods to determine the minimal inhibitory concentration (MIC) of antimicrobial substances. *Nat. Protoc.* **2008**, *3*, 163–175. [CrossRef] [PubMed]
66. Grey, P.A.; Kirov, S.M. Adherence to HEp-2 cells and enteropathogenic potential of *Aeromonas* spp. *Epidemiol. Infect.* **1993**, *110*, 279–287. [CrossRef] [PubMed]

© 2017 by the authors. Licensee MDPI, Basel, Switzerland. This article is an open access article distributed under the terms and conditions of the Creative Commons Attribution (CC BY) license (http://creativecommons.org/licenses/by/4.0/).

Communication

Chemical Constituents and Antifungal Activity of *Ficus hirta* Vahl. Fruits

Chunpeng Wan [1], Chuying Chen [1], Mingxi Li [1], Youxin Yang [1], Ming Chen [1] and Jinyin Chen [1,2,*]

[1] Jiangxi Key Laboratory for Postharvest Technology and Nondestructive Testing of Fruits & Vegetables; Collaborative Innovation Center of Post-Harvest Key Technology and Quality Safety of Fruits and Vegetables; College of Agronomy, Jiangxi Agricultural University, Nanchang 330045, China; chunpengwan@jxau.edu.cn (C.W.); ccy0728@126.com (C.C.); liming.xi@hotmail.com (M.L.); yangyouxin@jxau.edu.cn (Y.Y.); chenming@jxau.edu.cn (M.C.)

[2] Pingxiang University, Pingxiang 337055, China

* Correspondence: jinyinchen@126.com; Tel.: +86-791-8381-3058

Received: 1 September 2017; Accepted: 25 September 2017; Published: 27 September 2017

Abstract: Phytochemical investigation of *Ficus hirta* Vahl. (Moraceae) fruits led to isolate two carboline alkaloids (**1** and **2**), five sesquiterpenoids/norsesquiterpenoids (**3–7**), three flavonoids (**8–10**), and one phenylpropane-1,2-diol (**11**). Their structures were elucidated by the analysis of their 1D and 2D NMR, and HR-ESI-MS data. All of the isolates were isolated from this species for the first time, while compounds **2**, **4–6**, and **8–11** were firstly reported from the genus *Ficus*. Antifungal assay revealed that compound **8** (namely pinocembrin-7-*O*-β-D-glucoside), a major flavonoid compound present in the ethanol extract of *F. hirta* fruits, showed good antifungal activity against *Penicillium italicum*, the phytopathogen of citrus blue mold caused the majority rotten of citrus fruits.

Keywords: *Ficus hirta*; Moraceae; carboline alkaloids; sesquiterpenoids; flavonoids; antifungal

1. Introduction

The genus *Ficus* (Moraceae) contains more than 1000 species, most of them are distributed in tropical, sub-tropical, and Mediterranean regions [1]. There are around 98 species distributed in the South of China. *Ficus hirta* Vahl. is mainly distributed in Yunnan, Guizhou, Guangxi, Guangdong and Hainan province, China [1]. The fruits of *F. hirta* were used as medicine and food resource by the local people of Guangdong province, China. Previous chemical investigations on the *F. hirta* focused on its roots, which led to isolate and identify the predominant chemical constituents, flavonoids, and coumarins. To date, the total of 31 flavonoids [2–7] and 7 coumarins [2,3,5,8] have been reported from this species. Except the flavonoids and coumarins, there are some other compounds reported from this species, such as steroids [2,7] and benzoic acid derivatives [7].

Several studies on the pharmacological activities of *F. hirta* showed its antioxidation [9], cytotoxicity, and apoptosis of HeLa cells [10]; anti-inflammation and analgesia [11], antitussive and antiasthmatic [12], hepatoprotective [13], and radioresistance effects [14]. The fruits of *F. hirta* consumed as a plant-derived food that showed potential tonic effects [15]. Besides mentioned above, *F. hirta* also showed antibacterial activity against *Escherichia coli*, *Staphylococcus aureus* [16], and *Penicillium italicum*, a phytopathogenic cause of citrus blue mold resulted in the destructive fruit rotten of citrus. The fruits of *F. hirta* and several other medicinal plants were also used to control the phytopathogen in order to decrease the loss of citrus rotten [17,18]. The fruits of *F. hirta* showed promising antifungal activity against *P. italicum* and prolonged the Nanfeng mandarin preservation period [19], while the major antifungal constituents were not clear until now.

In order to continue our studies on isolation and identification of the antifungal compounds from plants. We have elucidated the antifungal constituents of *F. hirta* fruits. Fortunately, our previous studies have identified nine monosubstituted benzene derivatives from the extracts of *F. hirta* fruits and some of them showed good antifungal activities [17,18], while they were not the major antifungal constituents for their low content in the plant. In continuation, the current study was aimed to discover the major antifungal compounds with diverse structures from this species.

2. Results

As described previously, the ethanol extracts (FH) of the fruits of *F. hirta* and the fractions (FH1–FH4) fractionated by D101 macro resin column were evaluated for their antifungal activities against *P. italicum*. Fractions FH2–FH4 showed stronger antifungal activities in a concentration-dependent manner than that of FH crude extract [18]. The further isolation was focused on the fractions with potent antifungal activities to find more active compounds present in the fruits of *F. hirta*.

As a result, 11 compounds (**1–11**) (Figure 1) were isolated from those fractions, and their structures were elucidated based on the analysis of spectroscopic data (including HR-ESI-MS, ^{1}H-NMR, ^{13}C-NMR, and 2D NMR) and comparison of these data to previous published paper. The 11 compounds were determined as 1-methyl-1,2,3,4-tetrahydro-β-carboline-3-carboxylic acid (**1**) [20], methyl 1-methyl-1,2,3,4-tetrahydro-β-carboline-3-carboxylate (**2**) [21], vomifoliol (**3**) [22], dehydrovomifoliol (**4**) [22], icariside B$_2$ (**5**) [23], dihydrophaseic acid (**6**) [24], pubinernoid A (**7**) [25], pinocembrin-7-O-β-D-glucoside (**8**) [26], naringenin-7-O-β-D-glucoside (**9**) [27], eriodictyol-7-O-β-D-glucoside (**10**) [28], and 1-phenylpropane-1,2-diol (**11**) [29]. All of the isolates were isolated from this species for the first time, while compounds **2**, **4–6**, and **8–11** were first reported from the genus *Ficus*.

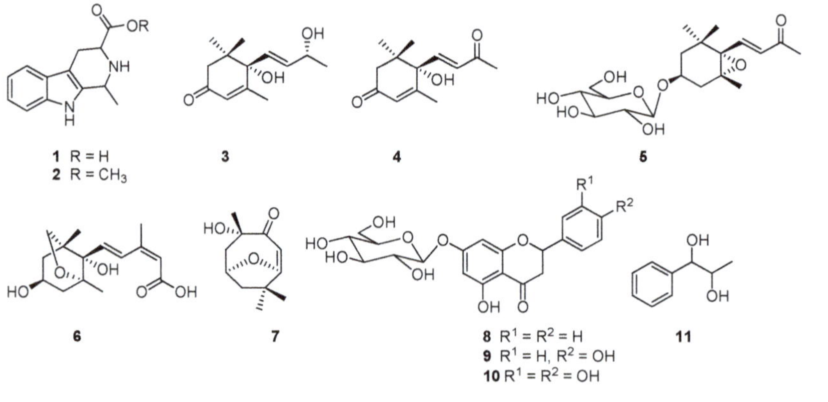

Figure 1. Structures of the compounds isolated from the fruits of *Ficus hirta*.

Compound **5**, was obtained as colorless amorphous solid, displayed a molecular formula of $C_{19}H_{30}O_8$ as determined by HRESIMS at *m/z* 409.1822 [M + Na]$^+$ (calcd. for $C_{19}H_{30}O_8Na$, 409.1838). In the ^{1}H-NMR spectrum, four tertiary methyls signals (δ_H 2.29, 1.21, 1.19, 0.96, each 3H, s), a *trans* double bond protons signal at δ_H 7.17 (1H, d, *J* = 15.8 Hz) and 6.18 (1H, d, *J* = 15.8 Hz), and an anomeric proton at δ_H 4.34 (1H, d, *J* = 7.0 Hz), as well as oxygen-bearing methine protons at δ_H 3.86 (1H, m) were observed. The ^{13}C-NMR and HSQC spectra revealed the presence of 19 carbon resonances, 6 of them were contributed to glucose. Analysis of the 1D and 2D NMR spectra data (including ^{1}H–^{1}H COSY, HSQC, HMBC) allowed for the establishment of the structure of **5**. The HSQC spectrum permitted the assignment of all of the protons to their bonding carbons. The ^{1}H–^{1}H COSY spectra (drawn with bold bonds in Figure 2) disclosed that compound **5** had three partial structure units (including a sugar moiety). Analysis of the HMBC spectrum then enabled the connectivity of these spin coupling fragments and the other functional groups. The HMBC correlations (Figure 2) from H$_3$-12 (H$_3$-13) to C-1, C-2, C-6, and C-13 (C-12); from H-11 to C-4, C-5, and C-6; from H-7 and H-8 to C-6; and from H$_3$-10 to C-8 and C-9, allowed the construction of the planar structure of aglycone. The glucopyranose was linked to C-3 by the HMBC correlation from H-1′ to C-3. Searching the structure with SCIFINDER revealed it has the same planar structure as the NMR data of **5** with those of icariside B$_2$ indicated they had the same stereochemistry. Therefore, compound **5** was determined as icariside B$_2$.

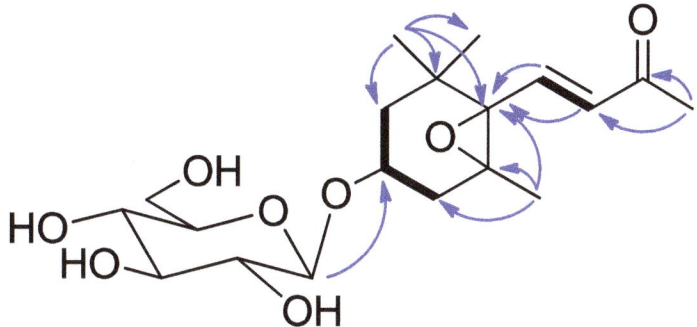

Figure 2. ^{1}H–^{1}H COSY (bold bonds) and HMBC (arrows) correlations of Compound **5**.

Compound **6**, had a molecular formula of C$_{15}$H$_{22}$O$_5$ as determined by HRESIMS at *m/z* 305.1349 [M + Na]$^+$ (calcd. for C$_{15}$H$_{22}$O$_5$Na, 305.1365). The ^{1}H-NMR spectrum showed a *trans* double bond signals at δ$_H$ 7.98 (1H, d, *J* = 15.8 Hz) and 6.52 (1H, d, *J* = 15.8 Hz), an olefinic proton signal at δ$_H$ 5.76 (1H, s), an oxygenated methylene signal at δ$_H$ 3.80 (1H, d, *J* = 7.4 Hz) and 3.70 (1H, d, *J* = 7.4 Hz), and three tertiary methyls signals (δ$_H$ 2.08, 1.15, 0.93, each 3H, s). The ^{13}C-NMR and HSQC spectra revealed the presence of 15 carbon signals, attributing to 3 methyls, 3 methylenes, 4 methines, and 5 quaternary carbons. Aforementioned data suggested that compound **6** was likely a sesquiterpenoid. Further analysis of 2D NMR (^{1}H–^{1}H COSY, HSQC, and HMBC) data allowed us to determine the structure of **6**. The ^{1}H–^{1}H COSY correlations of H-2/H-3, H-3/H-4, and H-7/H-8 indicated the presence of two structure units (drawn with bold bonds in Figure 3). The HMBC correlations (Figure 3) from H$_2$-12 to C-1, C-2, C-5, C-6, and C-13; from H$_3$-14 to C-4, C-5, and C-6; from H-7 and H-8 to C-6; from H$_3$-15 to C-8, C-9 and C-10; and from H-10 to C-11, constructed the planar structure of **6**. Compared the NMR data of **6** with those of dihydrophaseic acid revealed they had the same structure. Therefore, compound **6** was elucidated as dihydrophaseic acid.

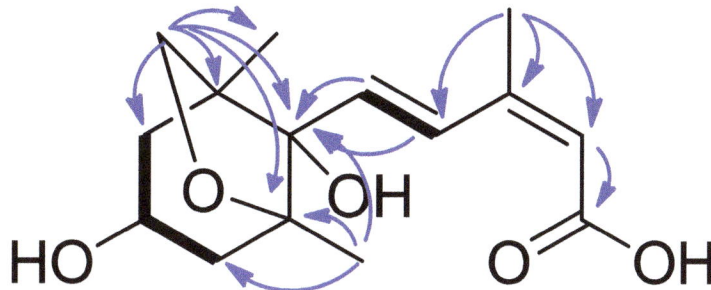

Figure 3. ^{1}H–^{1}H COSY (bold bonds) and HMBC (arrows) correlations of Compound **6**.

The antifungal activities of all of the isolates were tested at two concentrations (2.0 and 4.0 mg/mL). The results showed that none of them except compound **8** showed antifungal activity with the DIZs of 19.0 ± 0.5 mm and 24.0 ± 0.5 mm at 2.0 and 4.0 mg/mL, respectively, which are more powerful than that of FH (11.0 ± 0.6 mm at 2.0 mg/mL).

Moreover, compound **8** was also evaluated, its antifungal activity using mycelia growth method, the results are shown in Table 1. The inhibition rate was shown as concentration-dependent. Compound **8** exhibited more than 90% inhibitory effect against *P. italicum* at 400 μg/mL, while 100% inhibition rate was achieved at 800 μg/mL.

Table 1. Antifungal activity of Compound **8** tested by mycelia growth method

Concentration (μg/mL)	Inhibition Rate
25	13.70 ± 1.81
50	36.88 ± 1.08
100	56.10 ± 2.55
200	74.65 ± 1.61
400	92.05 ± 1.55
800	100

3. Discussion

The genus *Ficus* is characterized by flavonoids, coumarins, terpenoids (triterpenoids and sesquiterpenoids), and alkaloids [30]. In the current study, 11 compounds (**1–11**) were isolated and identified from the fruits of *F. hirta*, which were classified as carboline alkaloids (**1** and **2**), sesquiterpenoids/norsesquiterpenoids (**3–7**) and flavonoid glucosides (**8–10**). The structural classes of these isolates support the taxonomic placement of *F. hirta* in the genus *Ficus*. All of the chemical constituents are isolated from this species for the first time. Moreover, compounds **2, 4–6**, and **8–11** are firstly reported from the genus *Ficus*. Compound **1** has been isolated from *Ficus pumila* [17], which is the only carboline alkaloid reported from the genus *Ficus* before the current study. Five sesquiterpenoids/norsesquiterpenoids (**3–7**) could be further classified as megastigmanes (**3** and **5**), carotenoid sesquiterpenoid (**6**), and norsesquiterpenoid with 11C skeleton (**7**). Vomifoliol (**3**) has been exclusively obtained from the species of *Ficus platypoda* and *Ficus pumila* before our study [17,31].

The discovery of compounds **1** and **3** showed the relevance between this species and other *Ficus* species such as *F. platypoda* and *F. pumila*. Compound **6** is a derivative of plant hormone abscisic acid that can be classified as carotenoid sesquiterpenoid, which is different from other sesquiterpenoids isolated from the genus *Ficus* [32,33]. Compound **7** is a norsesquiterpenoids with 11C skeleton with different linkage with other 11C skeleton norsesquiterpenoids previously isolated from *Ficus microcarpa* [34,35]. Previously, two isomers of "Eriodictyol hexoside" and other flavonoids were tentative identified in the fruits of *F. carica* using HPLC-MS approaches [36]. The identification of eriodictyol-7-*O*-β-D-glucoside (**10**) in *F. hirta* confirmed these results and showed the relevance between this species and *F. carica*.

Pinocembrin-7-*O*-β-D-glucoside (**8**) showed good antifungal activity while the compounds **9** and **10** showed none activity, which indicated the antifungal activity of flavonoids maybe effected by the number of hydroxyl in the C loop. This is the first time the antifungal activity of compound **8** against *P. italicum* has been reported. However, some references have already revealed the antifungal activity of its aglycone, namely pinocembrin [37,38].

Overall, these results indicated that carboline alkaloid, megastigmanes, and flavonoids could be regarded as a chemotaxonomic marker of *F. hirta*. Also, while only one carotenoid sesquiterpenoid and a norsesquiterpenoid with 11C skeleton were identified herein, whether they may be regarded as a chemotaxonomic marker of *F. hirta* species remains to be established. Pinocembrin-7-*O*-β-D-glucoside (**8**) was the major antifungal constituent against *P. italicum* existed in the fruits of *Ficus hirta*.

4. Materials and Methods

4.1. Plant Material

The fruits of *F. hirta* were bought from Zhangshu medicinal market, Jiangxi Province, China, and authenticated by Prof. Shouran Zhou (College of Basic Medicine, Jiangxi University of Traditional Chinese Medicine). A voucher specimen (no. FH-201406) was deposited in the herbarium of Jiangxi Key Laboratory for Postharvest Technology and Nondestructive Testing of Fruits & Vegetables, Jiangxi Agricultural University (Nanchang, Jiangxi, China).

4.2. Equipment and Reagents

^{1}H- and ^{13}C-NMR spectral data were tested on a Varian 400 MHz Nuclear magnetic resonance spectrometer with Tetramethylsilane (TMS) as internal standard. HR-ESI-MS were detected on a TripleTOF™ 5600 LC/MS/MS (Applied Biosystems MDS, Foster City, CA, USA) mass spectrometer. Medium pressure liquid chromatography (MPLC) was carried out on a C-605 pump (BUCHI, Flawil, Switzerland) coupled with a reverse phase C18 column (3.6 × 46 cm). HPLC was conducted on a Hitachi Elite Chromaster system—consisting of a 5210 autosampler, 5110 pump, 5430 diode array detector, and 5310 column oven—which were operated by EZChrom Elite software. Luna C18 column (5 μm, 4.6 × 250 mm) for analysis and Luna C18 column (5 μm, 10 × 250 mm) for semi-preparative HPLC were purchased from Phenomenex Inc. (Torrance, CA, USA). The HPLC grade solvents were purchased from Sigma (Sigma, St. Louis, MO, USA). All analytical solvents were bought from Tansoole (Shanghai, China).

4.3. Extraction and Chromatography

The air dried fruits of *F. hirta* (4.9 kg) were ground and extracted using ultrasonic-assisted method with 95% ethanol (3 × 25 L) at 25 °C for 90 min. The extract were evaporated to remove ethanol solvent and yielded the dried ethanol extract (345.1 g), which was subjected to D101 macro rein column chromatography eluted with water, 30% ethanol (v/v), 50% ethanol, and 95% ethanol, respectively, to yield four fractions (FH1–FH4). Antifungal activity test indicated that three fractions (FH2–FH4) were the active fractions. Activity-guided isolation were performed accordingly, as follows.

The 30% ethanol fraction FH2 (113.6 g) was subjected to C_{18} silica gel column (3.6 × 46 cm) chromatography elution with MeOH/H$_2$O (MeOH/H$_2$O, 15/85 to 25/75, v/v) to yield five fractions (FH2a–FH2e). Fraction FH2c was separated on Sephadex LH-20 and eluted with MeOH to give six combined sub-fractions (FH2c1–FH2c6). Fraction FH2c2 was subjected to Sephadex LH-20 elution with MeOH to give five sub-fractions (FH2c2a–FH2c2e). Sub-fraction FH2c2c was purified using silica gel column chromatography using CH$_3$Cl-MeOH (10:1 to 1:1, v/v) for elution to give compound **1** (12.0 mg) and two sub-fractions (FH2c2c1 and FH2c2c2). Purification of FH2c2c1 with semi-preparative HPLC, eluting with MeOH-H$_2$O (0–25 min: 20:80 to 55:45; 25–26 min: 55:45 to 100:0; 26–27 min: 100:0; 27–28 min: 100:0 to 20:80; 28–35 min: 20:80; v/v, 3 mL/min), yielded compound **5** (2.4 mg). Fraction FH2c2d was subjected to silica gel column chromatography eluted with CH$_3$Cl-MeOH (100:1 to 1:1, v/v) to get seven fractions (FH2c2d1–FH2c2d7).

Purification of fraction FH-2C2d1 with semi-preparation HPLC, eluting with MeOH-H$_2$O (0–25 min: 20:80 to 68:32; 25–26 min: 68:32 to 100:0; 26–27 min: 100:0; 27–28 min: 100:0 to 30:70; 28–35 min: 30:70; v/v, 3 mL/min), yielded compound **4** (3.6 mg).

Fraction FH2c2d2 was purified by semi-preparative HPLC, eluting with MeOH-H$_2$O (0–20 min: 20:80 to 51:49; 20–21 min: 51:49 to 100:0; 21–22 min: 100:0; 22–23 min: 100:0 to 20:80; 23–30 min: 20:80; v/v, 3 mL/min), yielded compounds **7** (4.0 mg) and **11** (4.5 mg). Fraction FH2c2d5 was subjected on Sephadex LH-20 eluted with MeOH to give three sub-fractions (FH2c2d5a–FH2c2d5c). Purification of FH2c2d5b with semi-preparative HPLC, eluting with MeOH-H$_2$O (0–35 min: 30:70 to 34:66; 35–36 min: 34:66 to 100:0; 36–37 min: 100:0; 37–38 min: 100:0 to 30:70; 38–45 min: 30:70; v/v, 3 mL/min), yielded compounds **2** (3.2 mg) and **6** (8.2 mg). Fraction FH2c3 was purified by semi-preparative HPLC, eluting with MeOH-H$_2$O (0–21 min: 20:80 to 55:45; 21–22 min: 55:45 to 100:0; 22–23 min: 100:0; 23–24 min: 100:0 to 20:80; 24–31 min: 20:80; v/v, 3 mL/min), yielded compound **3** (9.7 mg). Fraction FH2d was separated on Sephadex LH-20 eluted with MeOH to give eleven combined sub-fractions (FH2d1–FH2d11). Fraction FH2d6 was recrystallized with methanol to yield compound **10** (15.4 mg).

The 50% ethanol fraction FH3 (35.5 g) was subjected to C$_{18}$ silica gel column chromatography eluted with MeOH/H$_2$O (MeOH/H$_2$O, 40/60 to 70/30, v/v) to yield six fractions FH3a–FH3f. Fraction FH3c was separated on Sephadex LH-20 eluted with MeOH to give eight combined sub-fractions (FH3c1–FH3c8). Fraction FH3c7 was recrystallized with methanol to yield compound **9** (13.2 mg).

The 95% ethanol eluted fraction FH4 (7.2 g) was separated over a column of Sephadex LH-20 eluted with MeOH to give eight combined sub-fractions (FH4a–FH4h). Fraction FH4e was recrystallized with methanol to yield compound **8** (21.5 mg).

4.4. Antifungal Activity Test

The antifungal activity of FH extracts and isolates against *P. italicum* was evaluated by the Oxford Cup method as described previously [17,18].

The antifungal activity of the pure compound **8** (pinocembrin-7-*O*-β-D-glucoside) against *P. italicum* were further examined by the mycelia growth method as described previously [17]. Briefly, the pure compound **8** were dissolved in 95% ethanol, and then added to the sterile PDA (potato dextrose agar) culture medium at the specified concentrations. The mixed media were then poured into plastic Petri dishes (90 mm). The agar-mycelial plugs (6 mm) infected with pathogens were incubated at the center of the Petri dishes sealed with parafilm and incubated in the dark. Mycelium colony growth diameters were measured when the fungal mycelium of the control group had completely covered the Petri dishes. All treatments were tested in six replicates. The inhibition of mycelial growth (IMG, %) was calculated as the following formula: IMG (%) = 100 × (dc − dt)/(dc − 6), where dc and dt were the mycelium diameters (mm) of the control and the treatment, respectively.

4.5. NMR and MS Data of Compounds **1–11**

The ^{1}H- and ^{13}C-NMR data of these compounds (**1–11**) were listed as follows.

Compound **1** HR-ESI-MS *m*/*z* 231.1130 [M + H]$^+$, ^{1}H-NMR (400 MHz, DMSO-d_6) δ: 4.52 (1, H d, *J* = 6.0 Hz, H-1), 3.64 (1H, dd, *J* = 4.5, 12.0 Hz, H-3), 2.78 (1H, m, H-4b), 7.45 (1H, d, *J* = 7.8 Hz, H-5), 7.01 (1H, t, *J* = 7.2 Hz, H-6), 7.10 (1H, t, *J* = 7.2 Hz, H-7), 7.34 (1H, d, *J* = 7.8 Hz, H-8), 3.18 (1H, dd, *J* = 4.5, 15.3 Hz, H-4a), 1.63 (3H, d, *J* = 6.2 Hz, CH$_3$). ^{13}C-NMR (100 MHz, DMSO-d_6) δ: 49.5 (C-1), 58.1 (C-3), 23.7 (C-4), 118.5 (C-5), 119.3 (C-6), 121.8 (C-7), 111.7 (C-8), 107.2 (C-4a), 126.6 (C-4b), 136.8 (C-8a), 132.7 (C-9), 17.4 (C-10), 170.0 (C-11).

Compound **2** HR-ESI-MS *m*/*z* 245.1284 [M + H]$^+$, ^{1}H-NMR (400 MHz, CD$_3$OD) δ: 4.60 (1H, dd, *J* = 5.3, 12.0 Hz, H-1), 3.95 (3H, s, OCH$_3$), 3.59 (1H, dd, *J* = 4.8, 11.2 Hz, H-3), 3.52 (1H, t, *J* = 6.0 Hz, H-4b), 7.50 (1H, d, *J* = 7.8 Hz, H-5), 7.07 (1H, t, *J* = 7.2 Hz, H-6), 7.16 (1H, t, *J* = 7.2 Hz, H-7), 7.37 (1H, d, *J* = 7.8 Hz, H-8), 3.14 (1H, m, H-4b), 1.78 (3H, d, *J* = 6.2 Hz, CH$_3$). ^{13}C-NMR (100 MHz, CD$_3$OD) δ: 50.2 (C-1), 55.6 (C-3), 22.3 (C-4), 117.7 (C-5), 119.4 (C-6), 122.3 (C-7), 110.0 (C-8), 104.6 (C-4a), 125.7 (C-4b), 137.2 (C-8a), 129.4 (C-9), 15.6 (C-10), 169.0 (C-11), 52.5 (OCH$_3$).

Compound **3** HR-ESI-MS *m*/*z* 247.1291 [M + Na]$^+$, ^{1}H-NMR (400 MHz, CD$_3$OD) δ: 1.00 (3H, s, CH$_3$-11), 1.03 (3H, s, CH$_3$-12), 1.24 (3H, d, *J* = 6.8 Hz, CH$_3$-10), 1.91 (3H, d, *J* = 1.2 Hz, CH$_3$-13), 2.16 (1H, d, *J* = 16.4 Hz, H-2a), 2.48 (1H, d, *J* =16.4 Hz, H-2b), 4.32 (1H, dq, *J* = 6.4, 6.4 Hz, H-9), 5.78 (1H, d, *J* = 16.0 Hz, H-7), 5.80 (1H, dd, *J* = 16.0, 6.4 Hz, H-8), 5.87 (1H, q, *J* = 1.4 Hz, H-4). ^{13}C-NMR (100 MHz, CD$_3$OD) δ: 41.0 (C-1), 49.3 (C-2), 199.8 (C-3), 125.7 (C-4), 166.0 (C-5), 79.1 (C-6), 128.5 (C-7), 135.5 (C-8), 67.2 (C-9), 18.1 (C-10), 23.1 (C-11), 22.4 (C-12), 22.0 (C-13).

Compound **4** HR-ESI-MS *m*/*z* 221.1164 [M − H]$^−$, ^{1}H-NMR (400 MHz, CD$_3$OD) δ: 1.01 (3H, s, CH$_3$-11), 1.06 (3H, s, CH$_3$-12), 2.31 (3H, s, CH$_3$-10), 1.90 (3H, d, *J* = 1.2 Hz, CH$_3$-13), 2.31 (1H, d, *J* = 16.4 Hz,

H-2a), 2.62 (1H, d, J = 16.4 Hz, H-2b), 6.98 (1H, d, J = 16.0 Hz, H-7), 6.43 (1H, dd, J = 16.0, 6.4 Hz, H-8), 5.93 (1H, s, H-4). ^{13}C-NMR (100 MHz, CD$_3$OD) δ: 40.6 (C-1), 49.1 (C-2), 189.9 (C-3), 126.6 (C-4), 163.2 (C-5), 78.6 (C-6), 146.9 (C-7), 130.3 (C-8), 199.2 (C-9), 26.2 (C-10), 22.1 (C-11), 23.3 (C-12), 17.7 (C-13).

Compound **5** HR-ESI-MS m/z 409.1822 [M + Na]$^+$, ^{1}H-NMR (CD$_3$OD, 400 MHz) δ: 7.17 (1H, d, J = 15.8 Hz, H-7), 6.18 (1H, d, J = 15.8 Hz, H-8), 4.34 (1H, d, J = 7.0 Hz, H-1′), 3.86 (1H, m, H-3), 3.12-3.83 (6H, Sugar H-2′, 3′, 4′, 5′, 6′), 2.42 (1H, m, H-4), 2.29 (3H, s, H-10), 1.81 (1H, dd, J = 8.2, 14.6 Hz, H-4), 1.73 (1H, m, H-2), 1.40 (1H, m, H-2), 1.21 (3H, s, H-13), 1.19 (3H, s, H-12), 0.96 (3H, s, H-11). ^{13}C-NMR (CD$_3$OD, 100 MHz) δ: 34.5 (C-1), 43.8 (C-2), 71.3 (C-3), 36.7 (C-4), 66.9 (C-5), 69.7 (C-6), 143.8 (C-7), 132.4 (C-8), 198.8 (C-9), 24.0 (C-10), 26.0 (C-11), 28.0 (C-12), 18.8 (C-13), 101.5 (C-1′), 73.7 (C-2′), 76.7 (C-3′), 70.2 (C-4′), 76.4 (C-5′), 61.3 (C-6′).

Compound **6** HR-ESI-MS m/z 305.1349 [M + Na]$^+$, ^{1}H-NMR (CD$_3$OD, 400 MHz) δ: 7.98 (1H, d, J = 15.8 Hz, H-4), 6.52 (1H, d, J = 15.8 Hz, H-5), 5.76 (1H, s, H-2), 0.93 (3H, s, H-13), 1.15 (3H, s, H-14), 2.08 (3H, s, H-15), 1.66 (1H, m, H-10b), 1.73 (1H, m, H-8b), 1.84 (1H, m, H-10a), 2.03 (1H, m, H-8a), 3.70 (1H, d, J = 7.4 Hz, H-12a), 3.80 (1H, d, J = 7.4 Hz, H-12b), 4.11 (1H, m, H-9). ^{13}C-NMR (CD$_3$OD, 100 MHz) δ: 168.1 (C-1), 117.8 (C-2), 150.1 (C-3), 130.4 (C-4), 133.8 (C-5), 81.8 (C-6), 86.3 (C-7), 44.6 (C-8), 64.6 (C-9), 43.1 (C-10), 75.8 (C-12), 14.9 (C-13), 18.2 (C-14), 19.8 (C-15).

Compound **7** HR-ESI-MS m/z 197.1155 [M + H]$^+$, ^{1}H-NMR (CD$_3$OD, 400 MHz) δ: 5.78 (1H, s, H-6), 4.10 (1H, m, H-2), 2.46 (1H, m, H-3β), 1.98 (1H, m, H-1β), 1.29 (1H, overlap, H-3α), 1.42 (1H, t, J = 11.6 Hz, H-1α), 1.59 (3H, s, H-11), 1.31 (3H, s, H-9), 1.28 (3H, s, H-10). ^{13}C-NMR (CD$_3$OD, 100 MHz) δ: 49.3 (C-1), 63.8 (C-2), 47.6 (overlap, C-3), 87.1 (C-4), 182.4 (C-5), 112.3 (C-6), 172.5 (C-7), 34.7 (C-8), 23.9 (C-9), 28.9 (C-10), 24.3 (C-11).

Compound **8** ESI-MS m/z 417.00 [M − H]$^-$, ^{1}H-NMR (600 MHz, DMSO-d_6) δ: 12.05 (5-OH), 7.55 (2H, d, J = 7.6 Hz, H-2′, 6′), 7.44 (3H, m, H-3′, 4′, 5′), 6.21 (1H, d, J = 1.5 Hz, H-8), 6.16 (1H, d, J = 1.5 Hz, H-6), 5.66 (1H, d, J = 12.9, H-2), 4.99 (1H, d, J = 7.4 Hz, H-1″), 3.66 (1H, d, J = 9.4 Hz, H-6″a), 3.15-3.45 (6H, H-3a, 2″, 3″, 4″, 5″, 6″b), 2.85 (1H, d, J = 16.7 Hz, H-3b). ^{13}C-NMR (150 MHz, DMSO-d_6) δ: 79.1 (C-2), 42.6 (C-3), 197.3 (C-4), 163.4 (C-5), 97.1 (C-6), 165.8 (C-7), 96.0 (C-8), 163.0 (C-9), 103.7 (C-10), 138.9 (C-1′), 127.2 (C-2′), 129.1 (C-3′), 129.1 (C-4′), 129.1 (C-5′), 127.2 (C-6′), 100.0 (C-1″), 73.5 (C-2″), 76.8 (C-3″), 69.9 (C-4″), 77.6 (C-5″), 61.0 (C-6″).

Compound **9** ESI-MS m/z 432.90 [M − H]$^-$, ^{1}H-NMR (600 MHz, DMSO-d_6) δ: 12.06 (5-OH), 7.33 (2H, d, J = 7.8 Hz, H-2′, 6′), 6.80 (3H, m, H-3′, 5′), 6.16 (1H, d, J = 1.5 Hz, H-8), 6.14 (1H, d, J = 1.5 Hz, H-6), 5.50 (1H, d, J = 12.7, H-2), 4.96 (1H, d, J = 7.6 Hz, H-1″), 3.67 (1H, d, J = 9.4 Hz, H-6″a), 3.14-3.46 (6H, H-3a, 2″, 3″, 4″, 5″, 6″b), 2.74 (1H, d, J = 16.9 Hz, H-3b). ^{13}C-NMR (150 MHz, DMSO-d_6) δ: 79.1 (C-2), 42.6 (C-3), 197.7 (C-4), 163.4 (C-5), 97.0 (C-6), 165.8 (C-7), 95.9 (C-8), 163.2 (C-9), 103.7 (C-10), 129.1 (C-1′), 128.9 (C-2′), 115.7 (C-3′), 158.3 (C-4′), 115.7 (C-5′), 128.9 (C-6′), 100.0 (C-1″), 73.5 (C-2″), 76.8 (C-3″), 69.9 (C-4″), 77.5 (C-5″), 61.0 (C-6″).

Compound **10** ESI-MS m/z 449.10 [M − H]$^-$, ^{1}H-NMR (600 MHz, CD$_3$OD) δ: 6.94 (1H, brs, H-5′), 6.80 (2H, brs, H-2′, 6′), 6.22 (1H, brs, H-8), 6.19 (1H, brs, H-6), 5.32 (1H, d, J = 12.6, H-2), 4.98 (1H, d, J = 7.4 Hz, H-1″), 3.89 (1H, d, J = 12.0 Hz, H-6″a), 3.70 (1H, dd, J = 5.3, 12.1 Hz, H-6″b), 3.40–3.48 (4H, H-3a, 2″, 3″, 4″), 3.13 (1H, m, H-5″), 2.75 (1H, d, J = 17.0 Hz, H-3b). ^{13}C-NMR (150 MHz, CD$_3$OD) δ: 79.3 (C-2), 42.8 (C-3), 197.1 (C-4), 163.5 (C-5), 96.6 (C-6), 165.6 (C-7), 95.5 (C-8), 163.2 (C-9), 103.5 (C-10), 130.1 (C-1′), 113.4 (C-2′), 145.1 (C-3′), 145.6 (C-4′), 114.9 (C-5′), 118.0 (C-6′), 99.8 (C-1″), 73.2 (C-2″), 76.4 (C-3″), 69.7 (C-4″), 76.8 (C-5″), 60.9 (C-6″).

Compound **11** HR-ESI-MS m/z 175.0713 [M + Na]$^+$, ^{1}H-NMR (CD$_3$OD, 400 MHz) δ: 7.26–7.35 (5H, m, H-2,3,4,5,6), 4.34 (1H, d, J = 7.1 Hz, H-7), 3.80 (1H, m, H-8), 0.95 (3H, d, J = 6.4 Hz, H-9).

5. Conclusions

Two carboline alkaloids (**1** and **2**), five sesquiterpenoids/norsesquiterpenoids (**3–7**) (three of which are megastigmanes), three flavonoids (**8–10**), and phenylpropane-1,2-diol (**11**) were isolated and identified from the fruits of *F. hirta* for the first time. Moreover, compounds **2**, **4–6**, and **8–11** were reported for the first time in the *Ficus* genus. Pinocembrin-7-O-β-D-glucoside (**8**) was the major antifungal constituent against *P. italicum* existed in the fruits of *Ficus hirta*. Chemotaxonomic analysis revealed that the carboline alkaloid, megastigmanes, and flavonoids could be regarded as a chemotaxonomic marker of *F. hirta*.

Acknowledgments: This project was supported by the National Natural Science Foundation of China (31500286) and the Natural Science Foundation of Jiangxi Province (20161BAB214167).

Author Contributions: Chunpeng Wan and Jinyin Chen conceived and designed the experiments; Chunpeng Wan, Chuying Chen, Mingxi Li, Youxin Yang and Ming Chen performed the experiments; Chunpeng Wan and Chuying Chen analyzed the data; Chunpeng Wan and Jinyin Chen wrote the paper.

Conflicts of Interest: The authors declare no conflict of interest.

References

1. Flora Compilation Committee of Chinese Academy of Science. *Flora of China*; Science Press: Beijing, China, 1998; Volume 23, pp. 67–160.
2. Li, C.; Bu, P.B.; Qiu, D.K.; Sun, Y.F. Chemical constituents from roots of *Ficus hirta*. *China J. Chin. Mater. Med.* **2006**, *31*, 131–133.
3. Ya, J.; Zhang, X.Q.; Wang, Y.; Li, Y.; Ye, W. Studies on flavonoids and coumarins in the roots of *Ficus hirta* Vahl. *Chem. Ind. For. Prod.* **2008**, *28*, 49–52.
4. Ya, J.; Zhang, X.Q.; Wang, G.C. Flavonoids from the roots of *Ficus hirta* Vahl. *Asia Chem. Lett.* **2009**, *13*, 21–26.
5. Ya, J.; Zhang, X.Q.; Wang, Y.; Zhang, Q.W.; Chen, J.X.; Ye, W.C. Two new phenolic compounds from the roots of *Ficus hirta*. *Nat. Prod. Res.* **2010**, *24*, 621–625. [CrossRef] [PubMed]
6. Zhao, L.P.; Di, B.; Feng, F. Chemical constituents from the roots of *Ficus hirta*. *Pharm. Clin. Res.* **2008**, *16*, 5–7.
7. Zheng, R.R.; Ya, J.; Wang, W.J.; Yang, H.B.; Zhang, Q.W.; Zhang, X.Q.; Ye, W.C. Chemical studies on roots of *Ficus hirta*. *China J. Chin. Mater. Med.* **2013**, *38*, 3696–3701.
8. Jiang, B.; Liu, Z.Q.; Zeng, Y.E.; Xu, H. Chemical constituents roots of *Ficus hirta*. *Chin. Tradit. Herb. Drugs* **2005**, *36*, 1141–1142.
9. Yi, T.; Chen, Q.; He, X.; So, S.; Lo, Y.; Fan, L.; Chen, H. Chemical quantification and antioxidant assay of four active components in *Ficus hirta* root using UPLC-PAD-MS fingerprinting combined with cluster analysis. *Chem. Cent. J.* **2013**, *7*, 115. [CrossRef] [PubMed]
10. Zeng, Y.; Liu, X.; Lv, Z.; Peng, Y.H. Effects of *Ficus hirta* Vahl.(Wuzhimaotao) extracts on growth inhibition of HeLa cells. *Exp. Toxicol. Pathol.* **2012**, *64*, 743–749. [CrossRef] [PubMed]
11. Zhou, T.N.; Wang, Y.; Tang, L.H.; Liu, D.D.; Hou, S.Z.; Deng, X.C.; Ye, M.R. Study of Radix *Fici hirtae* on anti-inflammatory analgesic and effect of acute liver injury of mice. *Pharm. Today* **2008**, *18*, 55–58.
12. Zhou, T.N.; Tang, L.H.; Huang, S.C.; Lu, D.D.; Wang, Y.; Liu, L.F.; Ye, M.R. Study on the antitussive and antiasthmatic effects of radix *Fici hirtae*. *J. Chin. Med. Mater.* **2009**, *32*, 571–574.
13. Jia, F.L.; Ruan, M. Radix *Fici hirtae* protective effect of aqueous extract on acute hepatic injury in mice induced by two methyl farmamide. *J. Chin. Med. Mater.* **2008**, *31*, 1364–1368.
14. Wang, X.P.; Duan, L.J.; Huang, X.; Cen, Y.W.; Li, G.F. protective role of aqueous extract from *Fici hirtae* radix for DNA damage of bone marrow cells by~(60) Co γ-ray in mice. *Chin. J. Mod. Appl. Pharm.* **2011**, *28*, 284–287.
15. Zhou, T.N.; Wang, Y.; Liu, D.D.; Tang, L.H.; Xiao, X.J.; Liu, L.F.; Ye, M.R. Experimental study on the tonic effect of different extracts from Radix *Fici hirtae*. *J. Chin. Med. Mater.* **2009**, *32*, 753–757.
16. Chen, Q.; Ye, S.X.; Yu, J. Antibacterial activity of Radix *Fici hirtae* by chromotest microassay. *Med. Plant* **2012**, *3*, 13–16.
17. Chen, C.; Wan, C.; Peng, X.; Chen, Y.; Chen, M.; Chen, J. Optimization of antifungal extracts from *Ficus hirta* fruits using response surface methodology and antifungal activity tests. *Molecules* **2015**, *20*, 19647–19649. [CrossRef] [PubMed]

18. Wan, C.; Han, J.; Chen, C.; Yao, L.; Chen, J.; Yuan, T. Monosubstituted benzene derivatives from fruits of *Ficus hirta* and their antifungal activity against phytopathogen *Penicillium italicum*. *J. Agric. Food Chem.* **2016**, *64*, 5621–5624. [CrossRef] [PubMed]
19. Chen, C.; Peng, X.; Zeng, R.; Chen, M.; Wan, C.; Chen, J. *Ficus hirta* fruits extract incorporated into an alginate-based edible coating for Nanfeng mandarin preservation. *Sci. Hortic.* **2016**, *202*, 41–48. [CrossRef]
20. Wei, W.; Fan, C.L.; Wang, G.Y.; Tang, H.J.; Wang, Y.; Ye, W.C. Chemical constituents from *Ficus pumila*. *Chin. Tradit. Herb. Drugs* **2014**, *45*, 615–621.
21. Zeng, Y.; Zhang, Y.; Weng, Q.; Hu, M.; Zhong, G. Cytotoxic and insecticidal activities of derivatives of harmine, a natural insecticidal component isolated from *Peganum harmala*. *Molecules* **2010**, *15*, 7775–7791. [CrossRef] [PubMed]
22. Yang, N.Y.; Duan, J.A.; Li, P.; Qian, S.H. Chemical constituents of *Glechoma longituba*. *Acta Pharm. Sin.* **2006**, *41*, 431–434.
23. Woo, K.W.; Lee, K.R. Phytochemical constituents of *Allium victorialis* var. platyphyllum. *Nat. Prod. Sci.* **2013**, *19*, 221–226.
24. Cai, L.; Liu, C.S.; Fu, X.W.; Shen, X.J.; Yin, T.P.; Yang, Y.B.; Ding, Z.T. Two new glucosides from the pellicle of the walnut (*Juglans regia*). *Nat. Prod. Bioprospect.* **2012**, *2*, 150–153. [CrossRef]
25. He, J.B.; Niu, Y.F.; Li, J.X.; Wang, L.B.; Zi, T.P.; Yu, S.; Tao, J. Studies on terpenoids from *Zygophyllum fabago*. *China J. Chin. Mater. Med.* **2015**, *40*, 4634–4638.
26. Zhao, Q.; Liu, F.; Li, Q.J.; Chen, W.P. Chemical constituents from flowers of *Rosa chinensis*. *Chin. Tradit. Herb. Drugs* **2012**, *43*, 1484–1488.
27. Ding, Y.X.; Guo, Y.J.; Ren, Y.L.; Dou, D.; Li, Q. Isolation of flavonoids from male flowers of *Eucommia ulmoides* and their anti-oxidantive activities. *Chin. Tradit. Herb. Drugs* **2014**, *45*, 323–327.
28. Deng, R.X.; Zhang, C.F.; Liu, P.; Duan, W.L.; Yin, W.P. Separation and identification of flavonoids from Chinese Fringetree Flowers (*Chionanthus retusa* Lindl et Paxt). *Food Sci.* **2014**, *35*, 74–78.
29. Mayorga, H.; Knapp, H.; Winterhalter, P.; Duque, C. Glycosidically bound flavor compounds of cape gooseberry (*Physalis peruviana* L.). *J. Agric. Food Chem.* **2001**, *49*, 1904–1908.
30. Fan, M.S.; Ye, G.; Huang, C.G. The advances of chemistry and pharmacological study of *Ficus* genus. *Nat. Prod. Res. Dev.* **2005**, *17*, 497–504.
31. El-Hela, A.; Mohammed, A.E.; Ragab, E.; Afifi, W. Chemical constituents and biological activity of *Ficus platypoda* (Miq.) leaves. *J. Biomed. Pharm. Res.* **2014**, *3*, 21–37.
32. Kitajima, J.; Kimizuka, K.; Tanak, Y. Three new sesquiterpenoid glucosides of *Ficus pumila* fruit. *Chem. Pharm. Bull.* **2000**, *48*, 77–80. [CrossRef] [PubMed]
33. Somwong, P.; Suttisri, R.; Buakeaw, A. New sesquiterpenes and phenolic compound from *Ficus foveolata*. *Fitoterapia* **2013**, *85*, 1–7. [CrossRef] [PubMed]
34. Li, Y.C.; Kuo, Y.H. A monoterpenoid and two simple phenols from heartwood of *Ficus microcarpa*. *Phytochemistry* **1998**, *49*, 2417–2419. [CrossRef]
35. Kuo, Y.H.; Li, Y.C. Three new compounds, ficusone, ficuspirolide, and ficusolide from the heartwood of *Ficus microcarpa*. *Chem. Pharm. Bull.* **1999**, *47*, 299–301. [CrossRef]
36. Ammar, S.; del Mar Contreras, M.; Belguith-Hadrich, O.; Bouaziz, M.; Segura-Carretero, A. New insights into the qualitative phenolic profile of *Ficus carica* L. fruits and leaves from Tunisia using ultra-high-performance liquid chromatography coupled to quadrupole-time-of-flight mass spectrometry and their antioxidant activity. *RSC Adv.* **2015**, *26*, 20035–20050. [CrossRef]
37. Yang, S.; Liu, L.; Li, D.; Xia, H.; Su, X.; Peng, L.; Pan, S. Use of active extracts of poplar buds against *Penicillium italicum* and possible modes of action. *Food Chem.* **2016**, *196*, 610–618. [CrossRef] [PubMed]
38. Peng, L.; Yang, S.; Cheng, Y.J.; Chen, F.; Pan, S.; Fan, G. Antifungal activity and action mode of pinocembrin from propolis against *Penicillium italicum*. *Food Sci. Biotechnol.* **2012**, *6*, 1533–1539. [CrossRef]

 © 2017 by the authors. Licensee MDPI, Basel, Switzerland. This article is an open access article distributed under the terms and conditions of the Creative Commons Attribution (CC BY) license (http://creativecommons.org/licenses/by/4.0/).

Article

Separation, Identification, and Antidiabetic Activity of Catechin Isolated from *Arbutus unedo* L. Plant Roots

Hanae Naceiri Mrabti [1,*], Nidal Jaradat [2,*], Ismail Fichtali [3], Wessal Ouedrhiri [4], Shehdeh Jodeh [5], Samar Ayesh [6], Yahia Cherrah [1] and My El Abbes Faouzi [1]

[1] Faculty of Medicine and Pharmacy, Laboratory of Pharmacology and Toxicology, Pharmacokinetics Team, Mohammed V University in Rabat, Rabat Institute, Rabat BP 6203, Morocco; y.cherrah1@um5s.net.ma (Y.C.); myafaouzi8@yahoo.fr (M.E.A.F.)
[2] Department of Pharmacy, Faculty of Medicine and Health Sciences, An-Najah National University, P.O. Box 7, 00970 Nablus, Palestine
[3] Laboratory of Applied Organic Chemistry, Faculty of Science and Technology, Sidi Mohamed Ben Abdellah University, Immouzer Road, 30050 Fez, Morocco; biocmb@gmail.com
[4] Laboratory of Medicinal and Aromatic Plants and Natural Substances, National Institute of Medicinal and Aromatic Plants-Taounate, Sidi Mohamed Ben Abdellah University, 30050 Fez, Morocco; wessal.ouedrhiri@gmail.com
[5] Department of Chemistry, An-Najah National University, P.O. Box 7, 00970 Nablus, Palestine; Sjodeh@hotmail.com
[6] Physical Science Department, Harold Washington College, 10 E. Lake Street, Chicago, IL 60601, USA; sayesh@ccc.edu
* Correspondence: Naceiri.mrabti.hanae@gmail.com (H.N.M.); nidaljaradat@najah.edu (N.J.); Tel./Fax: +97023345982 (N.J.)

Received: 10 March 2018; Accepted: 9 April 2018; Published: 12 April 2018

Abstract: Phytopharmaceuticals play an essential role in medicine, since the need to investigate highly effective and safe drugs for the treatment of diabetes mellitus disease remains a significant challenge for modern medicine. *Arbutus unedo* L. root has various therapeutic properties, and has been used widely in the traditional medicine as an antidiabetic agent. The current study aimed to isolate the pharmacologically active compound from *A. unedo* roots using accelerated solvent extraction technology, to determine its chemical structure using different instrumental analytical methods, and also to evaluate the α-glucosidase inhibitory activity. The roots of *A. unedo* were exhaustively extracted by high-pressure static extraction using the Zippertex® technology (Dionex-ASE, Paris, France), and the extract was mixed with XAD-16 resin to reach quantifiable amounts of active compounds which were identified by high-pressure liquid chromatography (HPLC), ^{1}H NMR (300 MHz), and ^{13}C NMR. The antidiabetic activity of the isolated compound was evaluated using the α-glucosidase inhibitory assay. The active compound was isolated, and its structure was identified as catechin using instrumental analysis. The results revealed that the isolated compound has potential α-glucosidase inhibitory activity with an IC_{50} value of 87.55 ± 2.23 µg/mL greater than acarbose. This was used as a positive control, which has an IC_{50} value of 199.53 ± 1.12 µg/mL. According to the results achieved, the roots of *A. unedo* were considered the best source of catechin and the Zippertex® technology method of extraction is the best method for isolation of this therapeutic active compound. In addition, the α-glucosidase inhibitory activity results confirmed the traditional use of *A. unedo* roots as an antidiabetic agent. Future clinical trials and investigations of antidiabetic and other pharmacological effects such as anticancer are required.

Keywords: *Arbutus unedo* L.; α-glucosidase; catechin; HPLC; NMR

1. Introduction

Diabetes mellitus is a disorder of carbohydrate metabolism characterized by the impaired ability of the body to produce or respond to insulin, and thereby maintain proper levels of glucose in the blood. This produces several devastating effects and complications, including neuropathy, nephropathy, retinopathy, hyperthyroidism, hypertension, arteriosclerosis, and many other serious diseases [1–3]. In fact, other complications are associated with the used antidiabetic drugs, as regular administration can lead to several adverse effects [4]. Accordingly, the achievement of sufficient control of hyperglycemia is difficult to reach with commercially available antidiabetic medications, thereby resulting in various and serious complications [5]. In fact, the investigation on medicinal plants usually started with extraction procedures which play a crucial role in the extraction outcomes, e.g., yields percentages and the quality of the produced phytochemicals. Nowadays, a wide range of technologies with different methods of extraction is available. Zippertex technology is one of the most efficient methods of extraction. It is a high-pressure static extractor which is considered the most efficient and convenient solid/liquid extraction device. It combines extraction and filtration steps, offering limpid highly concentrated extracts, ready for chemical and biological investigations. The combination of static pressure and heating favor the access of the solvent into the heart of the solid matter, and increases the solubilization of the target compounds. Zippertex offers the maximum qualitative and quantitative recovery, with the minimum required operations, solvents, handling, and time [6,7].

Medicinal plants offer a great opportunity to discover new natural therapeutic molecules. Some of these molecules may have beneficial effects on glucose homeostasis in diabetic patients without causing any undesirable effects currently observed in modern antidiabetic agents [5,8].

Arbutus unedo L. (Ericaceae family) is commonly known as wild strawberry, which is a perennial small tree native to the Mediterranean basin countries. It constitutes an important contribution to the nutritional culture and to the health promotion of Moroccan community [9,10].

However, in traditional medicine of many Mediterranean countries, *A. unedo* plant has been used widely with the employment of decoctions and infusions of all plant parts: fruits, leaves, barks, and roots [11,12]. For instance, the fruits are well known in folk medicine as an antiseptic, laxative, and diuretic [13,14], while the leaves are used as an astringent, urinary antiseptic, diuretic, antidiarrhea, and depurative. Moreover, recently, the leaves have been used to treat inflammatory diseases, hypertension, and diabetes [11].

Furthermore, *A. unedo* roots havebeen used traditionally to treat various gastrointestinal, urological, dermatologic, and cardiovascular diseases, as well as a diuretic, anti-inflammatory, and antidiabetic agent [15–18].

In fact, the *A. unedo* extract allowed us to identify several familial compounds, such as terpenoids, free quinine, and anthraquinone, which were the subject of our previous work and others, such as polyphenols, flavonoids, and tannins, which were quantified in the plant roots aqueous extract, thus, its antioxidant activity was also evaluated [10].

The current study aimed to isolate the therapeutic active molecule of *A. unedo* roots using accelerated solvent extraction Zippertex technology with XAD-16 resin, characterize its structure using different spectra methods and to evaluate its α-glucosidase inhibitory activity.

2. Results

The results of the present investigation showed that the extraction of *A. unedo* roots utilizing the Dionex-ASE (accelerated solvent extraction) prototype Zippertex produced a high percent of aqueous extract yield with 23.7%. The HPLC chromatogram of *A. unedo* rootsaqueous extract showed that this extract contained several phytochemical compounds, as shown in Figure 1A. This makes it difficult to identify the pharmacologically active principle, which encouraged us to look for another experimental method of separation, from which we can isolate a single therapeutic molecule. Satisfactory results were obtained using the XAD-16 resin. After adsorption on XAD-16 resin, the expected compound from the extract was separated using semi-preparative HPLC technique, and its chemical structure was identified as shown in Figure 1B.

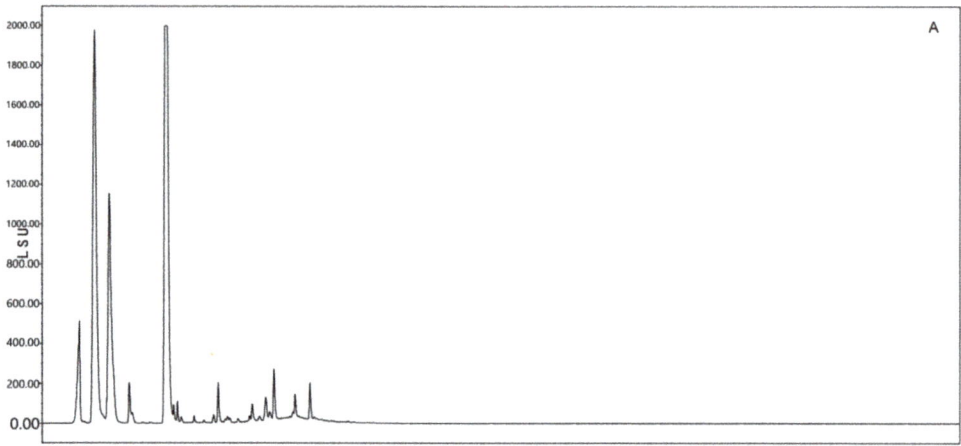

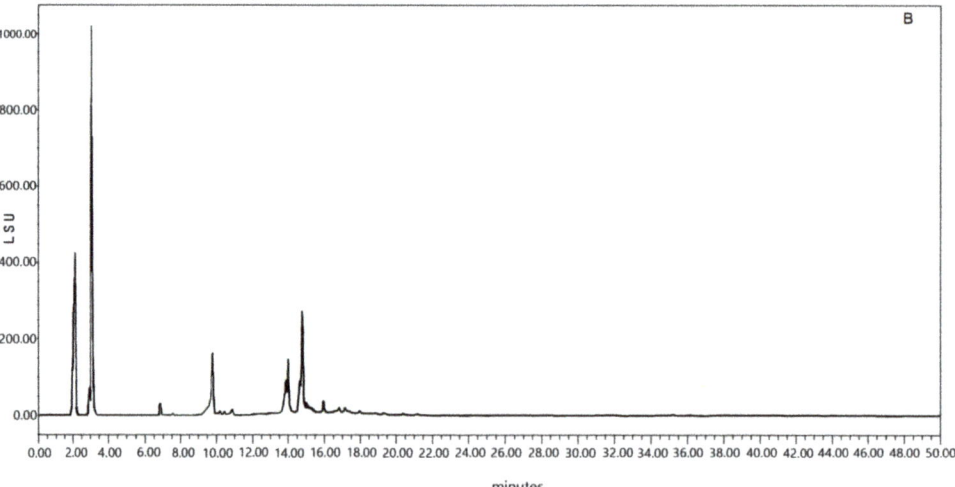

Figure 1. (**A**) HPLC analysis of *Arbutus unedo* roots aqueous extract obtained by Zippertex apparatus. (**B**) The aqueous extract of *A. unedo* roots adsorbed by the amberlite XAD-16.

The isolated compound showed interesting α-glucosidase inhibitory activity, results as shown in Table 1, in comparison with commercially used α-glucosidase inhibitory drug acarbose. These results encouraged us to identify and characterize the structure of this compound (Figure 2) using high-resolution mass spectrometry and NMR (^{1}H and ^{13}C) analyses. The negative-ion HRESIMS analysis gave the molecular formula $C_{15}H_{14}O_6$ on the basis of m/z 289.10 [M − H]$^-$ (calculated for $C_{15}H_{13}O_6$) containing nine unsaturation.

The ^{1}H and ^{13}C NMR spectra show the following signals:

^{1}H NMR (MeOD, 300 MHz): 2.46–2.54 (m, 1H, CH$_{2(2)}$); 2.81–2.88 (m, 1H, CH$_{2(2)}$); 3.29 (s, 1H, OH$_{(a)}$); 3.97 (dd, 1H, CH–OH$_{(1)}$, J = 7.5 Hz, J = 2.4 Hz); 4.57 (d, 1H, O–CH$_{(9)}$, J = 7.5 Hz); 4.85 (s, 4H, 4OH$_{(b,c,d,e)}$); 5.86 (d, 1H, CH$_{(5)}$, J = 2.4 Hz); 5.93 (d, 1H, CH$_{(7)}$, J = 2.1 Hz); 6.69–6.78 (m, 2H, 2CH$_{(14,15)}$); 6.84 (d, 1H, CH$_{(11)}$, J = 1.8 Hz).

^{13}C NMR (MeOD, 75 MHz): 27.09 (CH$_{2(2)}$); 67.40 (CH$_{(1)}$); 81.43 (CH$_{(9)}$); 94.17 (CH$_{(7)}$); 94.96 (CH$_{(5)}$); 99.49 (C$_{(3)}$); 113.89 (CH$_{(11)}$); 114.76 (CH$_{(14)}$); 118.69 (CH$_{(15)}$); 130.82 (C$_{(10)}$); 144.81 (C$_{(13)}$); 144.84 (C$_{(12)}$); 155.51 (C$_{(8)}$); 156.15 (C$_{(4)}$); 156.39 (C$_{(6)}$).

The ^{1}H and ^{13}C NMR spectra are shown in the Supplementary Material (Figures S1–S6).

The α-glucosidase inhibitory assay results showed that catechin inhibited α-glucosidase enzyme more than two folds of acarbose with IC$_{50}$ value 87.55 ± 2.23 µg/mL against 199.53 ± 1.12 µg/mL of acarbose, as represented in Table 1.

Table 1. The α-glucosidase inhibitory potency of catechin.

Compounds	α-Glucosidase Inhibitory Activity IC$_{50}$ (µg/mL), ±SD
Catechin	87.55 ± 2.23
Acarbose	199.53 ± 1.12

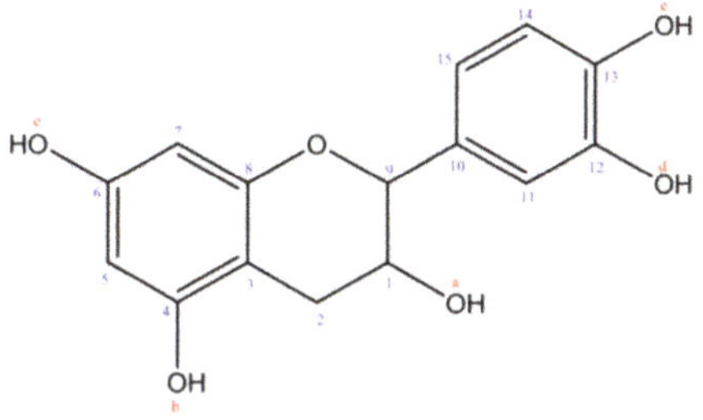

Figure 2. Catechin chemical structure isolated from A. unedo roots using XAD-16 resin.

3. Discussion

Due to problems associated with high processing temperatures and long processing times in conventional extraction methods, which can degrade or undergo undesirable oxidation processes, there is an essential need to promote development and application of alternative extraction techniques for phenolic compounds. Accelerated solvent extraction is a promising eco-friendly alternative providing exceptional separation and protection from degradation of unstable polyphenols. The current investigation revealed that extraction method of A. unedo roots utilizing of the Dionex-ASE (accelerated solvent extraction) prototype Zippertex produced a high yield of aqueous extract. Several conducted studies showed that the modern techniques, which have been replacing conventional ones, include accelerated solvent extraction, microwave-assisted extraction, supercritical fluid extraction, pressurized liquid extraction, and ultrasound-assisted extraction. These alternative techniques increased the extraction yields of polyphenols, reduced considerably the use of solvents, and accelerated the extraction process [19–21].

Flavonoid is a class of phytogenic polyphenolic compounds, including in many kinds of human diet, and has various physicochemical characters and chemical structures. The high-resolution mass spectrometry and NMR results revealed the presence of catechin in the roots of A. unedo extract, and these results are similar to those which were conducted by Junior et al. [22].

In fact, A. unedo fruits are considered as an alternative source of flavan-3-ols, in particular, catechin and its derivatives [2]. However, the inhibitory effect of catechin against α-glucosidase enzyme was evaluated in several previously conducted studies which initially demonstrated that catechin preferentially inhibited maltase rather than sucrase in an immobilized α-glucosidase inhibitory assay system [23–26]. This suggests that the α-glucosidase inhibition induced by catechin is closely associated with the presence of a free hydroxylgroupat the 3-position of this molecule [23]. Catechin is a very common and widely diffused metabolite in the plant kingdom. The traditional use of *A. unedo* as antidiabetic agent may be associated with other phytoconstituents, because many plants contain catechin. Therefore, the elevated biological property of this plant is due to the whole phytocomplex, and not only to just one molecule [27,28].

In addition, several conducted studies pointed out the interest of using catechin for many of health benefits, such as anticancer, antifungal, antioxidant, and antidiabetic purposes [29,30].

However, in a study established by Albuquerque et al. [21], they attempted the optimization of extraction of this molecule using different extraction methods, such as maceration and microwave techniques, and were capable of yielding 1.38 ± 0.1 and 1.70 ± 0.3 mg/g DW of catechin, respectively using the optimal extraction conditions. These extraction methods were found to be the most effective methods. The catechin yield from *A. unedo* root in the present study was of 95 mg of the plant dry weight. This high yield, which was obtained by using the Dionex-ASE (accelerated solvent extraction) prototype, Zippertex, in addition to the XAD-16 resin, which was used as an adequate option for an efficient cleanup step for purification and isolation of pure catechin compound from the plant root extract [31]. This yield was considered high in comparison with previously conducted extraction methods [21]. Moreover, the roots of *A. unedo* was considered the best source for isolation of the pharmacologically active molecule catechin in comparison with previously conducted isolation procedures of catechin from different sources [21]. However, many conducted research studies have demonstrated that the Zippertex method is friendly to the environment, with low cost and a low requirement for solvents, which decreased the time of extraction procedure and combined extraction/filtration steps [23].

The inhibitory effect of catechin against the α-glucosidase enzyme has been documented before; however, the advanced purification method used, with a greater percentage of yield, is novel.

To the best of the authors' knowledge, the current study is the first one which was carried out with the intent of isolation of the therapeutically active compound from the *A. unedo* plant roots, and to find out the best method for isolation of catechin from this plant. Moreover, this investigation approved the traditional antidiabetic use of *A. unedo* plant roots. Furthermore, in vivo trials are required to support this therapeutic use and to design a suitable dosage form to produce new drugs or supplements from *A. unedo* plant roots extract, to help in reducing blood glucose level and its complications.

4. Materials and Methods

4.1. Plant Material

The *A. unedo* roots were collected at Beni Mellal region, Morocco, in October 2016. The voucher specimen has been deposited in the Herbarium of the Botany Department at the Scientific Institute of Rabat/Morocco and then voucher specimen code is (RAB 101548). The roots were naturally dried in the shade at room temperature for 2–3 weeks.

4.2. Extraction and Isolation Procedures

The dry roots of *A. unedo* plantwere mechanically powdered, and 30 g of the plant material was extracted with 200 mL of water at 100 °C under static nitrogen pressure (100 bars) using the Dionex-ASE (accelerated solvent extraction) prototype Zippertex. The obtained aqueous extract was evaporated using a Rotavap at 100 °C, and lyophilized to give a brown powder which produced 7.12 g of yield (23.7%). After that, 1 g of the obtained extract was adsorbed by the amberlite XAD-16 (Sigma, Steinheim, Germany), which was previously washed with methanol. The pharmacological active molecule catechin was isolated from roots of *A. unedo* by a methanol-resin reagent, and the yield was 95 mg with 1.33% of the total aqueous extract [32].

4.3. Analytical Methods

4.3.1. Analytical HPLC

An Alliance® Waters W2695 HPLC chain (Waters Corporation, Milford, MA, USA) equipped with a Waters 2996 PDA detector (Waters Corporation, Milford, MA, USA) equipped with a Sunfire III C18 (4.6 mm × 150 mm) 3.5 µm (Waters) reverse phase column. This chromatographic system is coupled to a Waters 2424 light scattering (DEDL) detector. The HPLC system is controlled by Empower 3 software (Waters) (Waters Corporation, Milford, MA, USA). While ultrapure water (MilliQ), 0.1% formic acid/acetonitrile and 0.1% formic acid were the used solvents. In addition, the gradient was from 0 to 100% acetonitrile in 40 min and 10 min to 100% acetonitrile with flow rate of 0.7 mL/min [7].

4.3.2. HPLC Semi-Preparative

The pharmacological active molecule of *A. undo* roots was isolated by a methanol-resin reagent (95 mg) was identified by preparative HPLC chain equipped with a Waters 717 auto-sampler, a Waters 600 pump, a Waters 2998 DEDL, Waters 2420, PDA detector, and a Waters AF degasser. The column used is a Sunfire III C18 (10 mm× 250 mm) 5 µm. The HPLC system is controlled by Empower 2 software (Waters) (Waters Corporation, Milford, MA, USA). While ultrapure water (MilliQ), 0.1% formic acid/acetonitrile and 0.1% formic acid were the used solvents with a flow rate of 4 mL/min [7].

4.3.3. NMR and HRMS Analysis

The ^{1}H NMR (300 MHz) and ^{13}C NMR (75 MHz) spectra were recorded on Bruker spectrometer with chemical shift values (δ) given in part per million (ppm) relative to TMS (0.00 ppm) and using MeOD as a solvent, the coupling constants (J) are expressed in hertz (Hz) and singlet (s), doublet (d), and doublet of a doublet (dd) as well as the multiplet (m). The high-resolution mass spectra (HRMS) analysis was performed in a negative mode in full mass scan (*m/z* 100 to 600 amu) using a Thermo Scientific Orbitrap Mass Spectrometer Exactive equipped with a heated electrospray ionization source (HESI), and the used resolution was 1000 [22,33].

4.4. Enzyme Inhibition Assay

The α-glucosidase inhibition assay was performed according to the slightly modified method of Kee et al. [34], with some modifications. Briefly, α-glucosidase enzyme (0.1 U/mL) (Sigma-Aldrich, Lyon, France) and substrate *p*-nitrophenyl-α-D-glucopyranoside (*p*-NPG, 1 mM) (Sigma-Aldrich, Lyon, France) were dissolved in potassium phosphate buffer (0.1 M, pH 6.7), and all samples were dissolved in distilled water. The inhibitor (150 µL) was pre-incubated with the enzyme (100 µL) at 37 °C for 10 min, then a substrate (200 µL) was added to the reaction mixture. The enzymatic reaction was performed at 37 °C for 30 min. The reaction was then terminated by the addition of Na_2CO_3 (1 M, 1 mL) (Sigma-Aldrich, Lyon, France). All the samples were analyzed in triplicate with different concentrations to determine the IC_{50} values, and the absorbance was recorded at 405 nm. The α-glucosidase inhibitory activity was expressed as the percentage of inhibition, and the IC_{50} values were determined. Acarbose (Sigma-Aldrich, Lyon, France) was used as the positive control. The results were expressed as percentage inhibition and calculated using the following equation:

$$\text{Inhibition (\%)} = \frac{(Ac - Acb) - (As - Asb)}{Ac - Acb} \times 100$$

where Ac refers to the absorbance of the control (enzyme and buffer), Acb refers to the absorbance of the control blank (buffer without enzyme), As refers to the absorbance of the sample (enzyme and inhibitor), and Asb is the absorbance of the sample blank (inhibitor without enzyme).

4.5. Statistical Analysis

Determination of α-glucosidase inhibitory activity was carried out in triplicate for each sample. The obtained results were presented as means ± standard deviation (SD), and were then compared using an unpaired *t*-test.

5. Conclusions

The utilized Zippertex technology of extraction with XAD-16 resin had offered the maximum qualitative and quantitative recovery of catechin, with minimum operations, handling, minimum solvent, and time in comparison with other used methods of extraction and isolation. Moreover, the α-glucosidase inhibitory activity of catechin isolated from the *A. unedo* roots exhibited a potential effect in comparison with commercially available α-glucosidase inhibitory agent acarbose which explains their traditional use as a hypoglycemic plant. These findings further imply that catechin or *A. unedo* root extract separated by Zippertex technology can be used as a potential antidiabetic agent.

Supplementary Materials: The following are available online at http://www.mdpi.com/2223-7747/7/2/31/s1, Figure S1: ^{1}H NMR spectra analysis, Figure S2: ^{1}H NMR spectra analysis, Figure S3: ^{1}H NMR spectra analysis, Figure S4: ^{13}C NMR spectra analysis, Figure S5: ^{13}C NMR spectra analysis, Figure S6: ^{13}C NMR spectra analysis.

Acknowledgments: The authors wish to thank Mohammed V University in Rabat for its support to carry out this work.

Author Contributions: M.E.A.F., N.J. and H.N.M. designed and conceived the experiments; H.N.M. and W.O. performed the experiments; I.F. conducted the spectral NMR analysis; Y.C. contributed reagents and chemicals; N.J. wrote the paper; S.J. and S.A. revised the manuscript.

Conflicts of Interest: The authors declare that there are no conflicts of interest.

References

1. Wild, S.; Roglic, G.; Green, A.; Sicree, R.; King, H. Global prevalence of diabetes: Estimates for the year 2000 and projections for 2030. *Diabetes Care* **2004**, *27*, 1047–1053. [CrossRef] [PubMed]
2. American Diabetes Association. Standards of medical care in diabetes-2017 abridged for primary care providers. *Clin. Diabetes* **2017**, *35*, 5–26.
3. Crawford, K. Review of 2017 diabetes standards of care. *Nurs. Clin. N. Am.* **2017**, *52*, 621–663. [CrossRef] [PubMed]
4. Mazzotti, A.; Caletti, M.T.; Marchignoli, F.; Forlani, G.; Marchesini, G. Which treatment for type 2 diabetes associated with non-alcoholic fatty liver disease? *Dig. Liver Dis.* **2017**, *49*, 235–240. [CrossRef] [PubMed]
5. Eddouks, M.; Ouahidi, M.; Farid, O.; Moufid, A.; Khalidi, A.; Lemhadri, A. L'utilisation des plantes médicinales dans le traitement du diabète au maroc. *Phytothérapie* **2007**, *5*, 194–203. [CrossRef]
6. Bimakr, M.; Rahman, R.A.; Taip, F.S.; Ganjloo, A.; Salleh, L.M.; Selamat, J.; Hamid, A.; Zaidul, I. Comparison of different extraction methods for the extraction of major bioactive flavonoid compounds from spearmint (*Mentha spicata* L.) leaves. *Food Bioprod. Process.* **2011**, *89*, 67–72. [CrossRef]
7. Nothias, L.-F.L.; Boutet-Mercey, S.P.; Cachet, X.; De La Torre, E.; Laboureur, L.; Gallard, J.-F.O.; Retailleau, P.; Brunelle, A.; Dorrestein, P.C.; Costa, J. Environmentally friendly procedure based on supercritical fluid chromatography and tandem mass spectrometry molecular networking for the discovery of potent antiviral compounds from *Euphorbia semiperfoliata*. *J. Nat. Prod.* **2017**, *80*, 2620–2629. [CrossRef] [PubMed]
8. Barrajón-Catalán, E.; Herranz-López, M.; Joven, J.; Segura-Carretero, A.; Alonso-Villaverde, C.; Menéndez, J.A.; Micol, V. Molecular promiscuity of plant polyphenols in the management of age-related diseases: Far beyond their antioxidant properties. In *Oxidative Stress and Inflammation in Non-Communicable Diseases-Molecular Mechanisms and Perspectives in Therapeutics*; Springer: New York, NY, USA, 2014; pp. 141–159.
9. Mrabti, H.N.; Sayah, K.; Jaradat, N.; Kichou, F.; Ed-Dra, A.; Belarj, B.; Cherrah, Y.; Faouzi, M.E.A. Antidiabetic and protective effects of the aqueous extract of *Arbutus unedo* L. in streptozotocin-nicotinamide-induced diabetic mice. *J. Complement. Integr. Med.* **2018**. [CrossRef] [PubMed]

10. Mrabti, H.N.; Marmouzi, I.; Sayah, K.; Chemlal, L.; El Ouadi, Y.; Elmsellem, H.; Cherrah, Y.; Faouzi, M.A. *Arbutus unedo* L. aqueous extract is associated with in vitro and in vivo antioxidant activity. *J. Mater. Environ. Sci.* **2017**, *8*, 217–224.
11. Bento, I.; Pereira, J.A. *Arbutus unedo* L. and its benefits on human health. *J. Food Nutr. Res.* **2011**, *50*, 73–85.
12. Bnouham, M.; Merhfour, F.Z.; Ziyyat, A.; Aziz, M.; Legssyer, A.; Mekhfi, H. Antidiabetic effect of some medicinal plants of oriental morocco in neonatal non-insulin-dependent diabetes mellitus rats. *Hum. Exp. Toxicol.* **2010**, *29*, 865–871. [CrossRef] [PubMed]
13. Pallauf, K.; Rivas-Gonzalo, J.; Del Castillo, M.; Cano, M.; de Pascual-Teresa, S. Characterization of the antioxidant composition of strawberry tree (*Arbutus unedo* L.) fruits. *J. Food Compos. Anal.* **2008**, *21*, 273–281. [CrossRef]
14. Ruiz-Rodríguez, B.-M.; Morales, P.; Fernández-Ruiz, V.; Sánchez-Mata, M.-C.; Cámara, M.; Díez-Marqués, C.; Pardo-de-Santayana, M.; Molina, M.; Tardío, J. Valorization of wild strawberry-tree fruits (*Arbutus unedo* L.) through nutritional assessment and natural production data. *Food Res. Int.* **2011**, *44*, 1244–1253. [CrossRef]
15. Mariotto, S.; Esposito, E.; Di Paola, R.; Ciampa, A.; Mazzon, E.; de Prati, A.C.; Darra, E.; Vincenzi, S.; Cucinotta, G.; Caminiti, R. Protective effect of *Arbutus unedo* aqueous extract in carrageenan-induced lung inflammation in mice. *Pharmacol. Res.* **2008**, *57*, 110–124. [CrossRef] [PubMed]
16. Afkir, S.; Nguelefack, T.B.; Aziz, M. *Arbutus unedo* prevents cardiovascular and morphological alterations in l-name-induced hypertensive rats: Part i: Cardiovascular and renal hemodynamic effects of *Arbutus unedo* in L-name-induced hypertensive rats. *J. Ethnopharmacol.* **2008**, *116*, 288–295. [CrossRef] [PubMed]
17. Novais, M.; Santos, I.; Mendes, S.; Pinto-Gomes, C. Studies on pharmaceutical ethnobotany in Arrabida Natural Park (Portugal). *J. Ethnopharmacol.* **2004**, *93*, 183–195. [CrossRef] [PubMed]
18. Kivçak, B.; Mert, T.; Denizci, A. Antimicrobial activity of *Arbutus unedo* L. *J. Pharm. Sci.* **2001**, *26*, 125–128.
19. Arias, M.; Penichet, I.; Ysambertt, F.; Bauza, R.; Zougagh, M.; Ríos, Á. Fast supercritical fluid extraction of low-and high-density polyethylene additives: Comparison with conventional reflux and automatic soxhlet extraction. *J. Supercrit. Fluids* **2009**, *50*, 22–28. [CrossRef]
20. Luthria, D.L. Influence of experimental conditions on the extraction of phenolic compounds from parsley (*Petroselinum crispum*) flakes using a pressurized liquid extractor. *Food Chem.* **2008**, *107*, 745–752. [CrossRef]
21. Albuquerque, B.R.; Prieto, M.; Barreiro, M.F.; Rodrigues, A.; Curran, T.P.; Barros, L.; Ferreira, I.C. Catechin-based extract optimization obtained from *Arbutus unedo* L. Fruits using maceration/microwave/ ultrasound extraction techniques. *Ind. Crop. Prod.* **2017**, *95*, 404–415. [CrossRef]
22. Junior, O.V.; Dantas, J.H.; Barão, C.E.; Zanoelo, E.F.; Cardozo-Filho, L.; de Moraes, F.F. Formation of inclusion compounds of (+) catechin with β-cyclodextrin in different complexation media: Spectral, thermal and antioxidant properties. *J. Supercrit. Fluids* **2017**, *121*, 10–18. [CrossRef]
23. Matsui, T.; Tanaka, T.; Tamura, S.; Toshima, A.; Tamaya, K.; Miyata, Y.; Tanaka, K.; Matsumoto, K. α-glucosidase inhibitory profile of catechins and theaflavins. *J. Agric. Food Chem.* **2007**, *55*, 99–105. [CrossRef] [PubMed]
24. Bhandari, M.R.; Jong-Anurakkun, N.; Hong, G.; Kawabata, J. α-glucosidase and α-amylase inhibitory activities of nepalese medicinal herb pakhanbhed (*Bergenia ciliata*, haw.). *Food Chem.* **2008**, *106*, 247–252. [CrossRef]
25. Tadera, K.; Minami, Y.; Takamatsu, K.; Matsuoka, T. Inhibition of α-glucosidase and α-amylase by flavonoids. *J. Nutr. Sci. Vitaminol.* **2006**, *52*, 149–153. [CrossRef] [PubMed]
26. Justino, A.B.; Miranda, N.C.; Franco, R.R.; Martins, M.M.; da Silva, N.M.; Espindola, F.S. *Annona muricata* Linn. leaf as a source of antioxidant compounds with in vitro antidiabetic and inhibitory potential against α-amylase, α-glucosidase, lipase, non-enzymatic glycation and lipid peroxidation. *Biomed. Pharmacother.* **2018**, *100*, 83–92. [CrossRef] [PubMed]
27. Giovannini, D.; Gismondi, A.; Basso, A.; Canuti, L.; Braglia, R.; Canini, A.; Mariani, F.; Cappelli, G. *Lavandula angustifolia* Mill. Essential oil exerts antibacterial and anti-inflammatory effect in macrophage mediated immune response to *Staphylococcus aureus*. *Immunol. Investig.* **2016**, *45*, 11–28. [CrossRef] [PubMed]
28. Ettorre, A.; Frosali, S.; Andreassi, M.; Di Stefano, A. Lycopene phytocomplex, but not pure lycopene, is able to trigger apoptosis and improve the efficacy of photodynamic therapy in HL60 human leukemia cells. *Exp. Biol. Med.* **2010**, *235*, 1114–1125. [CrossRef] [PubMed]
29. Higdon, J.V.; Frei, B. Tea catechins and polyphenols: Health effects, metabolism, and antioxidant functions. *Crit. Rev. Food Sci. Nutr.* **2010**, *43*, 89–143. [CrossRef] [PubMed]

30. Malongane, F.; McGaw, L.J.; Mudau, F.N. The synergistic potential of various teas, herbs and therapeutic drugs in health improvement: A review. *J. Sci. Food Agric.* **2017**, *97*, 4679–4689. [CrossRef] [PubMed]
31. Selga, A.; Torres, J.L. Efficient preparation of catechin thio conjugates by one step extraction/depolymerization of pine (*Pinus pinaster*) bark procyanidins. *J. Agric. Food Chem.* **2005**, *53*, 7760–7765. [CrossRef] [PubMed]
32. Rabhi, C.; Arcile, G.; Cariel, L.; Lenoir, C.; Bignon, J.; Wdzieczak-Bakala, J.; Ouazzani, J. Antiangiogenic-like properties of fermented extracts of ayurvedic medicinal plants. *J. Med. Food* **2015**, *18*, 1065–1072. [CrossRef] [PubMed]
33. Fichtali, I.; Laaboudi, W.; Hadrami, E.E.; Aroussi, F.E.; Ben-Tama, A.; Benlemlih, M.; Stiriba, S. Synthesis, characterization and antimicrobial activity of novel benzophenone derived 1, 2, 3-triazoles. *J. Mater. Environ. Sci.* **2016**, *7*, 1633–1641.
34. Kee, K.T.; Koh, M.; Oong, L.X.; Ng, K. Screening culinary herbs for antioxidant and α-glucosidase inhibitory activities. *Int. J. Food Sci. Technol.* **2013**, *48*, 1884–1891. [CrossRef]

© 2018 by the authors. Licensee MDPI, Basel, Switzerland. This article is an open access article distributed under the terms and conditions of the Creative Commons Attribution (CC BY) license (http://creativecommons.org/licenses/by/4.0/).

Communication

First Report on the Ethnopharmacological Uses of Medicinal Plants by Monpa Tribe from the Zemithang Region of Arunachal Pradesh, Eastern Himalayas, India

Tamalika Chakraborty [1,2], Somidh Saha [3,4,5,*] and Narendra S. Bisht [3,6]

1. Institute of Ethnobiology, School of Studies in Botany, Jiwaji University, Gwalior 474011, India; tamalika.chakraborty@waldbau.uni-freiburg.de
2. Chair of Site Classification and Vegetation Science, Institute of Forest Sciences, University of Freiburg, Tennenbacherstr. 4, D-79106 Freiburg, Germany
3. Resource Survey and Management Division, Forest Research Institute, PO New Forest, Dehra Dun 248006, India; bishtnsifs@yahoo.com
4. Chair of Silviculture, Institute of Forest Sciences, University of Freiburg, Tennenbacherstr. 4, D-79106 Freiburg im Breisgau, Germany
5. Institute for Technology Assessment and Systems Analysis (ITAS), Karlsruhe Institute of Technology (KIT), Karlstr. 11, D-76133 Karlsruhe, Germany
6. Directorate of Extension, Indian Council of Forestry Research and Education, PO New Forest, Dehra Dun 248006, India
* Correspondence: somidh.saha@waldbau.uni-freiburg.de or somidhs@gmail.com; Tel.: +49-761-203-8627; Fax: +49-761-203-3781

Academic Editor: Milan S. Stankovic
Received: 28 November 2016; Accepted: 24 February 2017; Published: 2 March 2017

Abstract: The Himalayas are well known for high diversity and ethnobotanical uses of the region's medicinal plants. However, not all areas of the Himalayan regions are well studied. Studies on ethnobotanical uses of plants from the Eastern Himalayas are still lacking for many tribes. Past studies have primarily focused on listing plants' vernacular names and their traditional medicinal uses. However, studies on traditional ethnopharmacological practices on medicine preparation by mixing multiple plant products of different species has not yet been reported in published literature from the state of Arunachal Pradesh, India, Eastern Himalayas. In this study, we are reporting for the first time the ethnopharmacological uses of 24 medicines and their procedures of preparation, as well as listing 53 plant species used for these medicines by the Monpa tribe. Such documentations are done first time in Arunachal Pradesh region of India as per our knowledge. Our research emphasizes the urgent need to document traditional medicine preparation procedures from local healers before traditional knowledge of tribal people living in remote locations are forgotten in a rapidly transforming country like India.

Keywords: medicinal plants; traditional knowledge; Eastern Himalayas; mountain plants; ethnobotany; ethnopharmacology; bioprospecting

1. Introduction

The Himalayas are rich in diversity of medicinal plant species [1]. The culture of traditional healing of diseases using these plants is still prevalent among aboriginal mountain communities in the Himalayas. Arunachal Pradesh (approximately 84,000 km² in size), a state belonging to the Republic of India, is situated in the Eastern Himalayas. The entire state is declared as a "biodiversity hotspot" with 5000 endemic flowering plant species as well as very high faunal diversity [1,2]. Also, this state is the home to 28 major tribes and 110 sub-tribes and is considered to be one of the most splendidly variegated and multilingual tribal areas of the world [3]. The traditional wisdom of healing among mountain tribal communities is orally transferred from one generation to the next generation through traditional healers, spiritual gurus, and elderly or sometimes ordinary people. This traditional wisdom, if not properly documented, can be lost by rapid modernization and religious reformation among mountain communities in Arunachal Pradesh where traditional customary practices are often regarded as a symbol of *"backwardness"* and *"unscientific"* by the educated and younger generations. Nevertheless, plant-based traditional wisdom inherited and carried forward to generation after generation in traditional communities has become a recognized tool in the search for new sources of drugs and pharmaceuticals in modern medicine [4]. Therefore, field based ethnobotanical and ethnopharmacological surveys to list medicinal plants and their uses are still relevant and worth the effort in order to bring out new clues for the development of drugs to treat human diseases [5].

Before coming to our research objectives, we would like to briefly mention the state of the art of ethnopharmacological research in the Himalayas. There are plenty of research works on the listing of the traditional uses of medicinal plants from the Himalayas. A search with the terms "medicinal plants * Himalayas" yielded 163 peer-reviewed articles listed in ISI Web of Knowledge on 20 February 2017. However, out of those 163 articles, 19 articles were found from the Eastern Himalayas and only two were on the Monpa tribe (please see Materials and Methods section for a detailed sociocultural description of the Monpa tribe). Haridasan et al., in the seminal works produced in 1998 and 1990, comprehensively listed medicinal and edible plants of the Monpa tribe and other tribes of Arunachal Pradesh [2]. Recently, Namsa et al. (2011) listed 50 plant species and recorded their ethnobotanical uses among people of the Monpa tribe at the southern range of their habitation (i.e., Kalaktang circle of West Kameng district of Arunachal Pradesh) [6]. These two publications provided general descriptions of the plants, traditional uses of the plants to cure certain diseases, and traditional ways of consumption of these plants or plant parts (e.g., pills, syrups, decoctions, etc.). Nevertheless, no ethnopharmacological studies have yet reported how, and in what proportion, multiple plant parts from different species can be used to prepare specific ethnomedicines for healing of diseases among the Monpa tribes or any other tribes of the Eastern Himalayas as per our literature research as of 20 February 2017. In addition, the traditional knowledges of the people of the Monpa tribe residing at their northern habitation range (i.e., Zemithang circle of Tawang district of Arunachal Pradesh) are still not adequately documented due to the remoteness of the location.

Documentations of traditional ethnopharmacological know-hows are necessary for the preservation of traditional knowledges of Himalayan tribal communities. Such documentations could create interest among professional pharmacologists for the search of new medicines and motivate ethnologists to study high cultural diversity of the Eastern Himalayas of India. Those were the main motivations to carry out this research. This study aims to document traditional ethnopharmacological know-hows of medicinal drug making among Monpa people in the Zemithang region of the state of Arunachal Pradesh.

2. Results

Our study was a notable departure from the previous studies from the area that mostly documented and described the use of plant parts in individual plant species. We documented and described 24 ethnomedicines prepared by traditional healers based on 53 species (Table 1). The medicines were comprised of 53 plant species of medicinal plants belonging to 21 families (Table 2). These traditional medicines were most commonly used to heal a wide range of diseases such as arthritis, rheumatic pain, malaria, cough and cold, dysentery, etc. In addition, we recorded descriptions of medicines for the treatment of diseases such as epilepsy (*Pambrey*), herpes (*Bukbukpa-khaksa-chandongbra*), and oedema (*Darshek sheng nye putpoo*) that have rarely been reported in past studies. Our main result is presented in Table 1 which provides a list of ethnomedicines and their preparations by traditional ethnopharmacological techniques.

Table 1. List of 24 ethnomedicines used by the Zemithang Monpa people and the associated medicinal plants documented in this study.

Number	Name of the Ethnomedicines (in Monpa Tribal Language of Zemithang Dialect)	Type of Medicines	Name of Medicinal Plants Used for Ethnomedicines	Proportion of Used Plant Parts (*Bray* in local language is a Buddhist prayer bowl. It could be made of gold, silver, brass, copper, stone, or wood and is often used for religious offerings. 1 *bray* can contain approximately 900 g of grain).	Mode of Preparations	Medicinal Uses
1	Arkadamasisi	paste	*Crawfurdia speciosa* Wall.	1/4 bray of dried root + 1/8 bray of dried flower	dried roots and flower crushed together to prepare powder and then mixed with water to prepare paste	paste is applied externally for healing wounds
2	Baribama	decoction	*Aristolochia griffithii* Hook.f.	1/8 bray of raw washed roots	roots are boiled with water to prepare a decoction	decoction is taken as blood purifier and purgative
3	Blenga	pills	*Hedychium spicatum* Buch.-Ham.	1 bray of washed raw roots	raw roots are crushed and small round pills are prepared and sun dried	pills are taken orally for treatment of dysentery, chest pain, cough and cold
4	Bomdeng	paste	*Cirsium falconeri* Hook.f., *Cirsium verutum* D. Don and *Onopordum acanthium* L.	1/2 bray of washed raw root of *C. falconeri* + 1/2 bray of washed raw root of *C. verutum* + 3/4 bray of washed raw root of *O. acanthium*	raw roots are mixed together and crushed to prepare paste	paste is applied externally to treat arthritis
5	Bragen	syrup	*Bergenia stracheyi* Hook.f. & Thorns.	1 bray of washed fresh leaves	clear fresh leaves are crushed to prepare paste and mixed with 1/4 bray local millets wine to prepare syrup	syrup is taken for treating rheumatic pains
6	Bukbukpa-khaksa-chandongbra	paste	*Campanula latifolia* Linn., *Codonopsis clematidea* Schrenk. and *Codonopsis viridis* Wall.	1/2 bray of washed fresh leaves and 1/4 bray of fresh flowers of each species + 1/4 bray of conch powder + 1/4 bray of water	leaves and flowers are crushed together and mixed with conch powder and water to prepare paste	paste is applied externally to treat herpes
7	Chandoo-konghlin-bhor	powder	*Aconitum ferox* Wall. ex Ser., *Aconitum heterophyllum* Wall. ex Royle, *Aconitum hookeri* Stapf., *Geranium polyanthes* Edgeworth & J. D. Hooker, *Geranium wallichianum* D. Don and *Picrorhiza kurrooa* Royle ex Benth.	1 small dried root from each plants of *A. ferox, A. heterophyllum* and *A. hookeri* (total 5 g mixture of three plants) + 3 bray of dried root of *G. polyanthes* and *G. wallichianum* + 1 bray of dried root *P. kurrooa*	all ingredients are mixed together and crushed to prepare a powder	powder is taken orally to overcome poisoning effects
8	Chhalachhusar	syrup	*Meconopsis grandis* Prain and *Meconopsis paniculata* D. Don	1/2 bray of dry leaves from each plant + 1/2 bray of dy flowers from each plant	dried leaves and flowers are mixed together and crushed to prepare powder, and a small amount of powder (5 g) is mixed with 1 bray of water to prepare a syrup	syrup is taken to treat sexually transmitted diseases

Table 1. Cont.

Number	Name of the Ethnomedicines (in Monpa Tribal Language of Zemithang Dialect)	Type of Medicines	Name of Medicinal Plants Used for Ethnomedicines	Proportion of Used Plant Parts (Bray in local language is a Buddhist prayer bowl. It could be made of gold, silver, brass, copper, stone, or wood and is often used for religious offerings. 1 bray can contain approximately 900 g of grain).	Mode of Preparations	Medicinal Uses
9	Chhurchu doho keusheng	pills	Rheum australe D. Don, Rheum nobile Hook.f. & Thoms. and Bistorta affinis D. Don	1/2 bray of fresh roots from each species + 1/4 bray of dried flowers from each species	fresh roots and dried flowers are crushed together to make a paste, then small round pills are prepared and sun dried	pills are taken orally to overcome poisoning effects
10	Comrep	syrup	Rubus ellipticus Smith and Rubus paniculatus Smith	1/2 bray of fresh ripe fruits from each plant	roots are mixed together and crushed to prepare a thick syrup	syrup is used for treatment of cold and cough
11	Darshek sheng nye putpoo	decoction	Pieris formosa (Wallich) D. Don; Vaccinium nummularia Hook.f. & Thoms.	1/4 bray of fruits of P. formosa + 1/4 bray of fruits of V. nummularia + 1/2 bray of fresh roots of P. formosa + 1/2 bray of fresh roots of V. nummularia	mixture of all fresh fruits and roots along with water is boiled to prepare a decoction	decoction is taken to cure oedema
12	Dhamrep	paste	Fragaria nubicola Lindl., Geum elatum Wall. and Potentilla peduncularis D. Don.	1/2 bray of F. nubicola fresh fruits + 1/8 bray of dried roots of G. elatum + 1/4 bray of leaves of P. peduncularis	fresh fruits, leaves, and dried roots are crushed together to prepare a paste	paste is taken orally to treat cold, cough, and fever
13	Gin sheng	powder	Panax pseudoginseng Wall.	1/4 bray of dried rhizomes	dried rhizomes are crushed to prepare powder, which is taken with water	used for treating depression and fatigue
14	Karpo Chiito	paste	Iris clarkei Baker	1/4 bray of dried flower, leaves, stem, and root	dried flowers, leaves, stem parts, and roots are crushed together to prepare powder and mixed with local millets wine to prepare paste	paste is used externally to treat muscle pain
15	Lowa bur	pills	Lomatogonium carinthiacum (Wulfen) Rchb.	1/4 bray of dried roots	dried roots are crushed and small round pills are prepared and sun dried	pills are taken orally to treat cold, cough, and fever
16	Maraptang	pills	Houttuynia cordata Thunb.	1/4 bray of dried roots	dried roots are crushed and small round pills are prepared and sun dried	pills are taken orally for treatment of piles
17	Nyasheng jormu	paste and pills	Viscum articulatum Burm. f.	1/4 bray of fresh roots + 1/4 bray of fresh leaves + 1/4 bray of fresh stems	fresh roots, leaves, and parts of stem are crushed together to prepare paste; sometimes paste is used to prepare small round pills and sun dried	paste is used to join broken bones, treating pain from swelling of nerves and healing wounds; pills are used for treatment of infertility among women

136

Table 1. *Cont.*

Number	Name of the Ethnomedicines (in Monpa Tribal Language of Zemithang Dialect)	Type of Medicines	Name of Medicinal Plants Used for Ethnomedicines	Proportion of Used Plant Parts (*Bray* in local language is a Buddhist prayer bowl. It could be made of gold, silver, brass, copper, stone, or wood and is often used for religious offerings. 1 *bray* can contain approximately 900 g of grain).	Mode of Preparations	Medicinal Uses
18	Pambrey	mixture	*Anaphalis monocephala* DC., *Anaphalis triplinervis* Sims., *Gnaphalium hypoleucum* DC., *Leontopodium himalayanum* DC., *Leontopodium jacotianum* Beauv., *Tanacetum tibeticum* Hook.f. and *Tanacetum gracie* Hook.f. & Thoms.	1/2 bray of flowers from each of the plants	flowers are kept in a dark place for two days after plucking and then mixed together	used to treat epilepsy, mildly warm mixtures are applied on the bare head of the patient (two times a day) consecutively for 15 to 20 days
19	Pangen	pills	*Gentiana depressa* D. Don, *Gentiana ornata* Wallich ex G. Don, *Gentiana phyllocalyx* C. B. Clarke and *Gentiana tubiflora* Wallich ex G. Don.	1/4 bray of dried roots from each of the plants	dried roots are crushed and then mixed with 1/4 bray of local millet wine and 1/2 bray of water and small round pills are prepared and sun dried	pills are used to treat cough, cold, and headache
20	Rah-nya	decoction	*Smilacina purpurea* S. *oleracea* and *Polygonatum multiflorum* Allem.	1/4 bray of fresh roots from each of the plants	roots are boiled with water to prepare a decoction	is used for the treatment of malaria
21	Rambhoo tsarphakur	paste	*Morina longifolia* Wall., *Pterocephalus hookeri* (C. B. Clarke) Hock.	1/2 bray of dried flowers, 1/2 bray of fresh roots, 1/4 bray of fresh fruits of *M. longifolia* + 1/8 bray of dried flower, 1/2 bray of fresh roots, 1/2 bray of fresh fruits of *P. hookeri*	flowers, roots, and fruits of both plants are mixed together and crushed to prepare paste	paste is applied for healing chest pain
22	Trahm-Sheng	paste	*Corydalis cashmeriana* Royle.	1/4 bray of fresh leaves + 1/4 bray of fresh flower	fresh leaves and flowers are crushed to prepare paste	paste is applied for healing wounds
23	Wang La	powder	*Swertia chirayita* (Roxb. ex Flem.) Karst. and *Swertia hookeri* C. B. Clarke	1/2 bray of dried whole plants	dried whole plants are crushed to prepare powder, which is taken with water	powder is used to treat malaria, and is also used as a purgative and laxative
24	Whan	pills	*Lilium nepalense* D. Don	1/2 bray of dried roots	dried roots are crushed and mixed with water to prepare small round pills which are then sun dried	pills are used for treating gastritis and stomachic

Table 2. List of recorded plants used in Ethnomedicine.

Serial Number	Botanical Name	Family	Type
1	*Aconitum ferox* Wall. ex Ser.	Ranunculaceae	herb
2	*Aconitum heterophyllum* Wall. ex Royle	Ranunculaceae	herb
3	*Aconitum hookeri* Stapf.	Ranunculaceae	herb
4	*Anaphalis monocephala* DC.	Compositae	herb
5	*Anaphalis triplinervis* Sims.	Compositae	herb
6	*Aristolochia griffithii* Hook.f.	Aristolochiaceae	vine
7	*Bergenia stracheyi* Hook.f. & Thorns.	Saxifragaceae	herb
8	*Bistorta affinis* D. Don	Polygonaceae	herb
9	*Campanula latifolia* Linn.	Campanulaceae	herb
10	*Cirsium falconeri* Hook. f.	Asteraceae	herb
11	*Cirsium verutum* D. Don	Asteraceae	herb
12	*Codonopsis clematidea* Schrenk.	Campanulaceae	herb
13	*Codonopsis viridis* Wall.	Campanulaceae	herb
14	*Corydalis cashmeriana* Royle.	Papaveraceae	herb
15	*Crawfurdia speciosa* Wall.	Gentianaceae	herb
16	*Fragaria nubicola* Lindl.	Rosaceae	herb
17	*Gentiana depressa* D. Don	Gentianaceae	herb
18	*Gentiana ornata* Wallich ex G. Don	Gentianaceae	herb
19	*Gentiana phyllocalyx* C. B. Clarke	Gentianaceae	herb
20	*Gentiana tubiflora* Wallich ex G. Don.	Gentianaceae	herb
21	*Geranium polyanthes* Edgeworth & J. D. Hooker	Geraniaceae	herb
22	*Geranium wallichianum* D. Don	Geraniaceae	herb
23	*Geum elatum* Wall.	Rosaceae	herb
24	*Gnaphalium hypoleucum* DC.	Asteraceae	herb
25	*Hedychium spicatum* Buch.-Ham.	Zingiberaceae	herb
26	*Houttuynia cordata* Thunb.	Saururaceae	herb
27	*Iris clarkei* Baker	Iridaceae	herb
28	*Leontopodium himalayanum* DC.	Asteraceae	herb
29	*Leontopodium jacotianum* Beauv.	Asteraceae	herb
30	*Lilium nepalense* D. Don	Liliaceae	herb
31	*Lomatogonium carithiacum* (Wulfen) Rchb.	Gentianaceae	herb
32	*Meconopsis grandis* Prain	Papaveraceae	herb
33	*Meconopsis paniculata* D. Don	Papaveraceae	herb
34	*Morina longifolia* Wall.	Dipsacaceae	herb
35	*Onopordum acanthium* L.	Asteraceae	herb
36	*Panax pseudoginseng* Wall.	Araliaceae	herb
37	*Picrorhiza kurrooa* Royle ex Benth.	Scrophulariaceae	herb
38	*Pieris formosa* (Wallich) D. Don	Ericaceae	shrub
39	*Polygonatum multiflorum* Allem.	Convallariaceae	herb
40	*Potentilla peduncularis* D. Don.	Rosaceae	herb
41	*Pterocephalus hookeri* (C. B. Clarke) Hock.	Dipsacaceae	herb
42	*Rheum australe* D. Don	Polygonaceae	herb
43	*Rheum nobile* Hook.f. & Thoms.	Polygonaceae	herb
44	*Rubus ellipticus* Smith	Rosaceae	shrub
45	*Rubus paniculatus* Smith	Rosaceae	shrub
46	*Swertia chirayita* (Roxb. ex Flem.) Karst.	Gentianaceae	herb
47	*Smilacina oleracea* (Baker) Hook.f.	Liliaceae	herb
48	*Smilacina purpurea* (Wall.) H.Hara	Liliaceae	herb
49	*Swertia hookeri* C. B. Clarke	Gentianaceae	herb
50	*Tanacetum gracile* Hook.f. & Thoms.	Asteraceae	herb
51	*Tanacetum tibeticum* Hook.f.	Asteraceae	herb
52	*Vaccinium nummularia* Hook.f. & Thoms.	Ericaceae	shrub
53	*Viscum articulatum* Burm. f.	Viscaceae	shrub

3. Materials and Methods

3.1. Sociocultural Description of the People from the Monpa Tribe

The Monpa people are a Buddhist tribe belonging to the Mahayana (Tibetan–Lamaist) *Gelukpa* and *Nyngmapa* sect. The Monpa people are inhabitants of the western most districts of the Tawang and West Kameng regions of Arunachal Pradesh, India. Their main centers of habitation are in and around the administrative headquarters of Zemithang, Tawang, Dirang, and Kalaktang. Depending on the place of living and the geographical location of these centers, they are often called as Zemithang-Tawang or "Northern Monpas", Dirang or "Central Monpas", and Kalaktang or "Southern Monpas". The language used by Dirang and Kalaktang Monpa are different from that of Tawang Monpas. Dirang and Kalaktang Monpas use a dialect of Bhutanese Brokpa language, whereas Zemithang-Tawang Monpas use a dialect of Tibetan-Bhutanese Dakpa language. However, many other aspects of their life are quite similar. In Dakpa language, the name "Mon" and "Pa" signify the "Men of the Lower Country" or the inhabitants of southern regions to Tibet.

The Monpa villages are often situated on the slopes of the hills or in the valleys. A striking characteristic of the Monpa villages is the presence of a "*Gompa*" (Buddhist village monastery), often situated on the top of the hill and surrounded by prayer flags ("*phan*"), stone shrines ("*mane*"), and small chapels called "*chorten*" which are often found alongside the roads and foot-lanes. The houses are usually double or triple storeyed, and made mainly of locally sourced stone. Each house has a family chapel with a wooden, stone, or brass statue of the *Lord Buddha*.

The adornments and clothing are diverse and colourful. People cover their whole bodies with a variety of well-designed woolen garments. The women do the traditional spinning and weaving of the garments, as well as carpet making. The Monpa people can be recognized from a long distance owing to the attractive color of their clothing, which is a mellow strawberry red. The Monpa people love this color and dye their clothes themselves using the locally available natural dyes from diverse species of Rhododendrons and other plants. They love music and dance. Their musk-dances are very famous and attract a large number of tourists. The "Losar" or the Buddhist New Year is the most important festival celebrated among them, which is organized in February. Monpa villages could be located at a great distance by their high fluttering Buddhist prayer flags on which is printed in Tibetan script "*Om Mani Pame Hung*" which means "*Hail to him who is born as a Jewel in a Lotus*".

The Monpa people typically eat various types of locally grown vegetables, which are often cultivated by using tradtional methods [7]. Drinking yak milk, making homemade butter and dry cheese from yak milk (e.g., the famous "*churpi*" dry cheese), eating yak meat, pork, chicken, mutton, cultivation of multiple species of cereal and pulse through sustainable mountain agriculture based on tradtional ecological knowledges without any use of pesticides, herbicides, and chemical fertilizers are common practice [7]. Monogamy appears to be the form of marriage followed by the Tibetan Buddhist traditions. Tattooing is not typically observed among Monpa people, which is a stark contrast to the people from other tribes such as the *Nishi* and *Adi* in nearby districts. Information relating to the origin and migration of the Monpa people to their present habitat in Arunachal Pradesh is largely obscure. This is because written records on the history of Monpa people from the middle ages or beyond are very rare. Thus, it remains a matter of further anthropological and archeological research to find out the route and approximate time of their relationship with either the Tibetans or Bhutanese, or even with the people of Pan-Indian origin. When we visited Namshu village in Dirang region, the "*Gaobura*" or the village headman told us a folklore story about a marriage between a prince from Bhutan and a local Monpa girl from that village. The story indicates the Bhutanese influence among Dirang Monpa. The language of the Eastern Bhutan and Dirang areas are similar. Here we can quote from von Fürer-Haimendorf of Austria who was the most prominent anthropologist that ever worked with the tribes of this region [8]: "*THE REGION, WHICH ADJOINS TO THE WEST OF THE MOUNTAIN KINGDOM OF BHUTAN, DIFFERS FROM THE REST OF ARUNACHAL PRADESH BOTH TOPOGRAPHICALLY AND CULTURALLY. WHEREAS, ELSEWHERE NATURE AND THE TERRAIN HAD PREVENTED THE DEVELOPMENT OF CARAVAN ROUTES SUITABLE FOR PACK ANIMALS IN THE WESTERNMOST PART OF ARUNACHAL PRADESH. HOWEVER, WHERE THE CLIMATE AND GEOGRAPHICAL CONDITION WERE FAVORABLE, THE TRADE ROUTES WERE OPENED LINKING THE TERRITORY BOTH WITH TIBET AND THE PLAINS OF ASSAM OF INDIA. HENCE CONDITIONS ARE SIMILAR TO THOSE PREVAILING IN BHUTAN AND FURTHER WEST TO SIKKIM AND NEPAL. ALONG WITH THESE TRANS-HIMALAYAN TRADE ROUTES, TIBETAN CULTURAL ELEMENTS*

AND ULTIMATELY BUDDHIST MONKS AND NUNS INFILTRATED INTO THE MOUNTAIN REGION LYING BETWEEN THE EASTERNMOST OF BHUTAN AND THE SOUTHERN BORDER OF TIBET."

3.2. Study Area

The study area is located in the extreme north of the north-western Arunachal Pradesh. The areas of investigations are situated at the Lumla–Zemithang administrative circle of the Tawang district of Arunachal Pradesh (Figure 1). This region is situated along the bank of the river Namshyang Chu that flows through the area. The name, exact locations, and altitude of the three villages where the study took place are as follows: (1) village Kublaitang (27°37′070″ N, 91°41′618″ E, elevation 2224 m); (2) village Shakti (27°36′736″ N, 91°42′970″ E, elevation 2020 m); and (3) village Lumpho (27°43′140″ N, 091°43′069″ E, elevation 2550 m). The research areas fall under the middle Himalayan range of the Eastern Himalayas. The soil on the hills is moderately deep and moist, fertile loamy layer stained with humus. At places, shallow soils are not uncommon with underlying boulders and rocks. The subsoil at lower elevations consists of mostly boulders and pebbles superimposed by a layer of a sandy loam of various depths with layers of humus overtop. The relative humidity of this area varies from 30% to 80%. Southern aspects at low altitude areas are more humid than any other places in the region. The annual temperature in this area varies from −10 degrees Celsius to +15 degrees Celsius. The area typically receives 1500–1800 mm rainfall every year. The dry months are December, January, and February. The pre-monsoon rainfall starts from the end of the March. Highest rainfall is observed in June, July, and August [9]. The forest type of the research area is the Northeastern Himalayan subalpine mixed conifer forests. The top canopy of the forest consists of *Abies densa*, *Juniperus wallichiana*, *Illicium griffithi*, *Pinus wallichiana*, *Quercus* spp., and *Cupressus torulosa*. The secondary canopy layer mainly consists of *Rhododendron* spp., *Betula utilis*, *Pyrus aucuparia*, and *Salix wallichiana*. The trees of the forest ground storey are dominated by *Juniperus recurva*, *Cassiope fastigiata*, and *Rhododendron* spp. [10].

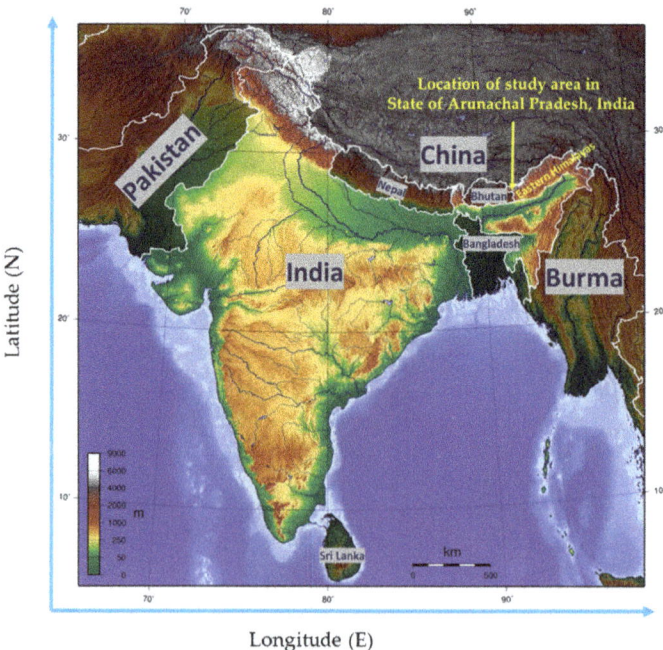

Figure 1. Location of the study area at the Zemithang region in the state of Arunachal Pradesh, India pointed by the yellow arrow (map not to scale).

3.3. Field Surveys

Field surveys were carried at the three sample villages of the Zemithang region of Tawang district. The research was carried out in three stages. In the first stage, ethnobotanical data were collected from the research area. At the second stage, ethnopharmacological information was collected from the same research area. The herbariums of the collected plant specimens were prepared and verified at the third stage of the research at the Forest Research Institute of Dehradun, India. The field identifications of the plants were mostly done by using field guide with colored photographs of the plants by Polunin and Stainton [11]. Some unidentified and partially identified plants from the field were brought to the specialists at the Forest Research Institute of Dehradun, India for full identification. The participatory transect walk, interview, and discussions with traditional healers were used for ethnobotanical data collection. The total number of participatory transects established were three for every village, resulting in a total of nine across all three villages. The length of each transect was 2 km from the center of the village to three different outward directions, depending on aspects of the village. We used three different groups for the transect walks. These groups were common village people including men and women, hunters, and traditional healers. Two walks with every group with different people were conducted. As such, the total number of transect walks per village was six, thereby totaling 18 transect walks across all three villages. This type of data collection design was followed for the robustness of ethnobotanical information. Apart from this technique for collecting plant specimens with ethnobotanical values, we used a structured questionnaire for interviews and group discussions regarding the ethnopharmacological techniques of medicine preparation for the collected plants. The people who participated in transect walks were not selected for questionnaire surveys in order to avoid repetition and establish a more general idea among larger population groups. The participatory transect walks were mostly carried out in spring and summer when a large flush of herbaceous plants grow in the forest, pasturelands, and meadows after the melting of winter snow. At the second stage of the research, ethnopharmacological information was collected from the high ranked monks and traditional healers who prepare medicine from plants for the healing of the tribal people. In each village, we selected at least three independent healers or monks for this purpose. After gathering the information, we performed a qualitative assessment for reaching a consensus among the respondents and rejected the conflicting responses. The basic information that was collected from these monks and traditional healers were regarding (1) the plants needed to make medicine; (2) the use of plant parts; (3) the different ratios of plant use; (4) the techniques of preparation; (5) the doses and prescription to the patients; and (6) the medicinal uses. The third stage of the research was carried out at the Resource Survey and Management Division of the Forest Research Institute, Dehradun, India. Taxonomical classification was performed with the help of the Botany Division of the Forest Research Institute, and identified plant specimens were confirmed by using the herbaria of the same division for comparison purposes. The specimens with detailed taxonomic information, name of collectors, and place and date of collections were finally deposited to The Course Coordinator of Postgraduate Programs, Forest Research Institute University (Dehra Dun, India) for future references. We had received permission from the local forest authorities in addition to having obtained consents from the traditional healers before doing this survey.

4. Discussion

The list of plant species and utilization of plant parts for different diseases documented in this study support a recent study carried out by Namsa et al. (2011) on the southern or Kalaktang Monpas [6]. The list we provided for the medicinal plants is not completely new to ethnobotanists, as it was already listed in old research works on medicinal plants of the Himalayas (see [12–14]). This proves that the plants we listed are already confirmed as "medicinal plants" by past researchers from the other parts of the Himalayas. However, the detailed ethnopharmacological descriptions or traditional ways of preparation for the herbal drugs and medicines were rarely documented. Due to this reason, a search with the terms "ethnopharmacology * Himalayas" yielded only three articles on 20 February 2017 in the ISI Web of Knowledge. For example, Gangwar et al. [15] worked on ethnopharmacological uses of *Mallotus philippinensis* Muell. Arg, and Stobdan et al. [16] did a similar work on *Hippophae rhamnoides* L. We found only one article, a study by Abbasi et al. [17], that was similar to our study and described ethnopharmacological knowledges of medicine preparation from a Himalayan region of the Pakistan Himalayas. Therefore, we emphasize that this is the first documented study on ethnopharmacology of a tribe from Arunachal Pradesh. We assume that in most cases modern pharmacologists and researchers start chemical assessments of the medicinal plants without giving much attention to the traditional ways of drug preparation by the tribal communities. This could be the reason behind the high number of studies on ethnobotany of medicinal plants from the Himalayas, but the comparatively minimal number of studies on ethnopharmacology. In South India, the Kani tribe uses similar approach for traditional medicine making [5], which supports the notion that tribal healers do use certain systematic techniques for drug preparation. The ethnopharmacological knowledge of traditional healers are generally transferred orally to the next generation, thus, making the knowledge vulnerable to being forgotten or lost.

In this context, we would like to provide a few examples of past pharmacological studies that had reported similar utilization of some medicinal plants listed in this study. Ghildiyal et al. (2012) showed that ethanolic extracts from *Hedychium spicatum* can inhibit respiratory as well as gastrointestinal disorders in rats and guinea pigs [18]. We showed in this study that the ethnomedicine *Blenga* prepared from the same plant was used for the treatments of dysentery and chest pain. In 2007, Nazir et al. extracted a drug called "Bergenin" from the species *Bergenia stracheyi* and proved that this drug can be used to treat arthritis in mice [19]. Interestingly, we found that an ethnomedicine named as *Bragen* (prepared from *Bergenia stracheyi* as well) was also used for the treatment of arthritis. Recently in 2014, Kumar et al. reported that the extracts of *Houttuynia cordata* can be used for the healing of hemorrhoids, and this species is frequently used in tradtional Tibetan and Chinese medicines [20]. We found that the ethnomedicine *Maraptang* prepared from *Houttuynia cordata* were also used by the tradtional Monpa healers for the treatment of piles which is a type of hemorrhoid. These examples mentioned above showed that the tradtional ethnomedicines used by the healers of Zemithang Monpa may have some potential to cure or manage some diseases. However, detailed pharmacological studies are needed to evaluate the potential of these medicines. A study by Witt et al. (2009) in Sikkim and Eastern Nepal (also part of the Eastern Himalayas) comprehensively listed 138 species of plants from tropical to alpine regions of the Himalayas used specifically in Tibetan medicine [21]. The majority of the species listed in our study were also reported by Witt et al., but detailed descriptions of the preparations for the ethnomedicines were not provided.

The results of this study should be interpreted very cautiously. The traditional ethnopharmacological knowledge of the Zemithang-Monpa tribe presented here for some diseases must not be treated as a general prescription under any circumstances, as scientific trials have not been undertaken nor the "traditional ethnomedicines" have ever been certified by any governmental authority such as the Central Drugs Standard Control Organization of India. There is also a high probability that the descriptions presented here may not be the same throughout the study region. Nevertheless, our main goal was not to certify or validate traditional medicines, but rather to document the uses and preparation of traditional medicines used by tribal people. The field method applied for data collection (i.e., participatory transect walk) also had some limitations. This method was helpful in remote regions where time and logistics are always a constraint of field work. Nevertheless, future research should establish more sample plots and cover larger regions in order to list more medicinal plants.

5. Conclusions

We have documented for the first time the vernacular names combined with ethnopharmacological preparations of ethnomedicines among Monpa tribes from the Zemithang region of Arunachal Pradesh, India. Past studies on ethnobotany in the Arunachal Pradesh, Eastern Himalayas, had listed uses of medicinal plants, however, we found that traditional healers use diverse species and plant parts in specific proportions for drug preparations. Our study illustrates the diversity of medicinal drug preparations and traditional knowledge that has passed through generation after generation of Monpa people. The ethnopharmacological documentation presented in this study should motivate researchers to carry out further scientific work on pharmacology, bioprospecting, and the cultivation of medicinal plants for the socioeconomic development in the region. Under ongoing warming of the Himalayas and mass migration of people from the mountain areas to cities, our study also highlights the need to document the traditional knowledge regarding the use of local flora and to develop strategies to conserve them before the traditional knowledges are lost or forgotten.

Acknowledgments: We sincerely thank the Divisional Forest Officers (DFOs) of Tawang Social Forestry Division and West Kameng Forest Division (Suneesh Buxy, M. Sambhu and Adukparon of Indian Forest Service) and R. C. Das (the Range Forest Officer, Lumla Forest Range, Arunachal Pradesh State Forest Service) for giving us permission and providing logistical support to do the field work. Without their support, this study could not have been done in this remote part of India. We thank Prema Khandu (village Lumpho) and Norbu (village Namshu) for all the supports provided during field data collection, and for organizing and participating in long field expeditions. We will never forget the support from Haridasan and Rao (State Forest Research Institute of Arunachal Pradesh, Itanagar), Saroj K. Barik (North Eastern Hill University, Shillong), and Mohammed Latif Khan (Northeastern Regional Institute of Science and Technology, Itanagar) in the identification of herbarium samples. We also acknowledge the cooperation provided by the Pijush K. Datta of WWF-India and Bibhab Talukdar of ATREE-Northeast Program and the IUCN-India during this research work. The article processing charge was funded by the open access publication fund of the Albert-Ludwigs-University of Freiburg.

Author Contributions: Tamalika Chakraborty and Somidh Saha conducted the field work, wrote the paper, and equally contributed to this work. Narendra Singh Bisht provided motivation, guidance, and supervision to carry out this research and acted as mentor for the graduate research works of Tamalika Chakraborty and Somidh Saha.

Conflicts of Interest: The authors declare no conflict of interest.

References

1. Borges, R.M. The frontiers of India's biological diversity. *Biotropica* **2005**, *37*, A1–A3.
2. Haridasan, K.; Bhuyan, L.R.; Deori, M.L. Wild edible plants of Arunachal Pradesh. *Arunachal For. News* **1990**, *18*, 1–8.
3. Adak, D.K. A morphometric study of the Thingbu-pa and population comparison with neighbouring Monpa tribes of Arunachal Pradesh, India. *Anthropologischer Anzeiger Bericht uber die Biologisch-Anthropologische Literatur* **2001**, *59*, 365–375. [PubMed]
4. Sharma, P.P.; Mujundar, A.M. Traditional knowledge on plants from Toranmal Plateau of Maharastra. *Indian J. Tradit. Knowl.* **2003**, *2*, 292–296.
5. Ayyanar, M.; Ignacimuthu, S. Traditional knowledge of kani tribals in Kouthalai of Tirunelveli hills, Tamil Nadu, India. *J. Ethnopharmacol.* **2005**, *102*, 246–255. [CrossRef] [PubMed]
6. Namsa, N.D.; Mandal, M.; Tangjang, S.; Mandal, S.C. Ethnobotany of the Monpa ethnic group at Arunachal Pradesh, India. *J. Ethnobiol. Ethnomed.* **2011**, *7*, 31. [CrossRef] [PubMed]
7. Saha, S.; Bisht, N.S. Role of traditional ecological knowledge in natural resource management among Monpas of north-western Arunachal Pradesh. *Indian For.* **2007**, *133*, 155–164.
8. Von Fürer-Haimendorf, C. *Tribes of India: The Struggle for Survival*; Oxford University Press: Delhi, India, 1982.
9. State Government of Arunachal Pradesh. *Working Plan: Tawang Social Forestry Division*; Arunachal Forest Department: Tawang, India, 2001.
10. Champion, S.H.G.; Seth, S.K. *A Revised Survey of the Forest Types of India*; Government of India: New Delhi, India, 1968.
11. Polunin, O.; Stainton, A. *Flowers of the Himalayas*; Oxford University Press: Delhi, India, 1997.
12. Haridasan, K.; Shukla, G.P.; Beniwal, B.S. Medicinal plants of Arunachal Pradesh. *SFRI Inf. Bull.* **1995**, *5*, 32.
13. Kala, C.P. *Medicinal Plants of Indian Trans-Himalaya: Focus on Tibetan Use of Medicinal Resources*; Bishen Singh Mahendra Pal Singh: Dehra Dun, India, 2003.

14. Kala, C.P. Status and conservation of rare and endangered medicinal plants in the Indian trans-Himalaya. *Biol. Conserv.* **2000**, *93*, 371–379. [CrossRef]
15. Gangwar, M.; Goel, R.K.; Nath, G. *Mallotus philippinensis* Muell. Arg (Euphorbiaceae): Ethnopharmacology and phytochemistry review. *BioMed Res. Int.* **2014**, *2014*, 213973. [CrossRef] [PubMed]
16. Stobdan, T.; Targais, K.; Lamo, D.; Srivastava, R.B. Judicious use of natural resources: A case study of traditional uses of seabuckthorn (*Hippophae rhamnoides* L.) in trans-Himalayan Ladakh, India. *Natl. Acad. Sci. Lett.* **2013**, *36*, 609–613. [CrossRef]
17. Abbasi, A.M.; Khan, M.A.; Zafar, M. Ethno-medicinal assessment of some selected wild edible fruits and vegetables of Lesser-Himalayas, Pakistan. *Pak. J. Bot.* **2013**, *45*, 215–222.
18. Ghildiyal, S.; Gautam, M.K.; Joshi, V.K.; Goel, R.K. Pharmacological evaluation of extracts of *Hedychium spicatum* (Ham-ex-Smith) rhizome. *Anc. Sci. Life* **2012**, *31*, 117–122.
19. Nazir, N.; Koul, S.; Qurishi, M.A.; Taneja, S.C.; Ahmad, S.F.; Bani, S.; Qazi, G.N. Immunomodulatory effect of bergenin and norbergenin against adjuvant-induced arthritis—A flow cytometric study. *J. Ethnopharmacol.* **2007**, *112*, 401–405. [CrossRef] [PubMed]
20. Kumar, M.; Prasad, S.K.; Hemalatha, S. A current update on the phytopharmacological aspects of *Houttuynia cordata* Thunb. *Pharmacogn. Rev.* **2014**, *8*, 22–35. [PubMed]
21. Witt, C.M.; Berling, N.E.; Rinpoche, N.T.; Cuomo, M.; Willich, S.N. Evaluation of medicinal plants as part of Tibetan medicine prospective observational study in Sikkim and Nepal. *J. Altern. Complement. Med.* **2009**, *15*, 59–65. [CrossRef]

© 2017 by the authors. Licensee MDPI, Basel, Switzerland. This article is an open access article distributed under the terms and conditions of the Creative Commons Attribution (CC BY) license (http://creativecommons.org/licenses/by/4.0/).

Article

An Investigation of the Antioxidant Capacity in Extracts from *Moringa oleifera* Plants Grown in Jamaica

Racquel J. Wright [1,2], Ken S. Lee [3], Hyacinth I. Hyacinth [4], Jacqueline M. Hibbert [5], Marvin E. Reid [2], Andrew O. Wheatley [1,6] and Helen N. Asemota [1,6,*]

1. Biotechnology Centre, University of the West Indies, Mona Kingston 8, Jamaica; racq.wright@gmail.com (R.J.W.); awheatley@mset.gov.jm (A.O.W.)
2. Caribbean Institute for Health Research, University of the West Indies, Mona Kingston 8, Jamaica; marvin.reid@uwimona.edu.jm
3. Department of Chemistry and Biochemistry, Jackson State University, Jackson, MS 39217, USA; ken.s.lee@jsums.edu
4. Department of Pediatrics, Aflac Cancer and Blood Disorder Center, Children's Healthcare of Atlanta and Emory University, Atlanta, GA 30322, USA; hhyacinth@emory.edu
5. Department of Microbiology, Biochemistry and Immunology, Morehouse School of Medicine, 720 Westview Drive SW, Atlanta, GA 30310, USA; jhibbert@msm.edu
6. Biochemistry Section, Department of Basic Medical Sciences, University of the West Indies, Mona Kingston 8, Jamaica
* Correspondence: helen.asemota@uwimona.edu.jm; Tel.: +1-876-977-1828

Academic Editor: Milan S. Stankovic
Received: 18 September 2017; Accepted: 28 September 2017; Published: 23 October 2017

Abstract: *Moringa oleifera* trees grow well in Jamaica and their parts are popularly used locally for various purposes and ailments. Antioxidant activities in *Moringa oleifera* samples from different parts of the world have different ranges. This study was initiated to determine the antioxidant activity of Moringa oleifera grown in Jamaica. Dried and milled Moringa oleifera leaves were extracted with ethanol/water (4:1) followed by a series of liquid–liquid extractions. The antioxidant capacities of all fractions were tested using a 2,2-diphenyl-1-picrylhydrazyl (DPPH) assay. IC_{50} values (the amount of antioxidant needed to reduce 50% of DPPH) were then determined and values for the extracts ranged from 177 to 4458 µg/mL. Extracts prepared using polar solvents had significantly higher antioxidant capacities than others and may have clinical applications in any disease characterized by a chronic state of oxidative stress, such as sickle cell anemia. Further work will involve the assessment of these extracts in a sickle cell model of oxidative stress.

Keywords: *Moringa oleifera*; DPPH; antioxidant activity; oxidative stress; sickle cell anemia

1. Introduction

The *Moringa oleifera* plant, which is also known as "Marengue" in Jamaica, is one of the 13 species in the *Moringa* genus. Other species include *Moringa stenopetala* and *Moringa ovalifolia*. *Moringa oleifera* is also called the Horseradish tree, and like horseradish it possesses a taproot system that supports the umbrella-like canopy of the trunk, leaves, and branches [1,2]. The leaves are tripinnate and grow with a fragile feather-like drooping crown. Fragrant flowers grow as spreading auxillary panicles and are yellowish white, and when fertilized produce pods resembling that of the common bean (*Phaseolus vulgaris*), commonly referred to as string beans or snap beans. Similar to stringbeans, the pods are initially green; however, as *Moringa* pods mature, they become brown and thicker [2]. In Jamaica, flowers and consequently pods are produced at an increased rate during the rainy season, although flowering and fruiting occurs throughout the year. The plant seems to prefer the drier climate of the southern areas in Jamaica but also grows in the north [3]. Several non-governmental organizations including the Food and Agricultural Organization (FAO), Educational Concerns for Hunger Organization (ECHO), Church World Service, and Trees for Life have endorsed *Moringa* as a nutritional gold mine for tropical areas due to its nutritional content and its ability to grow in tropical and drought affected areas [4,5].

Moringa oleifera is native to the Himalayas and grows well in sub-tropical and tropical regions of the world. It is widely used in ethnobotany and is thought to cure a variety of diseases [4]. *Moringa* has long been a part of Ayurvedic medicine in India and is understandably referred to as the "Miracle Tree". All parts of the *Moringa* plant are edible, with leaves and pods used most frequently. The leaves of the plant are utilized as a nutritional supplement, are thought to boost the immune system as well as energy levels, and are known to have anti-inflammatory as well as antioxidant properties. In Jamaica, *Moringa* is used in the preparation of hot and cold beverages (teas and juices) and as meals in many households. The interest in *Moringa* started in the 2000s, and this interest continues today. Various investigations have shown that *Moringa* contains antioxidants [4,6–8]. Antioxidants are useful in the management of oxidative stress in the body [9]. *Moringa* could potentially be used to improve the clinical condition of persons with oxidative stress conditions such as sickle cell anemia (SCA). Oxidative stress is the result of an imbalance between reactive oxidative species (ROS) and antioxidant components in the body. ROS can potentially damage cells in the body, destabilizing the cell integrity by reacting with cellular components [9]. Antioxidant components are designed to reduce ROS [9]. The body produces ROS as a part of its normal metabolic processes and consequently has biological mechanisms to counteract oxidation [10].

In healthy people, the antioxidant system usually restores balance easily by reducing ROS when formed [9]. People with SCA experience a relatively higher oxidant load due to factors including the increased frequency of erythrocyte destruction (which results in excessive free heme, a potential oxidant). People with SCA also seem to have an increased metabolic rate, which in turn increases the production of oxidants. This increased oxidant production results in an imbalance of oxidants compared to antioxidants causing oxidative stress [10,11]. Sickle cell anemics therefore have higher levels of ROS due to the inability of their antioxidant system to compensate for the abnormal free heme plasma levels. This results in inflammation and chronic organ damage [9,10].

In Jamaica, 1 in 300 persons are estimated to have SCA in Jamaica. People with SCA experience a variety of symptoms, the consequences of which could be serious and expensive for Jamaicans. Sickle cell anemics often experience sickling crises, for which the disease is named. Leg ulcers, splenomegaly, stroke, pulmonary hypertension, and other conditions may also occur. Inflammation is an underlying symptom associated with many conditions.

People with SCA are treated symptomatically. Symptoms include pain and infections, which are treated with analgesics and antibiotics, respectively. SCA is prevalent in African and Caribbean nations, developing countries with limited resources. *Moringa* is a tropical plant that grows easily in these countries and is therefore readily accessible [4]. The use of the plant would be a cost-effective way of combating SCA by reducing oxidative stress. There appears to be no evidence in the literature of antioxidant activity testing of the Jamaican grown *Moringa* plant. This study was initiated to determine the antioxidant activity of *Moringa* plants grown in Jamaica.

2. Results

Initial extraction with ethanol (E) resulted in 37% recovery. Further fractionation using various solvents, namely hexane, chloroform, butanol, and water was performed. In relation to the ethanol extract, the percentage recovery for the solvent extracts was as follows: hexane-E1 (48%), chloroform-E2 (1.75%), butanol-E3 (9.46%), and water-E5 (18.26%). DPPH reduction percentages are shown in Figure 1. The slope of the DPPH reduction percentage plot was used as an indicator of antioxidant capacity.

Moringa extracts E3 and E5 have higher slopes compared to E1 and E2 (Figure 1); therefore, E3 and E5 have higher antioxidant capacities compared to E1 and E2.

IC_{50} values, representing the sample concentration at which 50% of the DPPH radical has been reduced, were calculated from the DPPH reduction percentages. There is an inverse relation between IC_{50} values and antioxidant activities; this means that lower IC_{50} values indicate a higher antioxidant capacity. Extracts E2 and E1 had IC_{50} values of 1604 µg/mL and 4477 µg/mL, respectively, while extracts E, E3, E4, E5, and A had IC_{50} values of 832.8 µg/mL, 172.6 µg/mL, 1085 µg/mL, 516.9 µg/mL, and 1003 µg/mL, respectively. Based on the graph (Figure 1) showing the percentages of DPPH reduction, extracts E3 and E5 possessed more effective antioxidative capacity compared to the other extracts. Figure 2 also shows that the IC_{50} values of extracts E3 and E5 were lower compared to the other extracts.

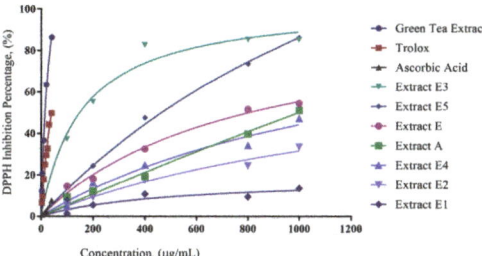

Figure 1. DPPH reduction percentage of Trolox, ascorbic acid, green tea, and the *Moringa oleifera* leaf extracts A, E, E1–5 after 30 minutes. DPPH Reduction Percentage gradients for extracts E3 and E5 (0.147; 0.082) were higher compared with the other extracts, particularly E1 and E2 (0.011; 0.029). E5 and E3 had gradients over 3–7 times greater than E2 and E1, respectively.

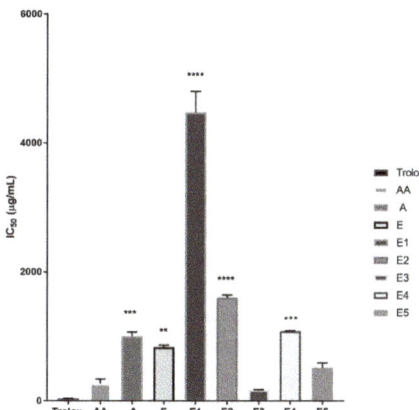

Figure 2. IC_{50} values of ascorbic acid (AA) compared with Trolox and *Moringa* leaf extracts A, E, and E1–E5. Statistical significance was calculated using one-way analysis of variance (ANOVA) followed by Dunnett's post-hoc test using GraphPad Prism statistical software; values of $p < 0.05$ were considered statistically significant. Significant values were denoted as follows: ** $p \leq 0.01$; *** $p \leq 0.001$; **** $p \leq 0.0001$. Ascorbic acid had an IC_{50} value of 272 µg/mL. Extracts E2 and E1 had IC_{50} values of 1604 µg/mL and 4477 µg/mL respectively, while extracts E, E3, E4, E5, and A had approximately 2–6 times lower IC_{50} values of 833 µg/mL, 173 µg/mL, 1085 µg/mL, 517 µg/mL, and 1003 µg/mL, respectively.

3. Discussion

Antioxidant activity of *Moringa oleifera* leaf extracts from Jamaica was assessed using the 2,2-diphenyl-1-picrylhydrazyl (DPPH) assay [6,12]. This seems to be the first study of antioxidant activity using *Moringa* sourced from Jamaica, as the literature tends to lack any such information. It should be noted that these extracts were produced from a direct ethanolic extraction of the *Moringa* leaves. Typically, when antioxidant principles are being extracted from samples or when antioxidant tests are being conducted, an alcoholic solvent, aqueous solvent, or a combination of both is used as the extracting agent [6,13–15].

The crude extract E was sequentially extracted using solvents of different polarities. This allowed for the sequential separation of compounds based on the affinity to the solvent. This further subdivision of extracts provided a clearer picture as to which solvents have the greater capacity to accumulate antioxidants as well as concentrating the antioxidant extracts to smaller fractions. Extracts E1 and E2 were produced using hexane and chloroform, respectively. Extracts E3 and E5 were produced using polar solvents, namely butanol and water, respectively. In Figure 1, the DPPH reduction percentage was calculated for samples and controls.

The IC_{50} values for each sample was calculated to show the concentration of sample needed to reduce 50% of the DPPH in the assay. Low IC_{50} values correspond to high antioxidant activity. The IC_{50} values show that E3 < E5 < E < A < E4 < E2 < E1 (Figure 2). Both E1 and E2 showed lower antioxidant activities than other extracts, with E1 being the lowest. This could be explained by the fact that hexane was the most non-polar solvent used in the experiment. Antioxidant compounds are known to accumulate in polar solvents and it is understandable that they would not be present in substantial amounts in an extract produced from hexane. Chloroform is more polar than hexane, so the increased antioxidant activity seen in extract E2, compared with extract E1 (hexane), is expected [16,17]. This data also suggests that liquid–liquid extraction of crude extract E with low polarity solvents, hexane and chloroform, was effective in separating fractions with high antioxidant potential, thus producing a more concentrated antioxidant fraction (the residual fraction after extraction with chloroform). This fraction when extracted further produced extracts E3 and E5, which had higher antioxidant capacity compared to the extracts.

Extract E3 had the highest antioxidant activity since it had the lowest IC_{50} value. It is notable that extract E3 possessed a similar IC_{50} value to ascorbic acid (vitamin C), a known antioxidant, which suggests that extract E3 could be a valuable antioxidant. Extract E had a lower antioxidant activity than extracts E3 and E5 based on the IC_{50} value; however, it still had a good antioxidant activity compared with ascorbic acid. These results suggest that the components responsible for antioxidant activity in *Moringa* leaves are polar compounds. The values for *Moringa* compared with the standard Trolox were not an exact match compared with values found in the literature; however, when the values were normalized as a ratio, the values correlated with results obtained by Chumark et al. (2008) [18]. Chumark et al. also showed that Trolox has a lower IC_{50} value compared to the *Moringa* extract, with *Moringa* being 35 times less potent than Trolox. In fact, it seems that the Jamaican *Moringa* had comparatively better antioxidant activity, as this study showed *Moringa* was only four times less potent than Trolox. At the time of this study, *Moringa* could be found in only a few regions of the island, and the material analyzed was from a single region [3]. Further studies are needed to test the *Moringa* leaves from other regions in Jamaica extracted under the same conditions to determine if the antioxidant activity differs from parish to parish, possibly depending on the nature of the soil. All leaves used in this assay were harvested at the same time. Researchers in Pakistan, Iqbal and Bhanger (2006), assessed the variations in the antioxidant activity of *Moringa* leaves in different seasons and sample locations and found that the antioxidant activity varied based on season and locations of the plants from which the samples were obtained [19]. However, it is important to note that Pakistan has a larger variation of climatic conditions compared to Jamaica. It is possible that the plants were affected by the cooler and/or drier conditions experienced there. In Jamaica, given the low variability in climatic conditions, it is possible that the Jamaican plants would not have significant differences in antioxidant activities based on the location and time of harvest. Additionally, it is also possible that Jamaican *Moringa* has adapted over the years to the moderate climate experienced in Jamaica compared with Pakistan's climate. Although the antioxidant activities of Jamaican *Moringa* leaves should be independent of the harvesting locations, this information is not known. This data would be significant if *Moringa* were to be exploited commercially. It may also be useful to investigate antioxidant activities of *Moringa* plants throughout the Caribbean, where the climatic conditions are generally similar but the nature of the soil may vary.

Siddharaj and Becker (2003) have shown that different agroclimatic conditions can result in different antioxidant capacities [6]. In his results, it was seen that India had the best antioxidant capacity compared to samples taken in Central America (Nicaragua) and Africa. This study, however, did not focus on the soil quality, which is another factor that could affect the quality of bioactive principles in plants. Forster and colleagues (2015) assessed the impact of sulfur (a soil factor) compared to water availability. In the field, water availability would typically be determined by rainfall, which is a climatic factor [20]. This study showed that the availability of sulfur had a positive impact on the presence of the beneficial bioactive principles in *Moringa*. There seems to be no direct study yet done on the effect of other soil factors such as pH, soil nutrient levels, and soil type on the antioxidant capacity in *Moringa* leaves. The studies mentioned above however, seem to allude to the premise that these factors could be responsible for differences between the antioxidant activity seen in different countries.

It should also be noted that the genotype of *Moringa* in Jamaica has not been elucidated. There is therefore no concrete evidence to indicate whether Jamaican *Moringa* is a different cultivar compared to those in Asia or Africa. *Moringa* was initially introduced to Jamaica in 1784; however, Asian and African migrants have entered Jamaica over the years, and it is quite possible that these immigrants may have brought different varieties or cultivars of *Moringa oleifera*. This means that there could be several cultivars of *Moringa oleifera* present in Jamaica as well. For example, a variety with completely white flowers and a different flavor profile has been noted in Jamaica.

Ndhlala and colleagues (2014) analyzed 13 *Moringa* cultivars from four regions (Thailand, Taiwan, United States of America, and South Africa) for antioxidant capacity. The data showed that the samples from Thailand had the best antioxidant capacity, whereas samples from South Africa (Silver Hill) had the lowest activity [21]. Ndhlala and others (2014) postulated among other things that climatic/environmental differences between the regions could be responsible for the variations seen [21]. *Moringa* has been present in Thailand for several decades, and it is possible that the antioxidant profile is due to the plant's adaptation to its environs [21]. Both Jamaica and Thailand have tropical climates; similarly, the environmental conditions under which the Jamaican *Moringa* plant is grown could have a positive effect on its antioxidant capacity. A comparison between samples from different agroclimatic origins would be beneficial to determine if this postulation is a valid claim. Analyses of soil factors would also be beneficial in providing a broader picture of the impact of environmental conditions on antioxidant capacity in *Moringa oleifera*.

4. Methods and Materials

4.1. Reagents

Methanol (HPLC grade), ethanol, butanol, hexane, and chloroform (ACS grade), ascorbic acid, and 2,2-diphenyl-1-picrylhydrazyl (DPPH) free radical were obtained from Sigma-Aldrich (Saint Louis, Missouri, USA). Green tea extract (pre-standardized to 50% epigallocatechin gallate) was obtained from HerbStoreUS (Walnut, CA, USA).

4.2. Extraction of Moringa oleifera Leaves

Moringa oleifera leaves were prepared according to the method in Luo et al. (2011) and Oyugi et al. (2009) with modifications [14,22]. The leaves were harvested, rinsed with clean water, and air-dried at room temperature overnight. The leaves were then placed in an oven at 40 °C and left until completely dry. The leaves were milled into a fine powder. The milled leaf samples (20 g) were subjected to soxhlet extraction using 80% ethanol: water mixture (250 mL). The resulting solution was dried in vacuo and stored at 4 °C until further analysis. This extract was called E. Different fractions were obtained by sequential washing with different solvents (Figure 3). Extract E was further purified by washing with non-polar and polar solvents. The dried extract E was re-suspended in 250 mL of water. This aqueous extract was subsequently extracted with 250 mL of hexane, chloroform, and butanol, respectively. The respective solvent was decanted after each extraction and dried in vacuo. Extracts were labeled as follows: E1: hexane; E2: chloroform; E3: butanol; E4: interphase between E5 and E3; E5: remaining water fraction. For the production of Sample A, double distilled water was added to the milled leaves, which were subsequently heated for 24 h with constant stirring. The decoction was filtered and the filtrate lyophilized. All samples were stored at 4 °C until further analysis.

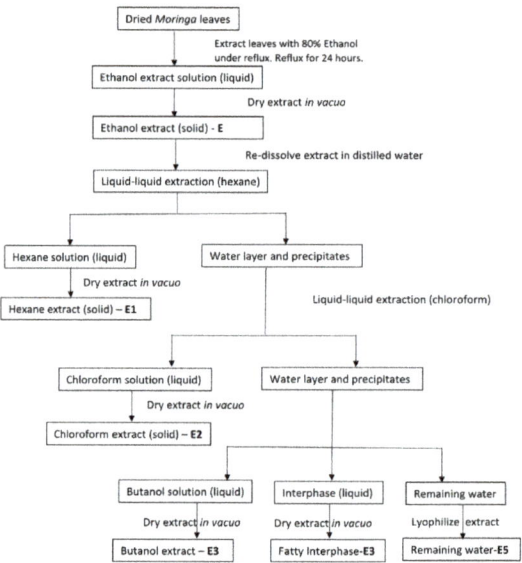

Figure 3. Schematic diagram of the ethanolic extraction process of *Moringa oleifera* leaves.

4.3. DPPH Radical Scavenging Assay

The antioxidant capacity of the *Moringa oleifera* leaf extracts were analyzed for reductive capacity using the 2,2-diphenyl-1-picrylhydrazyl (DPPH) assay [6,12]. DPPH is a stable free radical that absorbs strongly at 517 nm. When exposed to an antioxidant, it degrades to a light-yellow color that does not absorb at 517 nm. Therefore, the antioxidant activity of a substance can be analyzed based on the how strongly DPPH is absorbed at 517 nm. This assay was done using Trolox, a vitamin E analogue with well-documented antioxidant activity, as the main control for comparison with the extracts. Two additional controls, ascorbic acid and green tea extract, known chemical and herbal antioxidants, respectively, were used to compare the extracts' antioxidant capacities as they are typical antioxidants found in nature and are weaker than Trolox [15]. The green tea extract was pre-standardized to contain 50% epigallocatechin gallate, the main antioxidant in green tea.

4.3.1. Experimental Procedure

Controls and extracts E and E1–E5 were diluted to appropriate concentrations in methanol to ascertain the antioxidant capacity of all compounds in a linear range. The extracts were diluted to concentrations of 100, 200, 400, 800, and 1000 µg/mL. The standard (Trolox) was diluted to concentrations of 2.5, 5, 10, 15, 20, 25, 30, and 40 µg/mL. The controls (ascorbic acid and green tea extract) were diluted to concentrations of 2.5, 5, 10, 20, and 40 µg/mL; 0.1 mM DPPH in methanol was used and prepared fresh daily.

Controls and extracts were analyzed according to the method of Williams et al. (2006) with some modifications. The controls and extracts (40 µL each) were reacted with DPPH (200 µL each) in 96-well microplates for 30 min in the dark. A blank using methanol, in place of the sample, was also prepared and incubated with the samples. After incubation, the absorbance of samples was read using a Spectromax Pro 250 (Molecular Devices, Sunnnydale Ca.) microplate reader at 517 nm.

The mean optical density of the sample was used to calculate the DPPH reduction (inhibition), which is the percentage of DPPH that was neutralized by the antioxidants present in the added samples.

The equation below was used to calculate this percentage:

The DPPH reduction percentage is calculated as follows:

$$[(A_O - A_S)/A_O] \times 100 \tag{1}$$

where A_O is the blank absorbance, and A_S is the sample

The results were tabulated and presented in graphical format.

The IC_{50} values were calculated from the DPPH reduction percentages and referred to the concentration of sample (in µg/mL) required to reduce 50% of the DPPH present in the assay.

4.3.2. Statistical Analysis

Mean ± SEM was calculated from replicates of three or more for each sample. Significant differences between the values were calculated using Dunnett's post-hoc test with GraphPad Prism 5.0. Values of $p < 0.05$ was considered statistically significant.

Acknowledgments: This research was funded by grants from the Biotechnology Centre and Office of Graduate Studies University of the West Indies, Mona, Jamaica, the Eagle-i consortium, supported by NCRR (award #U24 RR 029825), as well as the National Institute on Minority Health and Health Disparities of the National Institutes of Health under Award Number G12MD007581. The content is solely the responsibility of the authors and does not represent the official views of the National Institutes of Health.

Author Contributions: Racquel Wright, Ken Lee, Jacqueline Hibbert, Andrew Wheatley and Helen Asemota conceived and designed the experiments. Racquel Wright performed the experiments. Andrew Wheatley, Hyacinth Hyacinth, Jacqueline Hibbert and Ken Lee supervised experimentation. Ken Lee, Andrew Wheatley and Jacqueline Hibbert contributed reagents/materials/analysis tools. Helen Asemota, Marvin Reid and Jacqueline Hibbert reviewed and revised the paper. Racquel Wright wrote the paper.

Conflicts of Interest: The authors declare no conflict of interest. The funding sponsors had no role in the design of the study; in the collection, analyses, or interpretation of data; in the writing of the manuscript; or in the decision to publish the results.

References

1. Adams, C.D. *Flowering Plants of Jamaica*; University of the West Indies: Mona, Jamaica, 1972; pp. 310–311.
2. Roloff, A.; Weisgerber, H.; Lang, H.; Stimm, B. *Moringa Oleifera Lam., 1785*; Enzyklopädie der Holzgewächse: Weinheim, Germany, 2009.
3. Grant, D. (Forestry Department, Ministry of Agriculture, St. Andrew, Jamaica). Personal communication, 2009.
4. Fahey, J.W. *Moringa oleifera*: A review of the Medical evidence for its nutritional, Therapeutic and prophylactic properties. Part 1. *Trees Life J.* **2005**, *1*, 1–21.
5. FAO Promotes *Moringa* Cultivation in Ethopia and Phillipines. 2014. Available online: http://www.intracen.org/itc/blogs/market-insider/FA-Promotes-Moringa-Cultivation-in-Ethiopia-and-Philippines/ (accessed on 30 August 2017).
6. Siddhuraju, P.; Becker, K. Antioxidant Properties of Various Solvent Extracts of Total Phenolic Constituents from Three Different Agroclimatic Origins of Drumstick Tree (*Moringa oleifera* Lam.) Leaves. *J. Agric. Food Chem.* **2003**, *51*, 2144–2155. [CrossRef] [PubMed]

7. Dahiru, D.; Onubiyi, J.A.; Umaru, H.A. Phytochemical screening and antiulcerogenic effect of *Moringa oleifera* aqueous leaf extract. *Afr. J. Tradit. Complement. Altern. Med.* **2006**, *3*, 70–75. [CrossRef]
8. Sreelatha, S.; Padma, P.R. Antioxidant Activity and Total Phenolic Content of *Moringa oleifera* Leaves in Two Stages of Maturity. *Plant Foods Hum. Nutr.* **2009**, *64*, 303–311. [CrossRef] [PubMed]
9. Wood, K.C.; Granger, D.N. Sickle cell disease: Role of reactive oxygen and nitrogen metabolites. *Clin. Exp. Pharmacol. Physiol.* **2007**, *34*, 926–932. [CrossRef] [PubMed]
10. Fasola, F.; Adedapo, K.; Anetor, J.; Kuti, M. Total antioxidants status and some hematological values in sickle cell disease patients in steady state. *J. Natl. Med. Assoc.* **2007**, *99*, 891–894. [PubMed]
11. Nagababu, E.; Fabry, M.E.; Nagel, R.L.; Rifkind, J.M. Heme degradation and oxidative stress in murine models for hemoglobinopathies: Thalassemia, sickle cell disease and hemoglobin C disease. *Blood Cells Mol. Dis.* **2008**, *41*, 60–66. [CrossRef] [PubMed]
12. Williams, L.; Hibbert, S.; Porter, R.; Bailey-Shaw, Y.; Green, C. Jamaican plants with in vitro anti-oxidant activity. In *Biologically Active Natural Products for the 21st Century*; Williams, L., Ed.; Research Signpost: Trivandrum, India, 2006; pp. 1–12.
13. Manian, R.; Anusuya, N.; Siddhuraju, P.; Manian, S. The antioxidant activity and free radical scavenging potential of two different solvent extracts of *Camellia sinensis* (L.) O. Kuntz, *Ficus bengalensis* L. and *Ficus racemosa* L. *Food Chem.* **2008**, *107*, 1000–1007. [CrossRef]
14. Luo, X.; Jiang, Y.; Fronczek, F.R.; Lin, C.; Izevbigie, E.B.; Lee, K.S. Isolation and structure determination of a sesquiterpene lactone (vernodalinol) from *Vernonia amygdalina* extracts. *Pharm. Boil.* **2011**, *49*, 464–470. [CrossRef] [PubMed]
15. Das, N.; Islam, M.E.; Jahan, N.; Islam, M.S.; Khan, A.; Islam, M.R.; Parvin, M.S. Antioxidant activities of ethanol extracts and fractions of Crescentia cujete leaves and stem bark and the involvement of phenolic compounds. *BMC Complement. Altern. Med.* **2014**, *14*, 1–9. [CrossRef] [PubMed]
16. Harborne, J.B. *Phytochemical Methods: A Guide to Modern Techniques of Plant Analysis*, 2nd ed.; Chapman and Hall: London, UK, 1972.
17. Trease, G.E.; Evans, W.C. *Pharmacognosy*, 10th ed.; Bailliere Tindall: London, UK, 1972.
18. Chumark, P.; Khunawat, P.; Sanvarinda, Y.; Phornchirasilp, S.; Morales, N.P.; Phivthong-ngam, L.; Klai-upsorn, S.P. The in vitro and ex vivo antioxidant properties, hypolipidaemic and antiatherosclerotic activities of water extract of *Moringa oleifera* Lam. leaves. *J. Ethnopharmacol.* **2008**, *116*, 439–446. [CrossRef] [PubMed]
19. Iqbal, S.; Bhanger, M.I. Effect of season and production location on antioxidant activity of *Moringa oleifera* leaves grown in Pakistan. *J. Food Compos. Anal.* **2006**, *19*, 544–551. [CrossRef]
20. Förster, N.; Ulrichs, C.; Schreiner, M.; Arndt, N.; Schmidt, R.; Mewis, I. Ecotype variability in growth and secondary metabolite profile in *Moringa oleifera*: Impact of sulfur and water availability. *J. Agric. Food Chem.* **2015**, *63*, 2852–2861. [CrossRef] [PubMed]
21. Ndhlala, A.R.; Mulaudzi, R.; Ncube, B.; Abdelgadir, H.A.; du Plooy, C.P.; van Staden, J. Antioxidant, antimicrobial and phytochemical variations in thirteen *Moringa oleifera* Lam. Cultivars. *Molecules* **2014**, *19*, 10480–10494. [CrossRef] [PubMed]
22. Oyugi, D.A.; Luo, X.; Lee, K.S.; Hill, B.; Izevbigie, E.B. Activity markers of the anti-breast carcinoma cell growth fractions of Vernonia amygdalina extracts. *Exp. Boil. Med.* **2009**, *234*, 410–417. [CrossRef] [PubMed]

© 2017 by the authors. Licensee MDPI, Basel, Switzerland. This article is an open access article distributed under the terms and conditions of the Creative Commons Attribution (CC BY) license (http://creativecommons.org/licenses/by/4.0/).

Article

Proanthocyanidin Characterization, Antioxidant and Cytotoxic Activities of Three Plants Commonly Used in Traditional Medicine in Costa Rica: *Petiveria alliaceae* L., *Phyllanthus niruri* L. and *Senna reticulata* Willd.

Mirtha Navarro [1], Ileana Moreira [2], Elizabeth Arnaez [2], Silvia Quesada [3], Gabriela Azofeifa [3], Diego Alvarado [4] and Maria J. Monagas [5],*

[1] Department of Chemistry, University of Costa Rica (UCR), Rodrigo Facio Campus, San Pedro Montes Oca, San Jose 2060, Costa Rica; mnavarro@codeti.org
[2] Department of Biology, Technological University of Costa Rica (TEC), Cartago 7050, Costa Rica; imoreira@itcr.ac.cr (I.M.); earnaez@itcr.ac.cr (E.A.)
[3] Department of Biochemistry, School of Medicine, University of Costa Rica (UCR), Rodrigo Facio Campus, San Pedro Montes Oca, San Jose 2060, Costa Rica; silvia.quesada@ucr.ac.cr (S.Q.); gabriela.azofeifacordero@ucr.ac.cr (G.A.)
[4] Department of Biology, University of Costa Rica (UCR), Rodrigo Facio Campus, San Pedro Montes Oca, San Jose 2060, Costa Rica; luis.alvaradocorella@ucr.ac.cr
[5] Institute of Food Science Research (CIAL), Spanish National Research Council (CSIC-UAM), C/Nicolas Cabrera 9, 28049 Madrid, Spain
* Correspondence: mjm@usp.org

Received: 5 September 2017; Accepted: 15 October 2017; Published: 19 October 2017

Abstract: The phenolic composition of aerial parts from *Petiveria alliaceae* L., *Phyllanthus niruri* L. and *Senna reticulata* Willd., species commonly used in Costa Rica as traditional medicines, was studied using UPLC-ESI-TQ-MS on enriched-phenolic extracts. Comparatively, higher values of total phenolic content (TPC), as measured by the Folin-Ciocalteau method, were observed for *P. niruri* extracts (328.8 gallic acid equivalents/g) than for *S. reticulata* (79.30 gallic acid equivalents/g) whereas *P. alliaceae* extract showed the lowest value (13.45 gallic acid equivalents/g). A total of 20 phenolic acids and proanthocyanidins were identified in the extracts, including hydroxybenzoic acids (benzoic, 4-hydroxybenzoic, gallic, prochatechuic, salicylic, syringic and vanillic acids); hydroxycinnamic acids (caffeic, ferulic, and p-coumaric acids); and flavan-3-ols monomers [(+)-catechin and (−)-epicatechin)]. Regarding proanthocyanidin oligomers, five procyanidin dimers (B1, B2, B3, B4, and B5) and one trimer (T2) are reported for the first time in *P. niruri*, as well as two propelargonidin dimers in *S. reticulata*. Additionally, *P. niruri* showed the highest antioxidant DPPH and ORAC values (IC$_{50}$ of 6.4 µg/mL and 6.5 mmol TE/g respectively), followed by *S. reticulata* (IC$_{50}$ of 72.9 µg/mL and 2.68 mmol TE/g respectively) and *P. alliaceae* extract (IC$_{50}$ >1000 µg/mL and 1.32 mmol TE/g respectively). Finally, cytotoxicity and selectivity on gastric AGS and colon SW20 adenocarcinoma cell lines were evaluated and the best values were also found for *P. niruri* (SI = 2.8), followed by *S. reticulata* (SI = 2.5). Therefore, these results suggest that extracts containing higher proanthocyanidin content also show higher bioactivities. Significant positive correlation was found between TPC and ORAC ($R^2 = 0.996$) as well as between phenolic content as measured by UPLC-DAD and ORAC ($R^2 = 0.990$). These findings show evidence for the first time of the diversity of phenolic acids in *P. alliaceae* and *S. reticulata*, and the presence of proanthocyanidins as minor components in latter species. Of particular relevance is the occurrence of proanthocyanidin oligomers in phenolic extracts from *P. niruri* and their potential bioactivity.

Keywords: *P. alliaceae*; *P. niruri*; *S. reticulata*; UPLC; TQ-ESI-MS; proanthocyanidins; mass spectrometry; antioxidant; cytotoxicity

1. Introduction

Characterization and quantification of secondary metabolites and their bioactivity studies are essential to increase the knowledge on plants with traditional medicinal uses towards efficient and safe utilization. Three plant species of special relevance because of their widespread utilization in traditional medicine in Costa Rica as anti-inflammatories are *Petiveria alliaceae*, *Phyllanthus niruri* and *Senna reticulata* [1]. *Senna reticulata* is a shrub belonging to the *Fabaceae* family, originally from Mesoamerica and South America (to Brazil), whose leaves and stems are used traditionally for rheumatism and skin conditions [2]. *Phyllanthus niruri* is a shrub native to America and tropical areas of India and China, which belongs to the *Phyllanthaceae* family, and has diuretic and hepatoprotective properties [3]. Finally, *Petiveria alliacea*, which belongs to the *Phytholacaceae* family, is distributed from the South of the United States to Brazil and its traditional uses include analgesic, anticoagulant and hypoglycemic properties [4].

Studies have attributed anti-inflammatory and other bioactivities mainly to anthraquinones present in *S. reticulate* [2], sulfur containing compounds in *P. alliaceae* [4], and lignans such as phyllanthosides in the case of *P. niruri*. More recent studies have shown the synergic effect of different families of phenolic compounds in bioactivity studies, for instance showing the important effects of proanthocyanidins on antioxidant and cytotoxic bioactivities [5,6]. Regarding phenolic compounds, mainly flavonoids such as kaempferol derivatives have been reported in *S. reticulata* [2] [7]; quercetin derivatives in *P. alliacea* [8] and *P. niruri* [9]; whereas caffeic acid derivatives and ellagitannins have also been reported in *P. niruri* [10]. Flavan-3-ols, including catechin and epicatechin monomers have been reported in *P. niruri* [11], whereas no detailed studies on proanthocyanidin oligomers have been performed in any of these three species.

Proanthocyanidins are condensed flavan-3-ols that constitute an important group of polyphenols because of their bioactivities, among others, ant-inflammatory, antioxidant and anti-cancer activities [12]. Despite the increasing number of studies on phenolics, the characterization of proanthocyanidins remains a complex task because of the need for high-end techniques such as High-Resolution Mass Spectrometry (HRMS). On the other hand, it has been argued that these bioactivities could be mediated by redox interaction, since the regulation on redox homeostasis has been implicated in the control of the transition from cell proliferation to cell differentiation and cell cycle progression in both plants and animals [13]. However, the mechanisms and factors that could affect these bioactivities remain to be elucidated [14], suggesting the importance of these studies.

Since proanthocyanidin composition of *P. alliaceae*, *P. niruri* and *S. reticulata* have been scarcely studied and because of findings demonstrating the synergic contribution of proanthocyanidins on plants whose bioactivities were attributed solely to other metabolites [5,15], the objective of this work was to obtain phenolic extracts from these three plant species and to characterize them UPLC-DAD-ESI-TQ-MS. Evaluation of the antioxidant activity through DPPH and ORAC methods, as well as the assessment of cytotoxicity in AGS adenocarcinoma gastric cells, SW620 adenocarcinoma colon cells, and Vero normal cells, was also carried out in the different extracts.

2. Results and Discussion

2.1. Phenolic Yield and Total Phenolic Contents

The extraction process described in the Materials and Methods section, allowed the phenolic enriched fractions to be obtained, as summarized in Table 1. *S. reticulata* presented the highest yield (6.53%) whereas *P. alliaceae* showed the lowest value (5.03%). The total phenolic contents (TPC)

shown in Table 1, also resulted in comparatively lower values for *P. alliaceae* extract (13.45 gallic acid equivalents/g dry extract) than *P. niruri* extract (328.80 gallic acid equivalents/g dry extract), which exhibited the highest values. These results agree with few reports indicating lower TPC values for an hydroalcoholic extract of *P. alliaceae* [16], and for an aqueous extract of *S. reticulata* [17]. However, higher TPC values, in the range of 263–270 gallic acid equivalents/g, which are slightly lower than our findings have been reported for ethanolic extracts of *P. niruri* [10,18]. Table 1 also summarizes the total proanthocyanidin (PRO) content for the different extracts. *P. niruri* showed the highest PRO content (322.23 cyanidin chloride equivalents/g dry extract) whereas no content was found in *P. alliaceae*. *S. reticulata* showed intermediate values for both TP and PRO contents. Thus, phenolic content varied according to plant species, the highest values for both TPC and PRO clearly corresponding to *P. niruri*.

Table 1. Extraction yield and total phenolic content.

Sample	Extraction Yield (%) [1]	Total Phenolic Content (TPC) (mg/g) [2,5]	Total Proanthocyanidin Contents (PRO) (mg/g) [3,5]
P. alliacea	5.03	13.45 [a] ± 0.46	nd [4]
P. niruri	5.58	328.80 [b] ± 13.41	322.93 [a] ± 11.12
S. reticulata	6.53	79.30 [c] ± 4.09	22.35 [b] ± 1.64

[1] g of extract/g of dry material expressed as %. [2] mg of gallic acid equivalent (GAE)/g extract. [3] mg of cyanidin chloride equivalent (CCE)/g extract. [4] nd = not detected. [5] Different superscript letters in the same column indicate differences are significant at $p < 0.05$.

2.2. Phenolic Profile by UPLC-DAD-ESI-TQ-MS Analysis

Table 2 summarizes the results of UPLC-DAD-ESI-TQ-MS analysis performed in the different extracts, as described in the Materials and Methods section. Also, Figure S1 and Table S1 (Supplementary Materials) show base chromatograms and main MS/MS parameters respectively. Among the 28 phenolic compounds screened, a total of 20 phenolic compounds were characterized and quantified, (benzoic, 4-hydroxybenzoic, gallic, protocatechuic, salicylic, syringic and vanillic acids); hydroxycinnamic acids (caffeic, ferulic, and p-coumaric acids); and flavan-3-ols monomers [(+)-catechin and (−)-epicatechin)], procyanidin dimers, propelargonidin dimers and procyanidin trimers). To our knowledge, our findings report for the first time the presence of proanthocyanidin oligomers (Figure 1) in extracts from *P. niruri* and *S. reticulata*.

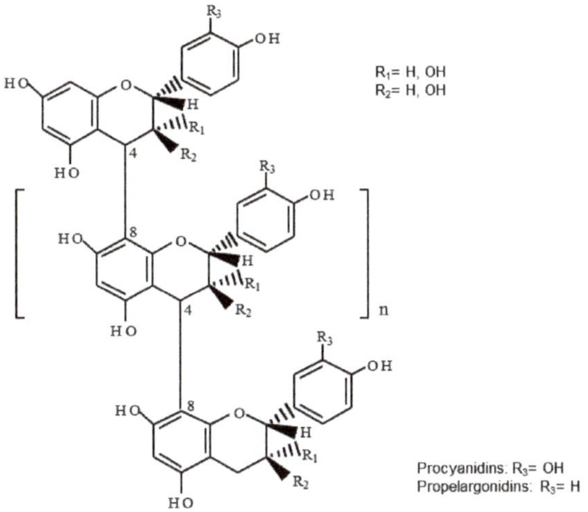

Figure 1. General chemical structure of B-type proanthocyanidins: procyanidins (composed by (epi) catechin units) and propelargonidins (composed by (epi) afzelechin units).

Table 2. Phenolic composition of *P. alliaceae*, *P. niruri* and *S. reticulata* extracts.

Compounds	P. alliaceae	P. niruri	S. reticulata
	Concentration (µg/g Extract)		
Hydroxybenzoic acids			
Benzoic acid	158.4 ± 7.5	nd	nd
Salicylic acid	175.9 ± 2.4	61.2 ± 1.3	16.7 ± 0.1
4-Hydroxybenzoic acid	28.1 ± 0.2	14.3 ± 0.1	80.9 ± 1.2
Protocatechuic acid	6.3 ± 0.2	192.4 ± 0.9	36.3 ± 0.3
Gallic acid	2.4 ± 0.0	763.3 ± 8.1	7.5 ± 0.3
Vanillic acid	12.1 ± 0.1	10.7 ± 0.2	49.6 ± 1.7
Syringic acid	9.2 ± 0.4	nd	25.0 ± 0.8
∑ Hydroxybenzoic acids	392.4	1041.9	216.0
Hydroxycinnamic acids			
p-Coumaric acid	31.6 ± 0.4	13.5 ± 0.8	39.0 ± 0.9
Caffeic acid	1.6 ± 0.0	25.0 ± 0.8	52.8 ± 0.6
Ferulic acid	47.5 ± 1.1	34.7 ± 1.1	372.5 ± 9.5
∑ Hydroxycinnamic acids	80.7	73.2	464.3
Flavan-3-ols: monomers			
(+)-Catechin	nd	186.6 ± 6.4	3.7 ± 0.1
(−)-Epicatechin	nd	331.7 ± 8.3	14.0 ± 0.1
∑ Monomers	nd	518.3	17.7
Flavan-3-ols: procyanidin dimers			
Procyanidin B1	nd	44.2 ± 1.5	nd
Procyanidin B2	nd	73.0 ± 3.2	nd
Procyanidin B3	nd	45.8 ± 1.6	nd
Procyanidin B4	nd	74.0 ± 1.5	nd
Procyanidin B5	nd	13.2 ± 0.3	nd
∑ Procyanidin dimers	nd	250.2	nd
Flavan-3-ols: propelargonidin dimers			
Propelargonidin dimer (5.03 min)	nd	nd	4.9 ± 0.1
Propelargonidin dimer (5.63 min)	nd	nd	5.9 ± 0.2
∑ Properlargonidin dimers	nd	nd	10.8
Flavan-3-ols: procyanidin trimers			
Trimer T2	nd	26.0 ± 0.6	nd
∑ Procyanidin trimers	nd	26.0	nd

nd—not detected.

In fact, five different procyanidin dimers, namely B1, B2, B3, B4 and B5 were found in *P. niruri* extract, two propelargonidin dimers (with retention times of 5.03 and 5.63 min) were detected in *S. reticulata*, whereas procyanidin trimer T was determined in *P. niruri*. On the other hand, flavan-3-ols monomers, namely (+)-catechin and (−)-epicatechin, were found in both *P. niruri* and *S. reticulata*, constituting to our knowledge, the first report for both monomers in this latter plant species. In contrast, proanthocyanidins were not detected in *P. alliaceae* extract.

However, *P. alliacea* was the richest extract in phenolic acids, particularly in hydroxybenzoic acids which constituted 82.9% of total phenolic content whereas hydroxycinnamic acids accounted for 17.1% of such content. *S. reticulata* exhibited the highest proportion of hydroxycinnamic acids (65.5%), followed by hydroxybenzoic acids (30.5%) and proanthocyanidin monomers (i.e., (+)-catechin and (−)-epicatechin) (2.5%). *S. reticulata* also showed two propelargonidin dimers (1.5%), which constitute an important group of compounds due to their particular bioactivities [6,19]. Finally, *P. niruri* presented the highest content of proanthocyanidins, representing 41.6% of total phenolic content, hydroxybenzoic acids representing the remaining 54.6% of the total content.

P. niruri also exhibited the highest structural diversity of proanthocyanidins. Besides the flavan-3-ol monomers (−)-epicatechin and (+)-catechin constituting 17.4% and 9.8%, respectively, procyanidin dimers accounted for 13.1% and procyanidin trimer T2 [(−)-epicatechin-(4β→8)-

(−)-epicatechin-(4β→8)-(+)-catechin] for 1.4%. These oligomers are important compounds because of their bioactivity as anti-inflammatory agents [20].

Procyanidin dimer B4 [(+)-(catechin-(4α→8)-(−)-epicatechin] (3.9%) was the most abundant in the extract of *P. niruri*, followed by B2 [(−)-epicatechin-(4α→8)-(−)-epicatechin] (3.8%), B3 [(+)-catechin-(4α→8)-(+)-catechin] (2.4%) and B1 [(−)-epicatechin-(4β→8)-(+)-catechin] (2.3%), while procyanidin B5 [(−)-epicatechin-(4β→6)-(−)-epicatechin] showed the lowest proportion (0.7%).

Regarding phenolic acids, salicylic acid (37.2%) and benzoic acid (33.5%) were the main components in *P. alliaceae*, followed by ferulic acid (10%), p-coumaric acid (6.7%) and 4-hydroxybenzoic acid (5.9%). In *P. niruri* extract, gallic acid is the single most abundant phenolic acid (40.0%), followed by protocatechuic acid (10.1%); whereas in *S. reticulata*, ferulic acid is the main component (52.6%) followed by 4-hydroxybenzoic acid (11.4%), caffeic acid (7.4%), vanillic acid (7.0%), p-coumaric acid (5.5%) and protocatechuic acid (5.1%).

The total UPLC contents were in agreement with the total phenolics (TPC) as measured by the Folin-Ciocalteau (Tables 1 and 2). For instance, *P. niruri* showed the higher contents by both methods (TPC of 328.8 mg GAE/g extract and UPLC value of 1909.9 µg/g extract), followed by *S. reticulata* which exhibited intermediate values (TPC of 72.3 mg GAE/g extract and UPLC value of 708.8 µg/g extract), and finally by *P. alliacea* (TPC of 13.45 mg GAE/g extract and UPLC value of 473.0 µg/g extract).

Finally, the total proanthocyanidin (PRO) contents was also in agreement with the UPLC findings. For instance, *P. niruri* showed the highest PRO value by both methods (322.93 mg CCE/g extract and UPLC value of 794.5 µg/g extract), followed by *S. reticulata* (PRO of 22.35 mg CCE/g extract and UPLC value of 25.6 µg/g extract), whereas no proanthocyanidin content were obtained by either method for *P. alliaceae*.

2.3. Antioxidant Activity

The DPPH and ORAC values are summarized in Table 3. In both antioxidant tests, values varied in the following order: *P. niruri* (IC_{50} (DPPH) 6.4 µg/mL and 6.5 mmol Trolox equivalents/g) > *S. reticulata* (IC_{50} (DPPH) 72.9 µg/mL and 2.68 mmol Trolox equivalents/mg) > *P. alliaceae* (IC_{50} (DPPH) > 1000 µg/mL and 1.32 mmol Trolox eq/mg).

Table 3. Total phenolic contents (UPLC-DAD-ESI-TQ-MS analysis) and antioxidant activity.

Sample	Total Phenolics UPLC [1] (µg/g Extract)	DPPH [2] IC50 (µg/mL)	ORAC [2] (mmol TE/mg Extract)
P. alliacea	473.0	>1000 [a]	1.32 [a] ± 0.11
P. niruri	1909.6	6.40 [b] ± 0.10	6.50 [b] ± 0.15
S. reticulata	708.8	72.90 [c] ± 1.10	2.68 [c] ± 0.28

[1] Σ = [hydroxybenzoic acids + hydroxycinnamic acids + flavan-3-ols monomers + procyanidin dimers + propelargonidin dimers + procyanidin trimers] contents (µg/g extract) (Table 2). [2] Different superscript letters in the same column indicate differences are significant at $p < 0.05$.

Regarding antioxidant values, our results are in agreement with other published results. For instance, low antioxidant values have also been reported for hydro-alcoholic extracts of *P. alliaceae* (DPPH, IC_{50} 255 µg/mL) [21]. For *S. reticulata*, values varied according to the method (TEAC, 0.03 mmoL TE/g; ORAC, 0.23 mmol TE/g) [17] and were lower when compared to ORAC values obtained herein. There is no report on DPPH for *S. reticulata*, but when comparing to other *Senna* species, IC_{50} values range between 89–424 µg/mL for ethanolic extracts of *S.gardneri*, *S. splendida*, *S.macranthera* and *S. trachypus* [22], thus a better result is obtained in our case for *S. reticulata* enriched-extract.

Finally, the higher antioxidant values found for *P. niruri* are in agreement with reports from the literature using DPPH method. For instance, studies on aqueous extracts report IC_{50} values of 15.3 µg/mL [3] and 6.85 µg/mL [18], whereas IC_{50} values of 9.1 µg/mL and 11.07 µg/mL were reported for methanolic [3] and ethanolic [18] extracts respectively. However, other studies reported lower antioxidant activity for a hydro-alcoholic extract (IC_{50} = 32.64 µg/mL) [10]. Therefore, our DPPH results for *P. niruri* are better than previous studies.

The difference in antioxidant values among *P. alliaceae*, *S. reticulata* and *P. niruri* extracts could be attributed to the differences in their phenolic content and distribution. Therefore, in order to investigate if the phenolic composition contributes to the antioxidant activity, a correlation analysis was carried out between DPPH and ORAC values, and the total phenolic contents (TPC, Table 1), as well as with the content by UPLC (Table 3). No correlation was found for DPPH, however, a significant and positive correlation was observed between TPC and ORAC values (R^2 = 0.996) and between UPLC contents and ORAC values (R^2 = 0.990). Therefore, our results are in agreement with previous studies reporting correlation between total polyphenolic contents and ORAC antioxidant activity [17].

2.4. Cytotoxicity of P. alliaceae, P. niruri and S. reticulata Extracts

Table 4 summarizes the IC_{50} values for the cytotoxicity of *P. alliaceae*, *P. niruri* and *S. reticulata* extracts on AGS human gastric adenocarcinoma, SW620 human colon adenocarcinoma and Vero monkey normal epithelial kidney cell lines. Also, Figure 2 shows dose-response curves. IC_{50} values indicate that there is no significant difference (one-way analysis of variance (ANOVA) followed by Tukey's post hoc test) between cytotoxicity values ($p < 0.05$) against gastric AGS adenocarcinoma cells and SW620 adenocarcinoma cells for *P. alliacea* and *S. reticulata* extracts. However, in the case of *P. niruri* extract, the ANOVA indicates that the cytotoxicity results are dependent on the cancer cell line.

Table 4. Cytotoxicity of extracts to gastric (AGS) and colon (SW620) adenocarcinoma cells as well as to control Vero cells.

Sample	IC_{50} (µg/mL)		
	AGS [1]	SW620 [1]	Vero [1]
P. alliacea [2]	106.5 [a,*] ± 7.9 (SI = 1.4)	108.4 [a,*] ± 4.7 (SI = 1.4)	151.5 [a,+] ± 3.3
P. niruri [2]	145.2 [b,*] ± 8.2 (SI = 2.2)	113.2 [a,+] ± 4.3 (SI = 2.8)	311.9 [b,◊] ± 24
S. reticulata [2]	208.4 [c,*] ± 8.9 (SI = 2.4)	202.5 [b,*] ± 9.1 (SI = 2.5)	>500 [c,+]

[1] Different superscript letters in the same column indicate differences are significant at $p < 0.05$. [2] Different superscript signs in the same row indicate differences are significant at $p < 0.05$.

Our results for the cytotoxicity of *P. alliacea* on both adenocarcinoma cell lines are similar to those reported for an ethanolic extract on Jurkat T cells [23]. Other studies indicate variability of results, with IC_{50} values ranging from 29 to 36 µg/mL for a C-18 chromatographic fraction of a hydro-alcoholic extract on erythroleukema and melanoma cell lines [24], and breast adenocarcinoma 4T1 cell line [25]. In contrast, no cytotoxic effect was observed for a methanolic extract on hepatic adenocarcinoma HepG2 [26].

For *S. reticulata* extracts, the results for the cytotoxicity are in agreement with a study reporting IC_{50} of 232.9 µg/mL for a methanolic extract on KB nasopharyngeal cells. However, no cytotoxicity was found in aqueous extracts [27]. Studies on other species of the genus *Senna*, including a methanolic extract of *S. covesii*, also reported no activity (IC_{50} > 800 µg/mL) on L929 tumor connective tissue cell line, HeLa cervix carcinoma and C3F6 lymphoma [28].

Finally, the cytotoxicity results obtained for the *P. niruri* extract on AGS tumor gastric cell lines and SW620 tumor colon cell lines, are similar to those found in the literature for methanolic and aqueous extracts on PC-3 prostate cancer cell lines (IC_{50} values of 117.7 and 155.0 µg/mL respectively), MeWo skin cancer cell lines (IC_{50} values of 153.3 and 193.3 µg/mL respectively) [29], and for an hydro-methanolic extract on MCF-7 human breast carcinoma cell lines (IC_{50} of 84.88 µg/mL) [30]. Similar to our cell line-dependence findings for *P. niruri* phenolic extracts, other studies performed on hydro-methanolic extracts of *P. amarus* and *P. virgatus* report lower cytotoxicity on Hep G2 hepatic carcinoma (IC_{50} > 250 µg/mL) [31].

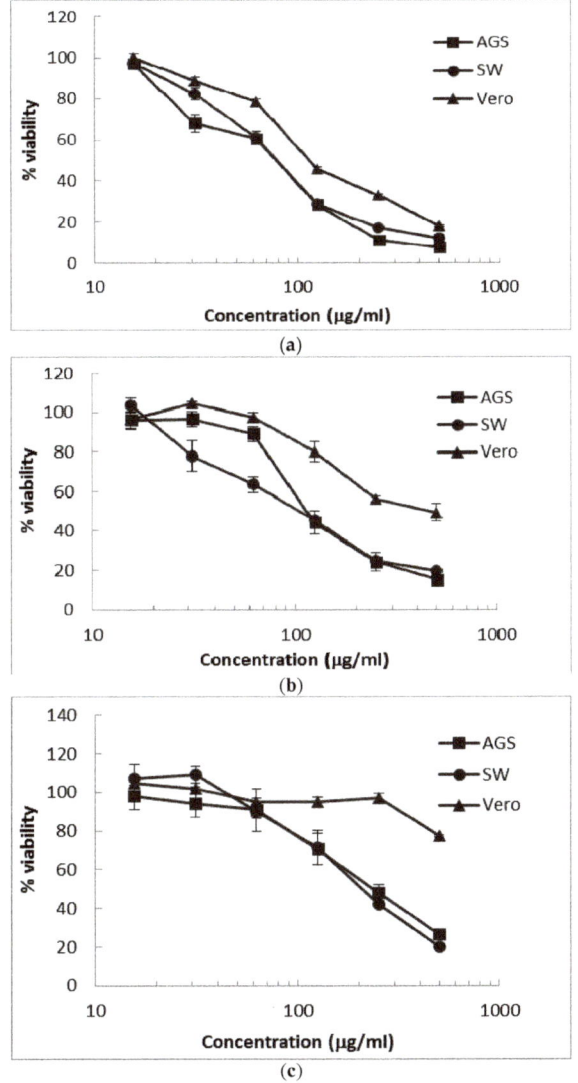

Figure 2. Cytotoxicity dose-response curves of extract treatment on tumoral cell lines. (**a**) *Petiveria alliacea*, (**b**) *P. niruri*, (**c**) *S. reticulate*. Results represent the mean ± SE of triplicates of one representative experiment of each cell line.

Concerning the selectivity of the cytotoxic activity of extracts against cancerous cell in normal Vero cells, our results indicate significant difference (ANOVA, $p < 0.05$) between IC_{50} values for both AGS and SW620 adenocarcinoma cell lines and normal Vero cells. When comparing selectivity index (SI), defined as the ratio of IC_{50} values of normal (Vero) cells to cancer cells (AGS or SW620), *P. alliaceae* extract showed the lowest values on both cell lines (SI = 1.4), while *P. niruri* extract showed the best selectivity result for SW620 colon cancer cells (SI = 2.8), which is in agreement with the selectivity range (SI ≥ 3) reported as promising for further anticancer studies [32,33]. In turn, *S. reticulata* extract exhibited a slightly lower value on SW620 colon cancer cells (SI = 2.5) and on AGS gastric cancer cells

(SI = 2.4), followed by *P. niruri* extract on SW620 cells (SI = 2.2). This variability has been reported in previous studies, showing selective cytotoxicity of a polyphenolic extract between gastric normal and cancer cells [34], whereas other studies report no selectivity on the cytotoxic effect in normal, as well as in cancer breast and prostate cell lines [35].

In fact, mechanisms related to the effect of polyphenols in cancer cells need to be elucidated, taking into account, for instance, the multiplicity of targets that can be reached [14] and the complexity of factors modulating cancer cell phenotypes [36]. However, there is enough evidence to support the potential anticancer effects of proanthocyanidins [12]. For instance, previous studies using proanthocyanidins from grape seeds showed a cytotoxic effect in cervix cancer [37], whereas cranberries polyphenols inhibited proliferation in prostate and colon cancer cells [38]. Also, other studies indicated the potential of propelargonidin dimers and procyanidin dimers and contents on higher and more selective cytotoxicity in gastric and colon cancer cell lines [6]. In addition, a report on another main component of *P. niruri* extract, gallic acid, indicates that this compound inhibits the growth of human hepatocellular carcinoma cells and induces apoptosis in these cell lines [39]. Since polyphenols could work in a synergistic manner, fractioning of *S. reticulata* and *P. niruri* extracts would contribute to further elucidate the structure-bioactivity relationship of these plant extracts.

3. Materials and Methods

3.1. Materials, Reagents and Solvents

Petiveria alliaceae, *Phyllanthus niruri* and *Senna reticulata* aerial plant material were acquired from local communities that grow and use the plants as traditional medicine, grouped as an Agricultural Producers Association (AMPALEC) in the Caribbean region. All plants were identified with the support of the Costa Rican National Herbarium and vouchers are deposited there. The material of each plant was rinsed in water and cut into small pieces. Subsequently, it was dried in a stove at 40°C until completely dry, and after being ground, it was preserved at −5 °C. Reagents such as AAPH, fluorescein, DPPH, and Trypsin-EDTA solution were provided by Sigma-Aldrich (St. Louis, MO, USA), while amphotericin B, penicillin-streptomycin, and Minimum essential Eagle's medium (MEM, 10% fetal bovine serum), were purchased from Life Technologies (Carlsbad, CA, USA). AGS human gastric adenocarcinoma, SW 620 human colorectal adenocarcinoma and Vero monkey normal epithelial kidney cell lines were obtained from American Type Culture Collection (ATCC, Rockville, MD, USA). DMSO was provided by Sigma-Aldrich (St. Louis, MO, USA), while MTBE, chloroform and methanol were purchased from Baker (Center Valley, PA, USA).

3.2. Phenolic Extracts from P. alliaceae, P. niruri and S. reticulata

The process followed for obtaining phenolic-enriched extracts was formerly described by our group [40]. Briefly, dried material from each plant was first extracted in a mixture of methyl ter-butyl ether (MTBE) and methanol (MeOH) 90:10 (v/v) at 25 °C during 30 min in ultrasound. Afterwards it was extracted for 24 h in order to obtain a non-polar extract of the samples. After filtration, the extraction was repeated once. The extracts were combined and the solvents evaporated in a rotavapor to dryness and subsequently washed with MeOH in order to extract any polyphenols. The residual plant material was extracted with MeOH at 25 °C during 30 min in ultrasound, and then extracted for 24 h. After filtration, the extraction was repeated twice. The three methanol extracts were combined with the previous MeOH washings and were evaporated in a rotavapor to dryness. Finally, the dried extract was washed with hexane, MTBE and chloroform consecutively in order to obtain a phenolic rich-extract.

3.3. Total Phenolic Content

The polyphenolic content was determined by a modification of the Folin-Ciocalteu (FC) method [41], whose reagent is composed of a mixture of phosphotungstic and phosphomolybdic acids. Each sample was dissolved in MeOH (0.1% HCl) and combined with 0.5 mL of FC reagent. Afterwards 10 mL of Na_2CO_3 (7.5%) were added and the volume was completed to 25 mL with water. Blanks were prepared in a similar way but using 0.5 mL of MeOH (0.1% HCl) instead of sample. The mixture was let standing in the dark for 1 h and then absorbance was measured at 750 nm. Values obtained were extrapolated in a gallic acid calibration curve. Total phenolic content was expressed as mg gallic acid equivalents (GAE)/g sample. Analyses were performed in triplicate.

3.4. Total Proanthocyanidin Content

The proanthocyanidin content was determined by a modification of the Bate-Smith method, which consists of the cleavage of the C–C interflavanic bond of proanthocyanidins in butanol-HCl through oxidative acid-catalysis [42]. Briefly, 0.2 mL of each sample were mixed with 20 mL of butanol/HCl (50:50) and 0.54 mM $FeSO_4$. The mixture was incubated at 90 °C for 1 h and after cooling, the volume was completed to 25 mL with butanol-HCl mixture. The absorbance was measured at 550 nm against a blank prepared in a similar way but without heating. The standard used was cyanidin chloride (which served to draw a calibration curve. Results were expressed as mg of cyanidin chloride equivalents (CAE)/g of extract.

3.5. UPLC-DAD-ESI-TQ-MS Analysis

The UPLC-MS system used to analyze the phenolic composition of the samples consisted of an UPLC coupled to an Acquity PDA eλ photodiode array detector (DAD) and an Acquity TQD tandem quadrupole mass spectrometer equipped with Z-spray electrospray interface (UPLC-DAD-ESI-TQ-MS) (Waters, Milford, MA, USA). The analyses were performed using a solution of 5 mg/mL of each extract in acetonitrile:H_2O (1:4). A volume of 2 µL was injected and a Waters® BEH C18 column (2.1 × 100 mm; 1.7 µm) was used. The elution consisted of a gradient composed of solvent A-water:acetic acid (98:2, v/v) and B-acetonitrile:acetic acid (98:2, v/v) [43]: 0–1.5 min: 0.1% B, 1.5–11.17 min: 0.1–16.3% B, 11.17–11.5 min: 16.3–18.4% B, 11.5–14 min: 18.4% B, 14–14.1 min: 18.4–99.9% B, 14.1–15.5 min: 99.9% B, 15.5–15.6 min: 0.1% B, 15.6–18 min: 0.1% B. The flow rate was 0.5 mL/min and DAD was recorded between 250–420 nm. ESI negative mode parameters included: source temperature, 130 °C; capillary voltage, 3 kV; desolvation temperature, 400 °C; cone gas (N_2) flow rate, 60 L/h; and desolvation gas (N_2) flow rate, 750 L/h. For quantification, MRM transitions were used, such as m/z 169/125 for gallic acid, m/z 289/245 for (+)-catechin and (−)-epicatechin, m/z 577/289 for procyanidin dimers, m/z 561/289 for propelargonidin dimers, and m/z 865/577 for procyanidin trimers. Commercial standards used were (−)-epicatechin, (+)-catechin, procyanidins B1 and B2. Assignment of procyanidins B3, B4 and B5 and procyanidin trimer T2 was performed with previously isolated standards and confirmed by MS/MS spectrum. Assignment of propelargonidins was performed through MS/MS spectrum at m/z 561 and quantification was performed on the calibration curve of procyanidin B1. The limit of detection (LOD) and limit of quantification (LOQ) are published elsewhere [43,44].

3.6. DPPH Radical-Scavenging Activity

A solution of 2,2-diphenyl-1-picrylhidrazyl (DPPH) (0.25 mM) was prepared using methanol as solvent. Next, 0.5 mL of this solution were mixed with 1 mL of extract at different concentrations, and incubated at 25° C in the dark for 30 min. DPPH absorbance was measured at 517 nm. Blanks were prepared for each concentration. The percentage of the radical-scavenging activity of the sample was plotted against its concentration to calculate IC_{50}, which is the amount of sample required to reach the 50% radical-scavenging activity. The samples were analyzed in triplicate.

3.7. ORAC Antioxidant Activity

Extracts (0.05 g) were mixed with 10 mL of methanol/HCl (1000:1, v/v) and sonicated for 5 min. Afterwards, the mixture was centrifuged and filtered. Fluorescein was used as fluorescence probe [45]. The reaction was performed in 75 mM phosphate buffer (pH 7.4) at 37 °C. The final assay mixture consisted of AAPH (12 mM), fluorescein (70 nM), and either Trolox (1–8 µM) or the extract at different concentrations. Fluorescence was recorded every minute for 98 min in black 96-well untreated microplates (Nunc, Denmark), using a Polarstar Galaxy plate reader (BMG Labtechnologies GmbH, Offenburg, Germany) with 485-P excitation and 520-P emission filters. Fluostar Galaxy software version 4.11-0 (BMG Labtechnologies GmbH, Offenburg, Germany) was used to measure fluorescence. Fluorescein was diluted from a stock solution (1.17 mM) in 75 mM phosphate buffer (pH 7.4), while AAPH and Trolox solutions were freshly prepared. All reaction mixtures were prepared in duplicate and three independent runs were completed for each extract. Fluorescence values obtained were normalized to the curve of the blank (no antioxidant). The area under the fluorescence decay curve (AUC) was calculated from the normalized curves, and the net AUC was then established. Subsequently, regression equation between antioxidant concentration and net AUC was obtained. Finally, ORAC value was estimated by dividing the slope of the latter equation by the slope of the Trolox line. ORAC values were expressed as mmol of Trolox equivalents (TE)/g of extract.

3.8. Evaluation of Cytotoxicity

The AGS, SW620 and Vero cells were grown in MEM (10% FBS) in the presence of glutamine (2 mmol/L), penicillin (100 IU/mL), streptomycin (100 µg/L) and amphotericin B (0.25 µg/m), at 37 °C, in a humidified atmosphere (5% CO_2) [5]. Briefly, 100 µL of 1.5×10^5 cells/mL (suspension) were seeded overnight into 96-well plates to reach 100% confluency. Subsequently, the cells were exposed for 48 h to 50 µL of extracts in concentrations varying 15–500 µg/mL in MEM (DMSO 0.1% v/v). Afterwards, the medium was eliminated, cells were washed with PBS (100 µL) and incubated with 100 µL of a MTT solution (0.5 mg/mL, final concentration) in PBS, for 2 h at 37 °C. Then, MTT was removed and the formazan crystals were dissolved in 100 µL of ethanol 95%. Absorbance was read at 570 nm in a microplate reader. DMSO was diluted in media in the same way as the extracts and incubated with the cells for 48 h to be used as control. Dose-response curves were established and IC_{50} was calculated. Extracts were tested in three independent experiments with different doses of extract analyzed in triplicate.

4. Conclusions

This study represents the first detailed MS analysis of phenolic-enriched extracts of *P. alliaceae*, *P. niruri* and *S. reticulata*, three species commonly used in traditional medicine in Costa Rica. Using different methods, including UPLC-DAD-ESI-TQ-MS techniques, results show diverse contents and distribution of 20 phenolic acids and proanthocyanidins among extracts. These findings constitute the first report on the diversity of phenolic acids in *P. alliaceae* and *S. reticulata*, and the presence of proanthocyanidins as minor components in this latter extract. In addition, five procyanidin dimers and one procyanidin trimer, were also detected for the first time in *P. niruri*. Further, significant positive correlation was found between total phenolic contents (TPC) and ORAC ($R^2 = 0.996$) antioxidant value as well as between UPLC contents and ORAC ($R^2 = 0.990$). *P. niruri* extract showed the highest antioxidant values in both DPPH and ORAC methods, as well as better cytotoxicity and selectivity on AGS gastric adenocarcinoma and SW620 colon adenocarcinoma cell lines in respect to normal cells. These results suggest that the high content of proanthocyanidins (41.6% of total phenolic content) found in *P. niruri* extracts could be responsible for the higher cytotoxicity and selectivity of the plant compared to the other two species from this study. Finally, the results show evidence of the potential health effects of *P. niruri* extracts on gut-related diseases, considering these polyphenols are metabolized by the gut [15,43]. Purification or fractioning of *P. niruri* phenolic extracts would be of interest to further evaluate their structure-bioactivity relationship.

Supplementary Materials: The following are available online at http://www.mdpi.com/2223-7747/6/4/50/s1, Figure S1: Chromatograms (UPLC-DAD) for polyphenolic compounds: (**a**) *P. alliaceae* extract, (**b**) *P. niruri* extract, (**c**) *S. reticulata* extract, Table S1: UPLC and MS/MS parameters for the identified polyphenols.

Acknowledgments: This work was partially funded by a grant from FEES-CONARE (Ref 115B0653). Authors also thank financial support from the Spanish National Research Council (CSIC), the University of Costa Rica and the Technological Institute of Costa Rica. Special thanks are due to Alonso Quesada from Costa Rican National Herbarium for his support with the vouchers.

Author Contributions: Mirtha Navarro and Maria J. Monagas participated in the conception and design of the study. Mirtha Navarro, Silvia Quesada, Gabriela Azofeifa and Diego Alvarado were involved in technical work and interpretation of data. Elizabeth Arnaez and Ileana Moreira participated in plant collection, identification and initial samples treatment. Mirtha Navarro and Maria J. Monagas drafted the manuscript that was revised and approved by all the authors.

Conflicts of Interest: The authors declare no conflicts of interest.

References

1. Arnaez, E.; Moreira, I.; Navarro, M. *Manejo Agroecológico de Nueve Especies de Plantas de uso Tradicional Cultivadas en Costa Rica*; FLACSO Latin American Institute: San Jose, Costa Rica, 2016; pp. 1–85.
2. Nunes dos Santos, R.; Vasconcelos Silva, M.G. Constituintes químicos do caule de *Senna reticulata* Willd. (Leguminoseae). *Quim. Nova* **2008**, *31*, 1979–1981. [CrossRef]
3. Harish, R.; Shivanandappa, T. Antioxidant activity and hepatoprotective potential of *Phyllanthus niruri*. *Food Chem.* **2006**, *95*, 180–185. [CrossRef]
4. Kim, S.; Kubec, R.; Musah, R.A. Antibacterial and antifungal activity of sulfur-containing compounds from *Petiveria alliacea* L. *J. Ethnopharmacol.* **2006**, *104*, 188–192. [CrossRef] [PubMed]
5. Navarro-Hoyos, M.; Lebrón-Aguilar, R.; Quintanilla-López, J.E.; Cueva, C.; Hevia, D.; Quesada, S.; Gabriela Azofeifa, G.; Moreno-Arribas, M.V.; Monagas, M.; Bartolomé, B. Proanthocyanidin Characterization and Bioactivity of Extracts from Different Parts of *Uncaria tomentosa* L. (Cat's Claw). *Antioxidants* **2017**, *6*, 12. [CrossRef] [PubMed]
6. Navarro, M.; Zamora, W.; Quesada, S.; Azofeifa, G.; Alvarado, D.; Monagas, M. Fractioning of Proanthocyanidins of *Uncaria tomentosa*. Composition and Structure-Bioactivity Relationship. *Antioxidants* **2017**, *6*, 60. [CrossRef] [PubMed]
7. Demirezer, L.O.; Karahan, N.; Ucakturk, E.; Kuruuzum-Uz, A.; Guvenalp, Z.; Kazaz, C. HPLC Fingerprinting of sennosides in laxative drugs with isolation of standard substances from some *Senna* Leaves. *Rec. Nat. Prod.* **2011**, *5*, 261–270.
8. Araújo-Luz, D.; Miranda-Pinheiro, A.; Lopes-Silva, M.; Chagas-Monteiro, M.; Prediger, R.D.; Ferraz-Maia, C.S.; Andrade-Fontes, E. Ethnobotany, phytochemistry and neuropharmacological effects of *Petiveria alliacea* L. (Phytolaccaceae): A review. *J. Ethnopharmacol.* **2016**, *185*, 182–201. [CrossRef]
9. Bagalkotkar, G.; Sagineedu, S.R.; Saad, M.S.; Stanslas, J. Phytochemicals from *Phyllanthus niruri* Linn. and their pharmacological properties: A review. *J. Pharm. Pharmacol.* **2006**, *58*, 1559–1570. [CrossRef] [PubMed]
10. Mahdi, E.; Noor, A.; Sakeena, M.; Abdullah, G.; Abdulkarim, M.; Sattar, M. Identification of phenolic compounds and assessment of in vitro antioxidants activity of 30% ethanolic extracts derived from two *Phyllanthus* species indigenous to Malaysia. *J. Pharm. Pharmacol.* **2011**, *5*, 1967–1978. [CrossRef]
11. Mediani, A.; Abas, F.; Khatib, A.; Tan, C.P.; Ismail, I.S.; Shaari, K.; Ismail, A.; Lajis, N.H. Relationship between metabolites composition and biological activities of *Phyllanthus niruri* extracts prepared by different drying methods and solvents extraction. *Plant Foods Hum. Nut.* **2015**, *70*, 184–192. [CrossRef] [PubMed]
12. Zhou, K.; Raffoul, J.J. Potential anticancer properties of grape antioxidants. *J. Oncol.* **2012**, *2012*, 803294. [CrossRef] [PubMed]
13. Considine, M.J.; Foyer, C.H. Redox Regulation of Plant Development. Antioxidants & Redox signaling. *Antioxid. Redox Signal.* **2014**, *21*, 1305–1326. [CrossRef] [PubMed]
14. Barrajon-Catalan, E.; Herranz-López, M.; Joven, J.; Segura-Carretero, A.; Alonso-Villaverde, C.; Menéndez, J.A.; Micol, V. *Oxidative Stress and Inflammation in Non-Communicable Diseases—Molecular Mechanisms and Perspectives in Therapeutics*, 1st ed.; Camps, J., Ed.; Springer International Publishing: Cham, Switzerland, 2014; pp. 141–159. ISBN 978-3-319-55857-8.
15. Monagas, M.; Urpi-Sarda, M.; Sanchez-Patán, F.; Llorach, R.; Garrido, I.; Gómez-Cordoves, C.; Andres-Lacueva, C.; Bartolome, B. Insights into the metabolism and microbial biotransformation of dietary flavan-3-ols and the bioactivity of their metabolites. *Food Funct.* **2010**, *1*, 233–253. [CrossRef] [PubMed]

16. Zaa, C.; Valdivia, M.; Marcelo, A. The anti-inflammatory and antioxidant effect of hydroalcoholic extract of *Petiveria alliacea*. *Rev. Peru. Biol.* **2012**, *19*, 329–334.
17. Lizcano, L.J.; Bakkali, F.; Ruiz-Larrea, B.; Ruiz-Sanz, J.I. Antioxidant activity and polyphenol content of aqueous extracts from Colombia Amazonian plants with medicinal use. *Food Chem.* **2010**, *119*, 1566–1570. [CrossRef]
18. Amin, Z.A.; Abdulla, M.A.; Ali, H.M.; Alshawsh, M.A.; Qadir, S.W. Assessment of In vitro antioxidant, antibacterial and immune activation potentials of aqueous and ethanol extracts of *Phyllanthus niruri*. *J. Sci. Food Agric.* **2012**, *92*, 1874–1877. [CrossRef] [PubMed]
19. Chang, E.; Lee, W.; Cho, S.; Choi, S. Proliferative effects of Flavan3-ols and Propelargonidins from Rhizomes of *Drynaria fortune* on MCF-7 and Osteoblastic Cells. *Arch. Pharm. Res.* **2003**, *26*, 620–630. [CrossRef] [PubMed]
20. Montagut, G.; Baiges, I.; Valls, J.; Terra, X.; Bas, J.; Vitrac, X.; Richard, T.; Mérillon, J.; Arola, L.; Blay, M.; et al. A trimer plus a dimer-gallate reproduce the bioactivity described for an extract of grape see procyanidins. *Food Chem.* **2009**, *116*, 265–270. [CrossRef]
21. Ramos, A.; Visozo, A.; Piloto, J.; García, A.; Rodríguez, C.A.; Rivero, R. Screening of antimutagenicity via antioxidant activity in Cuban medicinal plants. *J. Etnopharmacol.* **2003**, *87*, 241–246. [CrossRef]
22. Silva, G.A.; Monteiro, J.A.; Ferreira, E.B.; Fernandes, M.I.B.; Pessoa, C.; Sampaio, C.G.; Silva, M.G.V. Total phenolic content, antioxidant and anticancer activities of four species of *Senna* Mill. From northeast Brazil. *Int. J. Pharm. Pharm. Sci.* **2014**, *6*, 199–202.
23. Schmidt, C.; Fronza, M.; Goettert, M.; Geller, F.; Luik, S.; Flores, E.M.M.; Bittencourt, C.F.; Zanetti, G.D.; Heinzmann, B.M.; Laufer, S.; Merfort, I. Biological studies on Brazilian plants used in wound healing. *J. Etnopharmacol.* **2009**, *122*, 523–532. [CrossRef] [PubMed]
24. Urueña, C.; Cifuentes, C.; Castañeda, D.; Arango, A.; Kaur, P.; Asea, A.; Fiorentino, S. *Petiveria alliacea* extracts uses multiple mechanisms to inhibit growth of human and mouse tumor cells. *BMC Complement. Altern. Med.* **2008**, *8*, 60. [CrossRef]
25. Hernandez, J.F.; Urueña, C.P.; Cifuentes, M.C.; Sandoval, T.A.; Fiorentino, S. A *Petiveria alliacea* standardized fraction induces breast adenocarcinoma cell death modulating glycolytic metabolism. *J. Ethnopharmacol.* **2014**, *153*, 641–649. [CrossRef] [PubMed]
26. Ruffa, M.J.; Ferrar, G.; Wagner, M.L.; Calcagno, M.L.; Campos, R.H.; Cavallaro, L. Cytotoxic effect of argentine medicinal plant extracts on human hepatocellular carcinoma cell line. *J. Etnopharmacol.* **2002**, *79*, 335–339. [CrossRef]
27. Camacho, M.D.; Phillipson, J.D.; Croft, S.L.; Solis, P.N.; Marshall, S.J.; Ghazanfar, S.A. Screening of plants extracts for antiprotozoal and cytotoxic activities. *J. Ethnopharmacol.* **2003**, *89*, 185–191. [CrossRef]
28. Jiménez-Estrada, M.; Velásquez-Contreras, C.; Garibay-Escobar, A.; Sierras-Canchola, D.; Lapisco-Vásquez, R.; Ortiz-Sandoval, C.; Burgos-Hernández, A.; Robles-Zepeda, R. In vitro antioxidant and antiproliferative activities of plants of the ethnopharmacopeia from northwest of Mexico. *BMC Complement. Altern. Med.* **2013**, *13*, 12–19. [CrossRef] [PubMed]
29. Tang, Y.Q.; Jaganath, I.B.; Sekaran, S.D. *Phyllanthus* spp. induces selective growth inhibition of PC-3 and MeWo human cancer cells through modulation of cell cycle and induction of apoptosis. *PLoS ONE* **2010**, *5*, e12644. [CrossRef]
30. Jose, J.; Sudhakaran, S.; Sumesh-Kumar, T.M.; Jayadevi-Variyar, E.; Jayaraman, S. A comparative evaluation of anticancer activities of flavonoids isolated from *Mimosa pudica*, *Aloe vera* and *Phyllanthus niruri* against human breast carcinoma cell line (MCF-7) using MTT assay. *Int. J. Pharm. Pharm. Sci.* **2014**, *6*, 319–322. [CrossRef]
31. Poompachee, K.; Chudapongse, N. Comparison of the antioxidant and cytotoxic activities of *Phyllanthus virgatus* and *Phyllanthus amarus* extracts. *Med. Princ. Pract.* **2012**, *21*, 24–29. [CrossRef] [PubMed]
32. Mahavorasirikul, W.; Wiratchanee, M.; Vithoon, V.; Wanna, C.; Arunporn, I.; Kesara, N. Cytotoxic activity of Thai medicinal plants against human cholangiocarcinoma, laryngeal and hepatocarcinoma cells in vitro. *BMC Complement. Altern. Med.* **2010**, *10*, 1–8. [CrossRef] [PubMed]
33. Ramasamy, S.; Wahab, N.A.; Abidin, N.Z.; Manickam, S.; Zakaria, Z. Growth Inhibition of Human Gynecologic and Colon Cancer Cells by *Phyllanthus watsonii* through Apoptosis Induction. *PLoS ONE* **2012**, *7*, e34793. [CrossRef] [PubMed]

34. Ye, X.; Krohn, R.L.; Liu, W.; Joshi, S.S.; Kuszynski, C.A.; McGinn, T.R.; Bagchi, M.; Preuss, H.G.; Stohs, S.J.; Bagchi, D. The cytotoxic effects of a novel IH636 grape seed proanthocyanidin extract on cultured human cancer cells. *Mol. Cell. Biochem.* **1999**, *196*, 99–108. [CrossRef] [PubMed]
35. Weaver, J.; Briscoe, T.; Hou, M.; Goodman, C.; Kata, S.; Ross, H.; McDougall, G.; Stewart, D.; Riches, A. Strawberry polyphenols are equally cytotoxic to tumourigenic and normal human breast and prostate cell lines. *Int. J. Oncol.* **2009**, *34*, 777–786. [CrossRef] [PubMed]
36. Stoner, G.; Wang, L.; Casto, B. Laboratory and clinical studies of cancer chemoprevention by antioxidants in berries. *Carcinogenesis* **2008**, *29*, 1665–1674. [CrossRef] [PubMed]
37. Chen, Q.; Liu, X.F.; Zheng, P.S. Grape seed proanthocyanidins (GSPs) inhibit the growth of cervical cancer by inducing apoptosis mediated by the mitochondrial pathway. *PLoS ONE* **2014**, *9*, e107045. [CrossRef] [PubMed]
38. Seeram, N.; Lee, R.; Heber, D. Bioavailability of ellagic acid in human plasma after consumption of ellagitannins from pomegranate (*Punica granatum* L.) juice. *Clin. Chim. Acta* **2004**, *348*, 63–68. [CrossRef] [PubMed]
39. Sun, G.; Zhang, S.; Xie, Y.; Zhang, Z.; Zhao, W. Gallic acid as a selective anticancer agent that induces apoptosis in SMMC-7721 human hepatocellular carcinoma cells. *Oncol. Lett.* **2016**, *11*, 150–158. [CrossRef] [PubMed]
40. Navarro Hoyos, M.; Sánchez-Patán, F.; Murillo Masis, R.; Martín-Álvarez, P.J.; Zamora Ramirez, W.; Monagas, M.J.; Bartolomé, B. Phenolic Assesment of *Uncaria tomentosa* L. (Cat's Claw): Leaves, Stem, Bark and Wood Extracts. *Molecules* **2015**, *20*, 22703–22717. [CrossRef] [PubMed]
41. Singleton, V.; Rossi, J. Colorimetry of total phenolics with phosphomolybdic-phosphotungstic acid reagents. *Am. J. Enol. Vitic.* **1965**, *16*, 144–158.
42. Ribéreau-Gayon, P.; Stonestreet, E. Dósage des tannins du vin rouges et determination du leur structure. *Chem. Anal.* **1966**, *48*, 188–196.
43. Sánchez-Patan, F.; Monagas, M.; Moreno-Arribas, M.V.; Bartolome, B. Determination of microbial phenolic acids in human faeces by UPLC-ESI-TQ MS. *J. Agric. Food Chem.* **2011**, *59*, 2241–2247. [CrossRef] [PubMed]
44. Sánchez-Patan, F.; Cueva, C.; Monagas, M.; Walton, G.E.; Gibson, G.R.; Martin-Alvarez, P.J.; Moreno-Arribas, M.V.; Bartolome, B. Gut microbial catabolism of grape seed flavan-3-ols by human faecal microbiota. Targetted analysis of precursor compounds, intermediate metabolites and end-products. *Food Chem.* **2012**, *131*, 337–347. [CrossRef]
45. Davalos, A.; Gomez-Cordoves, C.; Bartolome, B. Extending applicability of the oxygen radical absorbance capacity (ORAC-Fluorescein) assay. *J. Agric. Food Chem.* **2004**, *52*, 48–54. [CrossRef] [PubMed]

 © 2017 by the authors. Licensee MDPI, Basel, Switzerland. This article is an open access article distributed under the terms and conditions of the Creative Commons Attribution (CC BY) license (http://creativecommons.org/licenses/by/4.0/).

Article

Comparative Evaluation of Polyphenol Contents and Antioxidant Activities between Ethanol Extracts of *Vitex negundo* and *Vitex trifolia* L. Leaves by Different Methods

Sarla Saklani [1], Abhay Prakash Mishra [1,*], Harish Chandra [2,*], Maria Stefanova Atanassova [3], Milan Stankovic [4], Bhawana Sati [5], Mohammad Ali Shariati [6,7], Manisha Nigam [8], Mohammad Usman Khan [9], Sergey Plygun [7], Hicham Elmsellem [10] and Hafiz Ansar Rasul Suleria [11,12]

1. Department of Pharmaceutical Chemistry, H. N. B. Garhwal (A Central) University, Srinagar Garhwal, Uttarakhand 246174, India; pharmachemhnbgu@gmail.com
2. High Altitude Plant Physiology Research Centre, H. N. B. Garhwal (A Central) University, Srinagar Garhwal, Uttarakhand 246174, India
3. Scientific Consulting, Chemical Engineering, University of Chemical Technology and Metallurgy (UCTM), Sofia 1734, Bulgaria; msatanassova@abv.bg
4. Department of Biology and Ecology, Faculty of Science, University of Kragujevac, Radoja Domanovića No. 12, Kragujevac 34000, Serbia; mstankovic@kg.ac.rs
5. Department of Pharmacy, Banasthali Vidyapeeth, Rajasthan 304022, India; bhawana.sati@gmail.com
6. Department of Scientific affairs, Kurks State Agricultural Academy, Kurks 305021, Russia; shariatymohammadali@gmail.com
7. All-Russian Research Institute of Phytopathology, Moscow, Bolshie Vyazemy 143050, Russia; rjoas@yandex.ru
8. Department of Biochemistry, H. N. B. Garhwal (A Central) University, Srinagar Garhwal, Uttarakhand 246174, India; anandmanisha23@gmail.com
9. Department of Biological Systems Engineering, Bio Product Sciences and Engineering Laboratory (BSEL), Washington State University, 2710 Crimson Way, Richland, WA 99354-1671, USA; engineer_usman_khan@yahoo.com
10. Laboratoire de Chimie Analytique Appliquée, Matériaux et Environnement (LC2AME), Faculté des Sciences, B.P. 717, Oujda 60000, Morocco; h.elmsellem@gmail.com
11. UQ Diamantina Institute, Translational Research Institute, Faculty of Medicine, The University of Queensland, 37 Kent Street Woolloongabba, Brisbane, QLD 4102, Australia; hafiz.suleria@uqconnect.edu.au
12. Department of Food, Nutrition, Dietetics & Health, Kansas State University, Manhattan, KS 66506, USA
* Correspondence: abhaypharmachemhnbgu@gmail.com (A.P.M.); hreesh5@gmail.com (H.C.); Tel.: +91-9452002557 (A.P.M.); +91-9456567555 (H.C.)

Received: 30 July 2017; Accepted: 25 September 2017; Published: 27 September 2017

Abstract: The in vitro antioxidant potential assay between ethanolic extracts of two species from the genus *Vitex* (*Vitex negundo* L. and *Vitex trifolia* L.) belonging to the Lamiaceae family were evaluated. The antioxidant properties of different extracts prepared from both plant species were evaluated by different methods. DPPH scavenging, nitric oxide scavenging, and β-carotene-linoleic acid and ferrous ion chelation methods were applied. The antioxidant activities of these two species were compared to standard antioxidants such as butylated hydroxytoluene (BHT), ascorbic acid, and Ethylene diamine tetra acetic acid (EDTA). Both species of *Vitex* showed significant antioxidant activity in all of the tested methods. As compared to *V. trifolia* L. (60.87–89.99%; 40.0–226.7 µg/mL), *V. negundo* has been found to hold higher antioxidant activity (62.6–94.22%; IC_{50} = 23.5–208.3 µg/mL) in all assays. In accordance with antioxidant activity, total polyphenol contents in *V. negundo* possessed greater phenolic (89.71 mg GAE/g dry weight of extract) and flavonoid content (63.11 mg QE/g dry weight of extract) as compared to that of *V. trifolia* (77.20 mg GAE/g and

57.41 mg QE/g dry weight of extract respectively). Our study revealed the significant correlation between the antioxidant activity and total phenolic and flavonoid contents of both plant species.

Keywords: phytochemicals; Nirgundi; chaste tree; antibacterial; free radical; scavenger; oxidative stress

1. Introduction

The genus *Vitex* (Lamiaceae) contains 270 species with diverse medicinal active constituents and properties. These species are predominantly trees and shrubs, found in tropical and subtropical regions. Other species hove fruits, seeds, and roots that are also important in traditional medicines. Some of *Vitex* species, including *Vitex negundo*, *Vitex glabrata*, *Vitex leucoxylon*, *Vitex penduncularis*, *Vitex pinnata*, and *Vitex trifolia*, are found in India [1]. These species are commonly used in traditional medicine to treat a wide range of ailments, such as depression, venereal diseases, asthma, allergy, skin diseases, snakebite, and body pains [2,3]. Many plants of the genus *Vitex* are used for their interesting biological activities, such as treatment of cough, wound healing, larvicidal, anti-HIV, anticancer, and trypanocidal [4–6].

V. negundo L. (*Lamiaceae*), known as Chinese chastetree, Sambhalu, or Nirgundi (in Ayurveda), grows gregariously in wastelands and is cultivated as a hedge-plant. The leaf extract of *V. negundo* is generally used as a grain preservation material to protect pulses against insects [7]. It contains many polyphenolic compounds, terpenoids, glycosidic iridoids, and alkaloids. Among its chemical constituents, it has several flavonoids such as casticin, orientin, isoorientin, luteolin, lutein-7-O-glucoside, corymbosin, gardenins A and B, 3-Odesmethylartemetin, 5-Odesmethylnobiletin, and 3',4',5,5',6,7,8-heptamethyoxyflavone. Interestingly, it is used conventionally for the treatment of eye-disease, toothache, inflammation, leukoderma, enlargement of the spleen, skin-ulcers, in catarrhal fever, rheumatoid arthritis, gonorrhea, and bronchitis. Moreover, it is also used as a tonic, vermifuge, lactagogue, emmenagogue, antioxidant, antibacterial, antipyretic, and antihistaminic agent. The oil of *V. negundo* has beneficial effect when applied to sinuses and scrofulous sores. Its extract has been reported to possess antitumor activity against Dalton ascites tumor cells in Swiss albino mice [8]. Lagundi tablets prepared from leaves of *V. negundo*, and commercially marketed as Ascof® (Rose Pharmacy, Mandaue, Philippines) are prescribed for the relief of mild to moderate bronchial asthma and cough [9].

V. trifolia L. (*Lamiaceae*) is commonly known as a chaste tree. It is a deciduous shrub found in tropical and subtropical regions. The plant species is native to Southeast Asia, Micronesia, Australia, and East Africa. This plant can be commonly found along the banks of water bodies like canals, rivers, and ponds. It is known to produce a variety of diterpenoids that display antioxidant, cytotoxic, and trypanosidal activities [10]. *V. trifolia* is conventionally consumed to improve memory, relieve pain, remove the bad taste in mouth, cure fever, and as a diabetes, amenorrhea, and cancer treatment. The flowers of *V. trifolia* mixed with honey are used to treat fever accompanied by vomiting and severe thirst. Additionally, it is used as an antibacterial, a sedative, and to treat rheumatism and the common cold in Asian countries [11–13].

Although all parts of *Vitex* species are used as medicament in different indigenous systems of medicine, the leaves are most potent for medicinal use. Hence, the basic aim of the present study was executed to explore the comparative account of the total polyphenolic contents and as well as the antioxidant activity for ethanolic extracts of *V. negundo* and *V. trifolia* (leaves) using a plethora of antioxidant assays.

2. Results

Phytochemical screening of the leaf extracts of *V. negundo* and *V. trifolia* revealed the presence of different phytochemicals, as summarized in Table 1. For both plants, a range of extracting solvents (petroleum ether, chloroform, ethyl acetate, ethanol, and water) were used. Out of tested solvents, ethanol was proven to be excellent for the extraction of phytochemicals as evident from the results (Table 1). Alkaloids were not detected in the petroleum ether extract of *V. negundo*. Water extract showed the presence of carbohydrates and tannins in both plants. Saponin was detected in ethanol, water, and petroleum ether extracts of *V. negundo*. However, in the case of *V. trifolia*, saponin was only detected in ethanol extracts. *V. trifolia* leave extract had proven to be a good source of flavonoids. The qualitative chemical screening test confirmed that the ethanol extract showed maximum phytoconstituents including flavonoids mostly responsible for antioxidant activity in the *V. negundo*.

Table 1. Qualitative screening of phytochemicals for selected plant extracts under different solvent systems.

Plant Part	Extract	‡ Carbo.	Alka.	Sapo.	Flav.	Phe.	Tan.	Terp.
Vitex negundo Leaves	Pet. Ether	−	−	−	+	−	−	+
	Chloroform	−	+	−	−	+	+	−
	Ethyl Acetate	−	−	−	+	+	+	+
	Ethanol	+	−	+	+	+	+	+
	Water	+	−	+	−	−	+	−
Vitex trifolia Leaves	Pet. Ether	−	−	+	−	−	−	−
	Chloroform	−	+	−	−	−	−	−
	Ethyl Acetate	−	+	−	+	+	+	+
	Ethanol	+	+	+	+	+	+	+
	Water	+	−	−	+	−	+	−

‡ Carbo. = Carbohydrates, Alka. = Alkaloids, Sapo. = Saponins, Flav. = Flavonoids, Phe. = Phenols, Tan. = Tannins, Terp. = Terpenoids; (+) = Presence, (−) = Absent.

2.1. Total Phenolic and Total Flavonoid Contents (TPC and TFC)

TPC in the ethanol extract of *V. negundo* and *V. trifolia* leaves extracts using the Folin-Ciocalteu reagent is expressed in terms of gallic acid equivalent or GAE (the standard curve equation: $y = 6.019x - 0.0186$, $r^2 = 0.989$) as mg GAE/g of extract. The concentrations of flavonoids are expressed in terms of quercetin equivalent (QE) (the standard curve equation: $y = 15.121x - 0.0472$, $r^2 = 0.986$), as mg QE/g of extract (Table 2). The ethanol extract of *V. negundo* leaves exhibited the higher content of total phenolics (89.71 mg GAE/g) and total flavonoids (63.11 mg QE/g) as compared to the ethanolic extract of *V. trifolia*, which have TPC (77.20 mg GAE/g) and TFC (57.41 mg QE/g).

Table 2. Secondary metabolite contents in *V. negundo* and *V. trifolia* leaves.

Parameter Analysed	*V. negundo*	*V. trifolia*
Total Phenolic Contents (mg GAE/g dry weight of extract)	89.71 ± 0.14	77.20 ± 0.22
Total Flavonoid Contents (mg QE/g dry weight of extract)	63.11 ± 0.31	57.41 ± 0.37

Each value is the average of three analyses ± standard deviation.

2.2. Antioxidant Activity

The antioxidant activity of ethanol extracts from both plant species is expressed in terms of percentage of inhibition (%) and IC_{50} values (µg/mL).

2.2.1. DPPH Free Radical-Scavenging Assay

To evaluate the scavenging effect of DPPH·+ in ethanol extract of *V. negundo* and *V. trifolia* leaves, DPPH·+ inhibition was investigated, and these results were shown as relative activities against control. The extracts constituted from the leaves of *V. negundo* and *V. trifolia* showed different antioxidant potential. Crude ethanol extract of the leaves of *V. negundo* and *V. trifolia* leaves and ascorbic acid (IC_{50} = 40.00 µg/mL) showed to have a potent antioxidant activity. A lower IC_{50} value indicates higher antioxidant potential. Both extracts have been shown to have significant DPPH radical scavenging activity (Figure 1). The *V. negundo* leaf ethanol extract was found to be the richest source of antioxidants among the samples investigated. The IC$_{50}$ value of the *V. negundo* leaf ethanol extract was found to be 77.09% (IC$_{50}$ = 70.20 µg/mL), which is lower than that of the *V. trifolia* leaf, which has a scavenging activity of 74.45% (IC$_{50}$ = 81.72 µg/mL). In addition, we compared the antioxidant potential of our samples with that of vitamin C (ascorbic acid). The same procedure was applied to vitamin C, and its IC$_{50}$ value was determined. Despite the scavenging activity of ascorbic acid (96.88%), a well-known antioxidant was fairly more prominent than that of extracts.

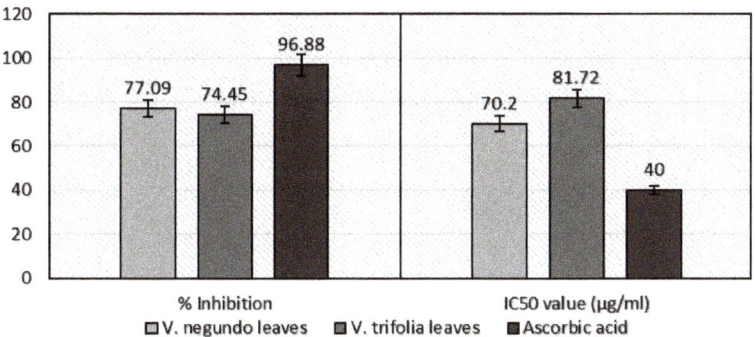

Figure 1. 2,2-Diphenyl-1-picrylhydrazyl (DPPH) assay of *V. negundo* and *V. trifolia* (leaves) ethanol extract.

2.2.2. β-Carotene-Linoleic Acid Assay

The inhibition extent of lipid oxidation by extracts (*V. negundo* and *V. trifolia* leaves) when compared to BHT showed significant activity (Figure 2). The higher antioxidant activity was observed in *V. negundo* leaves (68.66%) as compared to *V. trifolia* leaves (62.74%), with an IC$_{50}$ value of 208.3 µg/mL and 226.7 µg/mL, respectively. The antioxidant capacity of standard BHT was 92.19% at 195.74 µg/mL IC$_{50}$ value.

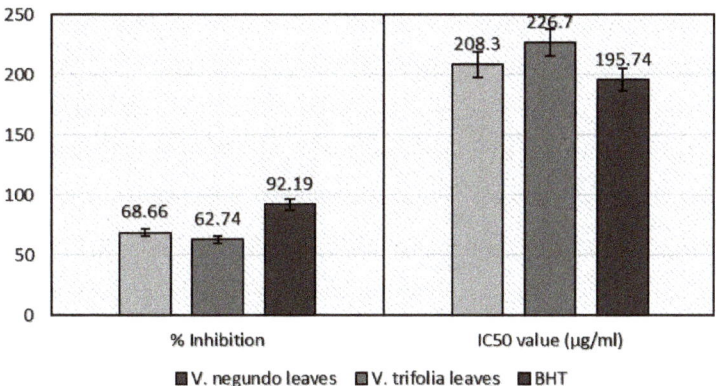

Figure 2. β-Carotene assay of *V. negundo* and *V. trifolia* (leaves) ethanol extract.

2.2.3. Nitric Oxide Radical Scavenging Assay

The current study proved that the extracts studied had comparable nitric oxide scavenging activity with the standard BHT (Figure 3). It was observed that the scavenging percentage of nitric oxide in the *V. negundo* leaves was 62.60% with an IC$_{50}$ value of 83.15 µg/mL, whereas in *V. trifolia* leaves was 60.87% over 92.78 µg/mL IC$_{50}$ value. An amount of 13.04 µg/mL BHT was needed to obtain 50% inhibition. The IC$_{50}$ value of the composed extracts was greater than that of the standard, which showed lower activity of extracts than the standard. Interestingly, in this assay, both plant extracts exhibited nitric oxide scavenging activity, which was moderately similar to each in terms of percentage.

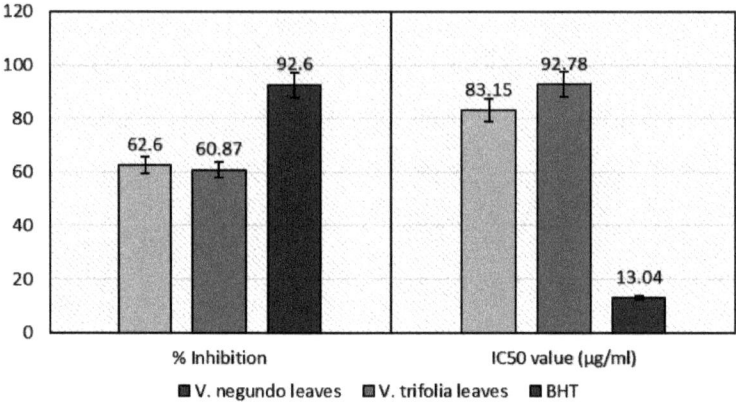

Figure 3. Nitric oxide (NO) scavenging assay of *V. negundo* and *V. trifolia* (leaves) ethanol extract.

2.2.4. Ferrous Ion Chelating Activity

EDTA is a well-known metal ion chelator, therefore, the chelating effect of *V. negundo* and *V. trifolia* leaves extracts was compared with it. Both extracts interfered with the formation of ferrous and ferrozine complex, suggesting that they had chelating activity. The strongest iron chelating activity of the extracts was noticed as 94.22% (IC$_{50}$ = 23.5 µg/mL) in *V. negundo* and 89.97% (IC$_{50}$ = 40.0 µg/mL) in *V. trifolia,* when compared with EDTA (98.78%, IC$_{50}$ = 6.03 µg/mL), as shown in Figure 4.

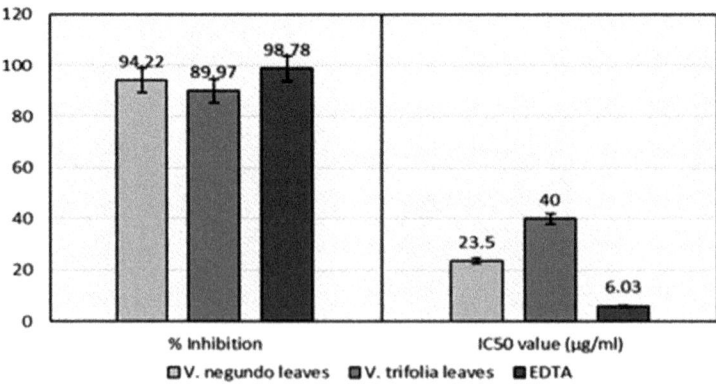

Figure 4. Ferrous ion chelating assay of *V. negundo* and *V. trifolia* (leaves) ethanol extract.

3. Discussion

Vitex species, an abundant herb/tree in the Indian subcontinent, possess great medicinal value. Therefore, it can be exploited for many herbal drugs therapeutics. The present study was carried out to compare the antioxidant potential of both species i.e., *V. negundo* and *V. trifolia*. Our study suggests that both plants have significant antioxidant potential, and both species can be exploited equally for preparation of Ayurvedic drugs or herbal drugs. Antioxidant properties imparting in any herbal preparation can be prescribed for premature skin aging for skin cancer, improving the immune system, removing free radicals from the body, eye health, troubles of memory, and so forth.

DPPH radical scavenging, β-carotene-linoleic acid assay, nitric oxide radical scavenging assay, Ferrous ion chelating activity, determination of total phenolic compounds, and determination of total flavonoid content of the ethanol extracts of the *V. negundo* and *V. trifolia* were examined in this study.

Significant variations were found in total polyphenolic contents of both *Vitex* species. The favorable properties resulting from the presence of TPC in the target species have been ascribed to their antioxidant activity. TPC may contribute directly to the antioxidative action mainly due to their redox properties, which can play an important role in absorbing and neutralizing free radicals, quenching singlet and triplet oxygen, or decomposing peroxides. Flavonoids are the most important natural phenolics and have a large number of biological and chemical properties, including radical scavenging. It has been suggested that up to 1.0 g polyphenolic compounds (from a diet rich in fruits and vegetables) ingested daily have inhibitory effects on mutagenesis and carcinogenesis in humans [14,15]. The presence of flavonoid, phenol, terpenoids, anthraquinones, carbohydrates, and steroids were also previously reported in *Vitex negundo* [16]. Total phenolic content of *V. negundo* was estimated to be 261 mg gram equivalents of catechol of *Vitex negundo*, and the total flavonoid content was expressed in Quercetin gram equivalents of 278 mg equivalents per gram of the extract of *V. negundo* [17]. The presence of phenolic compound in both species contributes to its antioxidant properties. The mechanism of phenolic content for imparting antioxidant properties was due to its neutralizing lipid free radicals and preventing decomposition of hydroperoxides into free radicals [18].

The results from the antioxidant analyses showed that both tested extracts might reach some confident level act as radical scavengers. The antioxidant activity of *V. negundo* and *V. trifolia* leaves extracts were determined using ethanol DPPH solution. This is a widely accepted technique for estimating free radical-scavenging activities of antioxidants. DPPH is a stable nitrogen-centered free radical, the color of which changes from violet to yellow upon reduction of ethanol solution of colored free radical DPPH by either the process of hydrogen or electron donation. The scavenging activity was measured as the decrease in absorbance of the samples versus DPPH standard solution [19,20]. In contrast to the lower IC_{50} DPPH value of methanolic, chloroform, ethyl acetate, and aqueous extract of *V. negundo* and *V. trifolia,* the ethanolic extract of both plants have higher IC_{50} DPPH value. However, in case of hexane extract IC_{50} DPPH, the value is slightly higher as compared to the ethanolic extract of our plants [21].

Antioxidant potential needs to be supported by diverse array of assays so as to recognize the distinctive biological activities of the complex assortment of secondary metabolites [14]. Therefore, the antioxidant activity of the extracts was tested by using the other three complementary systems, β-carotene-linoleic acid, nitric oxide radical scavenging and ferrous ion chelating activity.

In the β-carotene-linoleic acid assay, linoleic acid produces hydro-peroxides as free radicals and attacks the β-carotene molecules, resulting in the reduction in the absorbance at 470 nm. β-carotene in the systems undergoes rapid discoloration in the absence of antioxidant and vice versa in its presence. The presence of different antioxidants can delay the extent of β-carotene bleaching by neutralizing the linoleate free radical and other free radicals formed in the system. Thus, the degradation rate of β-carotene-linoleate depends on the antioxidant activity of the extracts. According to Boumerfeg et al. [22], the test of linoleic acid oxidation inhibition coupled with β-carotene, appears very useful as a mimetic model of lipid peroxidation in biological membranes. β-carotene-linoleic acid assay determines the inhibition ratios of linoleic acid oxidation as further methods to confirm the anti-lipoperoxidation effects of *V. negundo* and *V. trifolia*. Lower absorbance indicates a higher level of antioxidant activity. Interestingly, in this assay, both plant extracts exhibited nitric oxide scavenging activity, which was moderately similar to each other in respect of percentage.

In the in vitro nitric oxide radical scavenging assay, nitric oxide, which responds to macromolecules, may induce inflammation. It has been stated to show a key role in numerous inflammatory processes such as carcinomas, muscle sclerosis, arthritis, and ulcerative colitis [23]. The NO scavenging effect of ethanol extracts is shown in Figure 3. It was observed that the scavenging percentage of nitric oxide was higher in the ethanol extract of *V. negundo* leaves (62.60%) and lower in *V. trifolia* leaf extract (60.87%). So, it can be interpreted that the *V. negundo* leaves have greater potential to counteract the harmful effects of NO and other reactive nitrogen species than *V. trifolia* leaves. Therefore, *V. negundo* leaves extract showed a potent scavenger of nitric oxide and thus confirmed that the plant can also be used for the treatment of anti-inflammatory diseases caused by nitric oxide formation.

Ferrous ion chelating activity is characterized by the reduction of Fe^{3+} to Fe^{2+}. The method is used to assess the effectiveness of antioxidants for their electron transfer ability. An escalation in absorbance of the reaction mixture that changes color from yellow to blue indicates an increase in the reducing capacity due to increase in the formation of the complex. Unlike the DPPH assay, the iron chelating ability of Vitex extracts is more pronounced. Figure 4 shows the reductive proficiencies of ethanol extracts of *V. negundo* and *V. trifolia* leaves compared to EDTA. It can be perceived in Figure 4 that both ethanol extracts possess certain reducing capacity, but they were less effective than EDTA. *V. negundo* possesses better reducing power, in all applied concentrations, compared to *V. trifolia* [12].

Sengul et al. [24] reported the antioxidant capacity observed, on the one hand, was not solely from the phenolic contents, but could be due to the presence of some other phytochemicals, such as ascorbic acid, tocopherol, and pigments, as well as the synergistic effects among them, which also contribute to the total antioxidant capacity. On the other hand, total phenolic contents determined according to the Folin-Ciocalteu method is not an absolute measurement of a number of phenolic materials. Different types of polyphenolic compounds have different antioxidant activities, which is dependent on their structure. The extracts possibly contain different types of phenolic compounds, which have different antioxidant capacities.

4. Materials and Methods

4.1. Plant Material

Leaves of *V. negundo* and *V. trifolia* were collected from Lucknow, Uttar Pradesh, India in September 2014. Identification and authentication were carried out by the Botany Department, and the voucher specimens (PCHNBGU/2014/56 and PCHNBGU/2014/57) were deposited in the herbarium of our Pharmaceutical Chemistry Department, H. N. B. Garhwal (A Central) University, Srinagar Garhwal, Uttarakhand, India.

4.2. Chemicals and Reagents

2,2-Diphenyl-1-picrylhydrazyl (DPPH) and quercetin were purchased from Sigma Chemical Co. (St. Louis, MO, USA), while Ascorbic acid, Folin-Ciocalteu (FC) reagent, and ethanol were purchased from Thermo Fisher Scientific India Pvt. Ltd. Powai, Mumbai, India. Gallic acid, anhydrous sodium carbonate, aluminum chloride, and potassium acetate were purchased from Sisco Research Laboratory Pvt. Ltd. (Mumbai, India). All other chemicals and reagents obtained from S.D. Fine Chemicals Ltd., Mumbai, India.

4.3. Extraction Method

Leaves of *V. negundo* and *V. trifolia* were washed with running water and then with distilled water to remove dust and other contaminants. They were then shade dried at an average temperature of 40 °C for 84 h. Having dried, both plant materials were coarsely powdered with the help of an electric blender (Usha Pvt. Ltd., Gurgaon, India) and then passed through sieve no. 40 and stored in a closed container for further use. Different organic solvents (petroleum ether, chloroform, ethyl acetate, ethanol, and water) were used for the extraction of polar and non-polar organic compound. The powdered leaves (100 g) of *V. negundo* and *V. trifolia* were first extracted with petroleum ether using soxhlet apparatus (Borosil) for 72 h at room temperature and then successively extracted with chloroform, ethanol, and water. All extracts were concentrated and dried by using vacuum rotary evaporator (Popular Pvt. Ltd, Ambala, India) to evaporate solvents, while the concentrated extracts were kept in desiccators until further used.

4.4. Qualitative Phytochemical Screening

Phytochemical screening of active plant extracts was carried out according to the methods previously published by Mishra and Saklani [25], that is, the qualitative analysis of various phytochemicals such as alkaloids, tannins, saponins, total flavonoids and total phenols that could be responsible for antioxidant activity.

4.5. Determination of Total Phenolic Content (TPC)

TPC was determined using spectrophotometric method as described by Stankovic et al., 2012 [26]. In short, the reaction mixture was prepared by mixing 0.5 mL of ethanolic solution (1 mg/mL) of extract, 2.5 mL of 10% Folin-Ciocalteu's reagent dissolved in water and 2.5 mL 7.5% $NaHCO_3$. The samples were incubated at 45 °C for 15 min and absorbance was observed at 765 nm. The samples were prepared in triplicate, and the mean value of absorbance was obtained. Blank was concomitantly prepared with ethanol instead of the extract solution. The same procedure was repeated for the gallic acid, and the calibration line was constructed. The total phenolic content was expressed in terms of gallic acid equivalent (mg of GaA/g of extract).

4.6. Determination of Total Flavonoid Content (TFC)

TFC of the ethanolic leaf extract of both plants was measured using the aluminium chloride assay. Briefly, ethanol extract (10 mg) was dissolved in water (1 mL) in a test tube, to which 5% (*w/v*) $NaNO_2$ (60 µL) was added. After 5 min, a 10% (*w/v*) $AlCl_3$ solution (60 µL) was added. After 6 min, 1 M NaOH (400 µL) was added, and the total volume made up to 2 mL with distilled water. The solution was mixed well, and the absorbance was measured at 510 nm against a reagent blank. Concentrations were determined using a rutin standard curve. Mean total flavonoid contents ($n = 3$) were expressed as milligrams rutin equivalents (RE) per g (mg RE/g dry) [27].

4.7. Antioxidative Assay

4.7.1. Evaluation of DPPH Scavenging Activity

The ability of the plant extract to scavenge 2, 2-dyphenyl-2-picrylhydrazyl (DPPH) free radicals was assessed by the method described by Ćurčić et al. [28]. The stock solution of the plant extract was prepared in ethanol to achieve the concentration of 1 mg/mL. Diluted solutions (1 mL each) were mixed with DPPH (1 mL). After 30 min in darkness at room temperature (23 °C), the absorbance was recorded at 517 nm. The control samples contained all the reagents except the extract. The percentage inhibition was calculated using the following formula:

$$\% \text{ Inhibition} = (1 - A\text{ sample}/A\text{ control}) \times 100 \tag{1}$$

IC_{50} values were estimated from the % inhibition versus concentration sigmoidal curve using a non-linear regression analysis. The data were presented as mean values ± standard deviation ($n = 3$).

4.7.2. Nitric Oxide Radical-Scavenging Assay

The nitric oxide (NO) radical-scavenging activity of ethanol extracts were assayed according to Venkatachalam and Muthukrishnan, [29]. Briefly, the reaction mixture (5.0 mL) containing sodium nitroprusside (5 mM) in phosphate-buffered saline (pH 7.3), with or without the plant extract at different concentrations, was incubated at 25 °C for 3 h. The nitric oxide radical interacted with oxygen to produce the nitrite ion, which was assayed at 30-min intervals by mixing 1.0 mL incubation mixture with an equal amount of Griess reagent. The absorbance of the chromophore (purple azo dye) formed during the diazotization of nitrite ions with sulphanilamide and subsequent coupling with naphthyl ethylenediamine dihydrochloride was measured. The absorbance was measured at 546 nm by a spectrophotometer using BHT as the positive control. NO radical-scavenging activity (%) was calculated as follows:

$$\text{Scavenging activity (\%)} = (1 - \text{A sample}/\text{A control}) \times 100 \qquad (2)$$

4.7.3. β-Carotene-Linoleic Acid Assay

In this assay, antioxidant capacity is determined according to Katanic et al. [30] by measuring the inhibition of the volatile organic compounds, and the conjugated diene hydroperoxides arising from linoleic acid oxidation. A solution of β-carotene was prepared by dissolving β-carotene (2 mg) in chloroform (10 mL). The β-carotene-chloroform solution (2 mL) was pipetted into a round-bottomed flask, and chloroform was removed using a rotary evaporator at 40 °C for 5 min. Thereafter, 40 mg of linoleic acid, 400 mg of Tween 40 emulsifier, and 100 mL of distilled water were added to the flask with vigorous agitation to form an emulsion. The aliquots (4.8 mL) of this emulsion were added into test tubes containing different concentrations of sample solutions (0.2 mL), and the absorbance was immediately measured at 470 nm against a blank consisting of an emulsion without β-carotene. The tubes were incubated in a water bath at 50 °C. The absorbance was recorded at 20 min interval at 470 nm over a 60-min period using UV-visible spectrophotometer (Systronics India Ltd., Gujarat, India; Model No. AU-2701) at an initial time (t = 0). BHT was used as the reference compounds.

The degradation rate (dr) of the sample was calculated according to the first order kinetics as,

$$\text{dr of sample} = (\ln [A0/At])/t \qquad (3)$$

where ln = natural log; A0 = initial absorbance at time 0; At = absorbance at 20 min of incubation; t = 120 min; and dr = degradation rate. Antioxidant activity (AA) was expressed as percent of inhibition relative to the control by using the following equation:

$$\text{AA\%} = ([\text{dr control} - \text{dr sample}]/\text{dr control}) \times 100 \qquad (4)$$

4.7.4. Ferrous Ion Chelating Activity

The iron-chelating abilities of the *V. negundo* and *V. trifolia* leave extracts, and standards were estimated by the method of Robu et al. [31]. In brief, four dilutions in dimethylsulphoxide (DMSO) i.e., 20 mg/mL, 10 mg/mL, 5 mg/mL, and 2.5 mg/mL were prepared from the dried extracts. Briefly, 0.05 mL of each dilution were added to a 2.7 mL TRIS buffer (pH = 7.4). Thereafter, 0.05 mL of 2 mM $FeCl_2$ were added and vortexed for 15 s. At 30 s, the reaction was initiated by the addition of 5 mM of ferrozine (0.2 mL), and the mixture was shaken vigorously with the aid of cyclomixer (Remi Equipments Pvt. Ltd. Model No. CM-101, New Delhi, India) for 10 s. After 1 min beyond the addition of $FeCl_2$ solution, an absorbance of the solution was measured spectrophotometrically at 562 nm. The ability of extracts to chelate ferrous ion was calculated using the following formula:

$$\text{Chelating effect (\%)} = (1 - \text{A sample}/\text{A control}) \times 100 \qquad (5)$$

where A is the absorbance of the control and sample (extract or standard).

The IC_{50} value (μg/mL), which is the concentration of the extract/standard that chelate 50% of the ferrous ion, was calculated through linear interpolation between values above and below 50% activity.

5. Conclusions

In the present study, we have made an attempt to provide the comparative antioxidant potential of phytochemicals present in the two selected *Vitex* species (*V. negundo* and *V. trifolia*). The results indicate that considerable TPC and TFC presented in the *V. negundo* and *V. trifolia* leaf extracts could be an important source of antioxidant molecules. *V. negundo* shows polyphenolic content higher than *V. trifolia*. The tested *Vitex* extracts have a strong antioxidant activity against numerous oxidative systems in vitro. It was found that *V. negundo* has a more powerful antioxidant effect than *V. trifolia*. The antioxidant capacity of polyphenols is based on their molecular structure. Therefore, our result suggests that both plant species have potent antioxidant properties. However, *V. negundo* leaf extract as compared to *V. trifolia* possesses more antioxidant potential.

Acknowledgments: The authors are thankful to Head, Department of Pharmaceutical Sciences, H.N.B. Garhwal (A Central) University, Srinagar Garhwal, Uttarakhand, India, for providing facilities and encouragement to carry out this research.

Author Contributions: Sarla Saklani supervised this work, Abhay Prakash Mishra and Bhawana Sati did different antioxidant activities. Maria Stefanova Atanassova, Milan Stankovic, and Mohammad Ali Shariati helped in editing the manuscript. Harish Chandra and Manisha Nigam analyzed data and wrote the paper. Mohammad Usman Khan, Sergey Plygun and Hicham Elmsellem peer reviewed the manuscript and did possible corrections. Hafiz Ansar Rasul Suleria edited the whole manuscript and provided final suggestions to main authors. Abhay Prakash Mishra provided the final shape to the manuscript. All the authors read and approved the final manuscript.

Conflicts of Interest: The authors declare no conflict of interest regarding the publication of this article.

References

1. Kulkarni, L.A. Pharmacological review on *Vitex trifolia* Linn. (Verbaeaceae). *Pharmacologyonline* **2011**, *3*, 858–863.
2. Mary, R.N.I.; Meenashree, B.; Vasanthi, V.J. Screening of antibacterial activity and qualitative and quantitative analysis of phytochemicals in *Vitex trifolia*. *Int. J. Curr. Microbiol. Appl. Sci.* **2014**, *3*, 425–431.
3. Neuwinger, H.D. *African Traditional Medicine: A Dictionary of Plant Use and Applications with Supplement: Search System for Diseases*; Medpharm Scientific Publishers: Stuttgart, Germany, 2000.
4. Kannathasan, K.; Senthilkumar, A.; Chandrasekaran, M.; Venkatesalu, V. Differential larvicidal efficacy of four species of *Vitex* against *Culex quinquefasciatus* larvae. *Parasitol. Res.* **2007**, *101*, 1721–1723. [CrossRef] [PubMed]
5. Manjunatha, B.K.; Vidya, S.M.; Krishna, V.; Mankani, K.L.; Singh, S.D.J.; Manohara, Y.N. Comparative evaluation of wound healing potency of *Vitex trifolia* L. and *Vitex altissima* L. *Phytother. Res.* **2007**, *21*, 457–461. [CrossRef] [PubMed]
6. Anandan, R.; Jayakar, B.; Karar, B.; Babuji, S.; Manavalan, R.; Kumar, R.S. Effect of ethanol extract of flowers of *Vitex trifolia* Linn. on CCL4 induced hepatic injury in rats. *Pak. J. Pharm. Sci.* **2009**, *22*, 391–394. [PubMed]
7. Lakshmanashetty, R.H.; Nagaraj, V.B.; Hiremath, M.G.; Kumar, V. In Vitro antioxidant activity of *Vitex negundo* L. leaf extracts. *Chiang Mai J. Sci.* **2010**, *37*, 489–497.
8. Chitra, V.; Sharma, S.; Kayande, N. Evaluation of anticancer activity of *Vitex negundo* in experimental animals: An in vitro & in vivo study. *Int. J. PharmTech Res.* **2009**, *1*, 1485–1489.
9. Tiwari, O.P.; Yamini, B.T. Antioxidant properties of different fractions of *Vitex negundo* Linn. *Food Chem.* **2007**, *100*, 1170–1176. [CrossRef]
10. Shaba, P.; Pandey, N.N.; Sharma, O.P.; Rao, J.R.; Singh, R.K. In Vitro trypanosidal and cytotoxicity effects of methanolic extract of *Vitex negundo* leaves against trypanosoma evansi. In Proceedings of the 15th Congress of FAVA-OIE Joint Symposium on Emerging Diseases, Bangkok, Thailand, 27–30 October 2008; pp. O43–O44.
11. Hossain, M.M.; Paul, N.; Sohrab, M.H.; Rahman, E.; Rashid, M.A. Antibacterial activity of *Vitex trifolia*. *Fitoterapia* **2001**, *72*, 695–697. [CrossRef]
12. Kiuchi, F.; Matsuo, K.; Ito, M.; Qul, T.K.; Honda, G. New norditerpenoids with trypanocidal activity from *Vitex trifolia*. *Chem. Pharm. Bull.* **2004**, *52*, 1492–1494. [CrossRef] [PubMed]
13. Tiwari, N.; Thakur, J.; Saika, D.; Gupta, M.M. Antitubercular diterpenoids from *Vitex trifolia*. *Phytomedicine* **2013**, *20*, 605–610. [CrossRef] [PubMed]

14. Jagtap, S.S.; Stpute, R.A.; Rahatgaonakar, A.M.; Lanjewar, K.R. Phytochemical screening, antioxidant, antimicrobial and quantitative multi-elemental analysis of *Habenaria longicorniculata* J. Graham. *J. Acad. Ind. Res.* **2014**, *3*, 108–117.
15. Lazaro, M.L. Distribution and biological activities of the flavonoid Luteolin. *Mini Rev. Med. Chem.* **2009**, *9*, 31–59. [CrossRef]
16. Sastri, B.N. The wealth of India, a Dictionary of Raw material and Industrial Products. *Publ. Inf. Dir.* **1950**, *5*, 285–293.
17. Prasanna, K.; Karthikeyan, V. Preliminary phytochemical, total phenolics and flavonoid content analysis of *Vitex negundo* and *Calatropis gigantea* leaf ethanolic extracts. *J. Chem. Pharm. Res.* **2015**, *7*, 282–285.
18. Li, B.B.; Smith, B.; Hossain, M.M. Extraction of phenolics from citrus peels: I. Solvent extraction method. *Sep. Purif. Technol.* **2006**, *48*, 182–188. [CrossRef]
19. Ruan, Z.P.; Zhang, L.L.; Lin, Y.M. Evaluation of the antioxidant activity of *Syzygium cumini* leaves. *Molecules* **2008**, *13*, 2545–2556. [CrossRef] [PubMed]
20. Jakovjevic, Z.D.; Stankovic, S.M.; Topusovic, D.M. Seasonal variability of *Chelidonium majus* L. secondary metabolites content and antioxidant activity. *EXCLI J.* **2013**, *12*, 260–268.
21. Shah, S.; Dhanani, T.; Kumar, S. Comparative evaluation of antioxidant potential of extracts of *Vitex negundo, Vitex trifolia, Terminalia bellerica, Terminalia chebula, Embelica officinalis* and *Asparagus racemosus*. *Innov. Pharm. Pharmacother.* **2013**, *1*, 44–53.
22. Boumerfeg, S.; Baghiani, A.; Djarmouni, M.; Ameni, D.; Adjadi, M.; Belkhiri, F.; Charef, N.; Khennoue, S.; Arrar, L. Inhibitory activity on xanthine oxidase and antioxidant properties of *Teucrium polium* L. extracts. *Chin. Med.* **2012**, *3*, 30–41. [CrossRef]
23. Russo, D.; Valentao, P.; Andrade, B.; Fernandez, E.C.; Milella, L. Evaluation of antioxidant, antidiabetic and anticholinesterase activities of *Smallanthus sonchifolius* landraces and correlation with their phytochemical profiles. *Int. J. Mol. Sci.* **2015**, *16*, 17696–17718. [CrossRef] [PubMed]
24. Sengul, M.; Yildiz, H.; Gungor, N.; Cetin, B.; Eser, Z.; Ercisli, S. Total phenolic content, antioxidant and antimicrobial activities of some medicinal plants. *Pak. J. Pharm. Sci.* **2009**, *22*, 102–106. [PubMed]
25. Mishra, A.P.; Saklani, S. *Satyrium nepalense*: A rare medicinal orchid of western Himalaya (India); phytochemical screening, antimicrobial evaluation and conservation studies. *Indones. J. Pharm.* **2012**, *23*, 162–170. [CrossRef]
26. Stankovic, M.S.; Nicforovic, N.; Mihailovic, V.; Topusovic, M.; Solujiv, S. Antioxidant activity, total phenolic content and flavonoid concentrations of different plant parts of *Teucrium polium* L. subsp. polium. *Acta Soc. Bot. Pol.* **2012**, *81*, 117–122. [CrossRef]
27. Mishra, A.P.; Saklani, S.; Chandra, S.; Tiwari, P. Total phenolics, flavonoids and antioxidant evaluation in the leaves of *Argyreia nervosa* Burm. *Int. J. Pharm. Sci. Rev. Res.* **2015**, *32*, 34–37.
28. Curic, M.G.; Stankovic, M.S.; Radojevic, I.D.; Stefanovic, O.D.; Comic, L.R.; Topuzovic, M.D.; Dacic, D.S.; Markovic, S.D. Biological effects, total phenolic content and flavonoid concentrations of fragrant yellow onion (*Allium flavum* L.). *Med. Chem.* **2012**, *8*, 46–51. [CrossRef]
29. Venkatachalam, U.; Muthukrishnan, S. Free radical scavenging activity of ethanolic extract of *Desmodium gangeticum*. *J. Acute Med.* **2012**, *2*, 36–42. [CrossRef]
30. Katanic, J.; Mihailovic, V.; Stanmovic, N.; Boroja, T.; Mladenovic, M.; Solujic, S.; Stankovic, M.S.; Vrvic, M.M. Dropwort (*Filipendula hexapetala* Gilib.): Potential role as antioxidant and antimicrobial agent. *EXCLI J.* **2015**, *14*, 1–20. [CrossRef] [PubMed]
31. Robu, S.; Aprotosoaie, A.C.; Miron, A.; Cioanca, O.; Stanescu, U.; Hancianu, M. In Vitro Antioxidant activity of ethanolic extracts from some *Lavandula* species cultivated in Romania. *Farmacia* **2012**, *60*, 394–401.

© 2017 by the authors. Licensee MDPI, Basel, Switzerland. This article is an open access article distributed under the terms and conditions of the Creative Commons Attribution (CC BY) license (http://creativecommons.org/licenses/by/4.0/).

Communication

Variability of Secondary Metabolites of the Species *Cichorium intybus* L. from Different Habitats

Nenad M. Zlatić * and Milan S. Stanković

Department of Biology and Ecology, Faculty of Science, University of Kragujevac, 34000 Kragujevac, Serbia; mstankovic@kg.ac.rs
* Correspondence: nzlatic@gmx.com; Tel.: +381-34-336223

Received: 26 July 2017; Accepted: 8 September 2017; Published: 11 September 2017

Abstract: The principal aim of this paper is to show the influence of soil characteristics on the quantitative variability of secondary metabolites. Analysis of phenolic content, flavonoid concentrations, and the antioxidant activity was performed using the ethanol and ethyl acetate plant extracts of the species *Cichorium intybus* L. (Asteraceae). The samples were collected from one saline habitat and two non-saline habitats. The values of phenolic content from the samples taken from the saline habitat ranged from 119.83 to 120.83 mg GA/g and from non-saline habitats from 92.44 to 115.10 mg GA/g. The amount of flavonoids in the samples from the saline locality varied between 144.36 and 317.62 mg Ru/g and from non-saline localities between 86.03 and 273.07 mg Ru/g. The IC_{50} values of antioxidant activity in the samples from the saline habitat ranged from 87.64 to 117.73 µg/mL and from 101.44 to 125.76 µg/mL in the samples from non-saline habitats. The results confirmed that soil types represent a significant influence on the quantitative content of secondary metabolites. The greatest concentrations of phenols and flavonoids and the highest level of antioxidant activity were found in the samples from saline soil. This further corroborates the importance of saline soil as an ecological factor, as it is proven to give rise to increased biosynthesis of secondary metabolites and related antioxidant activity.

Keywords: saline habitats; secondary metabolites; adaptation; different solvents

1. Introduction

The abiotic factor with the most harmful effect upon the productivity and growth of plants is soil salinity. Increased concentrations of salt in soil exert a negative influence on plants due to the following reasons: water absorption is more difficult, the growth of the plant may be hindered, and plant metabolism as well as both physiological and chemical processes may be disturbed to a significant extent [1]. In order to successfully adapt to saline soil, plants develop a series of specific mechanisms which enable them to respond appropriately to salinity-induced stress [2]. The mechanisms develop at molecular, cellular, metabolic, and physiological levels [3].

Salinity-induced stress causes the production of reactive oxygen species (ROS), such as hydrogen peroxide, hydroxyl radicals, and superoxide anions in a variety of cells resulting in oxidative stress [4]. Increased production of ROS leads to oxidative damage of cellular membranes, proteins, carbohydrates, and DNA [5]. Plants activate enzymatic and nonenzymatic antioxidant systems for the purpose of protection from the negative influence of free radicals. The activation of enzymatic systems includes superoxide dismutase, catalase, and glutathione reductase, whereas carotenoids, flavonoids, and other phenolic compounds belong to nonenzymatic systems of protection [4]. Different ecological factors that directly influence the plant primary metabolism have different effects. Correspondingly, the differentiation of secondary metabolism occurs. The primary function of secondary metabolites is their participation in the process of plant adaptation to the effects of ecological factors [6]. From an evolutionary viewpoint, biosynthesis of secondary metabolites in plants was considered as an ecophysiological response of a plant to different influences of abiotic factors in a certain habitat, including salinity [7]. Secondary metabolites participate in the process of plant adaptation to the ecological conditions of environment and their quantity in plant organs varies depending on both the abiotic and biotic factors that a plant is exposed to [8].

A great number of investigated medicinal plants that are used in the fields of pharmacy and medicine belong to the family Asteraceae. Within this plant family there are several important genera, which abound in medicinal plant species, and such genera are: *Artemisia, Achillea, Inula,* and *Matricaria*. These genera show the intense biological activity applied in the treatment of respiratory, digestive, cardiovascular, and some other types of diseases [9]. Apart from therapeutic effects, the active metabolites of the species that belong to this family manifest antimicrobe, antiviral, antiproliferative, and antioxidant activity in both in vitro and in vivo conditions [10,11].

The genus *Cichorium* encompasses approximately nine species. Among the best known medicinal species that belong to the genus *Cichorium* are: *C. intibus, C. spinosum,* and *C. dubium*. Chicory (*C. intybus* L., Asteraceae) is a perennial herbaceous plant up to 120 cm high. Its stem is erect and branched. The lower leaves are curled, whereas the upper ones are bare and lanceolate. The flowers are blue in colour and grow either individually or in groups. The fruit is achene, smallish and ovoid in shape. *C. intybus* populates meadows, saline habitats, areas by the road, the edges of forests, and the territories of Europe, West Asia, and North Africa. Due to the limited distribution and wider application, the successful growth of this species is enabled on territories outside its initial area [12].

Cichorium intybus is well known in the folk and modern medicine for its anticancer [13], antioxidant [14], antidiabetic [15], anti-inflammatory [16], antimicrobial [17], anthelmintic [18], analgesic [19], cardiovascular [20], gastroprotective [21], hepatoprotective [22], immunological [23], reproductive effects [24], wound healing abilities [25], and many other pharmacological applications [17,26]. It is also important to mention the numerous active substances of this species: alkaloids, coumarins, caffeic acid derivatives, sesquiterpene lactones, flavonoids, terpenoids, and volatile compounds [26].

Secondary metabolites, apart from other adaptive mechanisms, have significance in relation to the reaction of plants to the chemical stress induced by increased quantity of salt in soil in the process of neutralisation of consequences of toxicity for the purpose of ecophysiological adaptation. Taking the previously stated fact into account, the species *C. intybus* was selected as a plant suitable for analysis, with its suitability stemming from the fact that it grows in habitats with both normal mineral regime and with saline soils. Accordingly, by means of sampling the species from one habitat with saline soil and two habitats with normal mineral regime as well as by analysing both the quantity of secondary metabolites from the group of phenolic compounds and their antioxidant activity, a comparison was performed in order to determine their significance in terms of adaptation to toxic effects of salt.

2. Results and Discussion

Total phenolic content, flavonoid concentration, and antioxidant activity in vitro were determined using ethanol and ethyl acetate extracts of whole *C. intybus* plant from saline and non-saline habitats. In order to extract the active substances of different polarity, the extraction solvents of different polarity were used.

2.1. The Total Quantity of Phenolic Compounds

The results of the analysis of the total quantity of phenolic compounds, flavonoid concentration, and antioxidant activity in ethanol and ethyl acetate extracts of the aboveground plant parts of the species *C. intybus* are shown in Table 1.

Table 1. Total phenolic content, flavonoid concentration, and antioxidant activity of the analysed species in ethanol and ethyl acetate extracts.

Locality	Type of Analysis					
	Total Phenolic Content (mg of GA/g of Extract)		Flavonoid Content (mg of Ru/g of Extract)		Antioxidant Activity IC_{50} (µg/mL)	
	Type of Extract					
	Ethanol	Ethyl Acetate	Ethanol	Ethyl Acetate	Ethanol	Ethyl Acetate
Oblačinska slatina	120.83 ± 1.02	119.83 ± 1.34	144.36 ± 0.83	317.62 ± 2.04	117.73 ± 1.71	87.64 ± 1.90
Ivanjica	95.53 ± 0.97	115.10 ± 1.50	129.00 ± 1.18	273.07 ± 1.56	120.53 ± 2.21	101.44 ± 1.53
Kragujevac	92.44 ± 1.12	96.55 ± 1.45	86.03 ± 0.59	176.09 ± 1.39	121.05 ± 1.66	125.76 ± 2.33

Each value is the average of three analyses ± standard deviation.

The total quantity of phenolic compounds in ethanol extracts ranged from 92.44 to 120.83 mg GA/g of extract, whereas the values in the ethyl acetate extracts varied from 96.55 to 119.83 mg GA/g of extract. The results showed that the greatest quantity of phenolic compounds in the ethanol extracts was found in the sample from the locality Oblačinska slatina (120.83 mg GA/g), while the smallest quantity was observed in the sample collected in Kragujevac (92.44 mg GA/g). Similarly, the highest concentration of phenolic compounds in the ethyl acetate extracts was measured in the samples from the locality Oblačinska slatina (119.83 mg GA/g), whereas the lowest concentration was found in the samples collected in Kragujevac (96.55 mg GA/g). In the extracts obtained using moderate and lower solvents (ethanol and ethyl acetate) the order of concentrations for phenolic compounds is: Oblačinska slatina > Ivanjica > Kragujevac.

The analyses showed that the extract obtained using solvents of lower polarity contained the highest concentration of active compounds. The total quantity of phenolic compounds was greater in the ethanol extracts made from the solvent of moderate polarity and the samples collected from the locality of Oblačinska slatina (120.83 mg GA/g of extract). Consequently, it was shown that a certain group of phenolic compounds is important for the adaptation to saline habitats [27].

The total quantity of phenolic compounds in the aboveground plant parts of the species *C. intybus* varied depending on the type of locality from which the plant material was taken. The substrate of Oblačinska slatina contains high concentrations of salt. Contrary to this, the substrates of Ivanjica and Kragujevac contain low salt concentrations. Due to this difference in salt concentrations, greater quantities of phenolic compounds (120.83 mg GA/g) were measured in the plant extract of the sample from Oblačinska slatina. The obtained results for the quantity of phenolic compounds of the species *C. intibus* are in accordance with the previously recorded values [28]. It has been confirmed that the quantity of polyphenols in plants belonging to the species *Mentha pulegium* increases due to the stress caused by greater concentrations of salt in ground substrate [29]. Similarly, the examination of the species *Nigella sativa* demonstrated the increase in the total quantity of phenols due to saline treatment [30]. Salinity-induced increases in the concentration of the group of phenolic acids (protocatechuic, chlorogenic, and caffeic acids) has been observed in the species *Matricaria chamomilla* [31].

The plants sampled in Ivanjica contained higher concentrations of phenolic compounds in comparison with the samples from Kragujevac. These localities differ in altitude, as Ivanjica is located at an altitude of 997 m and Kragujevac at an altitude of 194 m. Previously performed studies showed that the plant species in habitats with increased intensity of light contain phenolic compounds, which have considerable influence on the adaptational abilities of plants populating these habitats. The increased quantity of phenolic compounds in plants plays a protective role from ultraviolet-B radiation, which is more intense at higher altitudes [32,33].

2.2. The Total Quantity of Flavonoids

The results of the analysis of the total quantity of flavonoids in ethanol and ethyl acetate extracts of aboveground plant parts of the species *C. intybus* are shown in Table 1. The total quantity of flavonoids in the ethanol extracts ranged from 86.03 to 144.36 mg Ru/g of extract, whereas the quantity in the ethyl acetate extracts varied from 176.09 to 317.62 mg Ru/g of extract. The results showed that the greatest quantity of flavonoids in the ethanol extracts was found in the sample from the locality Oblačinska slatina (144.36 mg Ru/g), while the smallest quantity was observed in the sample collected in Kragujevac (86.03 mg Ru/g). Similarly, the highest concentration of flavonoids in the ethyl acetate extracts was measured in the sample from the locality Oblačinska slatina (317.62 mg Ru/g), whereas the lowest concentration was found in the sample collected in Kragujevac (176.09 mg Ru/g). In the extracts obtained using moderate and lower solvents (ethanol and ethyl acetate) the order of concentrations for phenolic compounds is: Oblačinska slatina > Ivanjica > Kragujevac.

The concentration of flavonoids in plant extracts depends on the polarity of solvents used in the extract preparation. Based on the obtained values for concentration of flavonoids, the highest concentration of these compounds is observed in the extracts obtained using solvents of low polarity. High concentrations of flavonoids in ethyl acetate extracts may be attributed to their high solubility in this solvent [34]. Accordingly, the analysis of whole plant extracts indicated that ethyl acetate was the most effective solvent for the extraction of flavonoids from the species *C. intybus* and that therefore solvents with low and moderate polarity should be used to this end.

Apart from solvent polarity, what proved to be relevant in terms of the quantity of flavonoids is the type of locality from which samples were taken. The greater quantity of flavonoids was observed in the extract made from the samples collected from the locality of Oblačinska slatina (317.62 mg Ru/g) the substrate of which contains high concentrations of salt. Previous studies confirmed that certain plants from saline habitats synthetize greater concentrations of flavonoids as a response of the secondary metabolism to the adaptation of species to the increased level of salt in the substrate [35,36].

The concentration of flavonoids differs in the localities of Ivanjica and Kragujevac, with Ivanjica having higher concentrations of flavonoids in both types of extracts. This study revealed a difference in the quantity of flavonoids in the samples collected at different altitudes. Increases in altitude may cause increases in the plant production of flavonoids. UV-B radiation increases the production of flavonoids in barley [37]. The greater intensity of light in a habitat is accompanied by a larger accumulation of secondary metabolites in certain plant organs [32].

2.3. Antioxidant Activity

The obtained values for antioxidant activity of the ethanol and ethyl acetate extracts made from the aboveground plant parts of the species *C. intybus* are shown in Table 1. Antioxidant activity of the ethanol extracts from different localities ranged from 117.73 to 121.05 μg/mL, while the antioxidant activity of the ethyl acetate extracts varied from 87.64 to 125.76 μg/mL. The results showed that the highest level of antioxidant activity of both the ethanol and ethyl acetate extracts was observed in the samples from the locality Oblačinska slatina, with values of 87.64 μg/mL and 117.73 μg/mL for each type of extract, respectively. The lowest values of both the ethanol and ethyl acetate extracts were obtained from the samples collected in Kragujevac, with values of 121.05 μg/mL and 125.76 μg/mL for each type of extract, respectively. In the extracts obtained using moderate and lower solvents (ethanol and ethyl acetate) the order of antioxidant activity is: Oblačinska slatina > Ivanjica > Kragujevac.

The extraction of antioxidant substances of different chemical structure was achieved using solvents of different polarity. The largest capacity to neutralize DPPH (2,2-dyphenyl-1-picrylhydrazyl) radicals was measured in the ethyl acetate extract from the saline habitat Oblačinska slatina, which neutralized 50% of free radicals at concentrations of 87.64 μg/mL. It is well known that the phenolic content of plants may contribute directly to their antioxidant activity [34] due to its role in scavenging free radicals.

The plants sampled from the locality Oblačinska slatina had the highest concentrations of phenols and flavonoids. This indicates that secondary metabolites of the phenolics group in *C. intybus* are the key active substances for the expression of antioxidant activity. Earlier studies on major secondary metabolites in upper parts of *C. intybus* demonstrated the presence of sesquiterpene lactones as well as caffeic acids derivates (chicoric acid, chlorogenic acid, isochlorogenic acid, dicaffeoyl tartaric acid). Significant biological activity of these phenol acids and flavonoids has already been confirmed for both in vitro and in vivo model systems. The antioxidant activity has been reported as well [26].

Numerous studies have shown increased activity of antioxidant enzymes in plants located in habitats with saline substrates. Neutralisation of free radicals in the species *Phragmites karka* increases when the plant is exposed to higher concentrations of salt [38]. Similar results have been obtained for the species *Hordeum vulgare* [39].

In comparison with the locality of Kragujevac, Ivanjica stands out in terms of its greater ability to neutralise free radicals. This is attributable to the differences in the type of habitat and altitude. These results are in accordance with the findings that the extracts of the samples collected at higher altitudes have greater antioxidant capacity than the extracts of the samples taken at lower altitudes [40,41].

2.4. Correlation between Phenolic Compounds, Flavonoids, and Antioxidant Activity

The obtained results for the total quantity of phenolic compounds and flavonoids, as well as for the level of antioxidant activity of the ethanol and ethyl acetate extracts of the species *C. intybus* were statistically analysed for the purpose of determining the degree of correlation. The values of correlation point to the significant link between the analysed parameters.

The correlation between the total quantity of phenolic compounds, flavonoids, and the values of antioxidant activity measured in both ethanol and ethyl acetate extracts is established in *C. intybus* (Table 2). It can be noticed that an increase in the phenol content of extract decreases the IC_{50} value, i.e., increases their scavenging DPPH free-radical activity (negative correlation). This is important in the light of the fact that the lower IC_{50} values imply higher levels of antioxidant activity. Similar conclusions have been reached in other studies [42,43].

Table 2. The coefficient of correlation ($p < 0.05$) between the total phenolic compounds (TPC), total flavonoids (TF) and the values for antioxidant activity (AA).

(r)	AA Ethanol	AA Ethyl Acetate
TPC Ethanol	−0.999 *	−0.835
TPC Ethyl acetate	−0.760	−0.985
TF Ethanol	−0.800	−0.994
TF Ethyl acetate	−0.832	−0.999 *

* Correlation is significant at the 0.05 level.

3. Materials and Methods

3.1. Plant Material

Plants were sampled from natural populations found in different localities with saline and non-saline substrate. The plant material was sampled from three different localities: one saline (Oblačinska slatina, Serbia) and two non-saline ones (Ivanjica and Kragujevac, Serbia). The sampling was performed in the phase of flowering in August 2014 (Table 3). The collected samples were identified in the Department of Biology and Ecology of the Faculty of Science in Kragujevac. Aboveground plant parts were dried in a dark room at room temperature. Dry samples were ground in a blender and thus kept in dark vials until analysis.

Table 3. The basic characteristics of the localities in which the species *C. intybus* was sampled.

Locality	Type of Habitat	Altitude	Latitude and Longitude
Oblačinska slatina	Meadow, Hygrophilous habitat	285 m	43°18'17.76" N 21°41'0.340" E
Ivanjica	Meadow, Mesophilous habitat	997 m	43°28'32.55" N 20°10'29.11" E
Kragujevac	Meadow, Mesophilous habitat	194 m	44°01'31.18" N 20°54'50.62" E

3.2. Preparation of Plant Extracts

Dried plant material (10 g) was put into Erlenmeyer flasks filled with 200 mL of ethanol or ethyl acetate and thus left at room temperature. The extract was filtrated after 48 h using Whatman No. 1 filter paper. The plant extracts were condensed to dry on a rotary vacuum at 40 °C. The obtained extracts were placed in sterile containers and kept in the fridge at 4 °C. The total quantity of phenolic compounds, flavonoids, as well as the level of antioxidant activity were determined in the extract concentration of 1 mg/mL using ethanol or ethyl acetate as solvents.

3.3. Chemicals

Organic solvents and sodium hydrogen carbonate were purchased from Zorka pharma Sabac, Serbia. 2,2-dyphenyl-1-picrylhydrazyl (DPPH) was obtained from Sigma Chemicals Co., St. Louis, MO, USA. Folin-Ciocalteu phenol reagent, 3-tert-butyl-4-hydroxyanisole (BHA), and aluminium chloride hexahydrate ($AlCl_3 \cdot 6H_2O$) were purchased from Fluka Chemie AG, Buchs, Switzerland. All other solvents and chemicals were of analytical grade.

3.4. Determination of Total Phenolic Contents

The total phenolic content was determined using the spectrophotometric method [44]. First, 0.5 mL of methanol solution (1 mg/mL) of extract, 2.5 mL of 10% Folin-Ciocalteu's reagent dissolved in water, and 2.5 mL 7.5% $NaHCO_3$ were used to obtain the reaction mixture. Then, the samples were incubated at 45 °C for 15 min. The absorbance was measured at λ_{max} = 765 nm. The samples were prepared in triplicate and the mean value of absorbance was obtained. The blank was prepared with methanol solution. The same procedure was repeated for the gallic acid and the calibration curve was constructed. The total phenolic content was expressed as gallic acid equivalent (mg of GA/g).

3.5. Determination of Flavonoid Concentrations

The concentration of flavonoids was determined using the spectrophotometric method [45]. The sample contained 1 mL of methanol solution of the extract in the concentration of 1 mg/mL and 1 mL of 2% $AlCl_3$ solution dissolved in methanol. The samples were incubated at ambient temperature for an hour. The absorbance was measured at λ_{max} = 415 nm. The samples were prepared in triplicate and the mean value of absorbance was obtained. The same procedure was repeated for the rutin and the calibration line was constructed. Concentration of flavonoids in extracts was expressed in terms of rutin equivalent (mg of Ru/g).

3.6. Evaluation of DPPH Scavenging Activity

The efficiency of the plant extract to neutralise DPPH (1,1-diphenyl-2-picrylhydrazyl radical) free radicals was determined using the spectrophotometric method previously described [46] with adequate modifications [47]. The plant extract was dissolved in methanol to obtain the concentration 1 mg/mL. Dilutions were made to obtain concentrations of 500, 250, 125, 62.5, 31.25, 15.62, 7.81, 3.90, 1.99, and 0.97 mg/mL. Diluted solutions (1 mL each) were mixed with 1 mL of DPPH methanolic solution

(80 mg/mL). The absorbance was recorded at 517 nm. The control samples contained methanol and DPPH reagents. The percentage inhibition was calculated using the equation: % inhibition = 100 × (A of control − A of sample)/A of control)), whilst IC_{50} values were estimated based on the sigmoidal curve presenting the dependence of the percent of DPPH scavenging on sample concentration. Antioxidant activity was expressed as the half-maximal inhibitory concentration (IC_{50} values in mg/mL). In the presented results, antioxidant efficiency of the extract increased with decreasing IC_{50} values. The data were presented as mean values ± standard deviation ($n = 3$).

3.7. Statistical Analysis

All experimental measurements were carried out in triplicate and are expressed as the average of three analyses ± standard deviation. Results were analysed statistically using IBM, SPSS, Statistics, ver. 19, Armonk, NY: IBM Corp. Pearson's correlation coefficient (r) was used to evaluate relationships between contents and antioxidant properties of chicory extracts.

4. Conclusions

The results of this research show that there is a significant difference in the quantity of secondary metabolites and their activity in the species *C. intybus*, which populates both saline and non-saline habitats. The total quantity of phenolic compounds and flavonoids increases due to the presence of salt in the substrate. The level of antioxidant activity was higher in the samples taken from saline habitats, and the result implies that this is a mechanism of plant adaptation to the increased concentrations of salt in the substrate. Plants may adapt to the stressful conditions in the habitat by means of synthesis regulation and accumulation of secondary metabolites. The plants tolerant to salt stress are sources of secondary metabolites and may be highly applicable in pharmaceutical and food industries.

Acknowledgments: This investigation was supported by the Ministry of Education, Science and Technological Development of the Republic of Serbia III41010. The authors acknowledge Ana Vučićević for manuscript lecturing.

Author Contributions: Nenad M. Zlatić partly conducted field work, organised and performed experiments, planned the effective presentation of data, and wrote the paper; Milan S. Stanković proposed the theme, partly conducted the field work, and provided guidance and supervision to organise the experiments and analyse the data; both authors revised the paper in accordance with the instructions.

Conflicts of Interest: The authors declare no conflict of interest.

References

1. Munns, R. Comparative physiology of salt and water stress. *Plant Cell Environ.* **2002**, *25*, 239–250. [CrossRef] [PubMed]
2. Bohnert, J.H.; Shen, B. Transformation and compatible solutes. *Sci. Hortic.* **1999**, *78*, 237–260. [CrossRef]
3. Gupta, B.; Huang, B. Mechanism of salinity tolerance in plants: Physiological, biochemical, and molecular characterization. *Int. J. Genom.* **2014**. [CrossRef] [PubMed]
4. Apel, K.; Hirt, H. Reactive oxygen species: Metabolism, oxidative stress, and signal transduction. *Annu. Rev. Plant Biol.* **2004**, *55*, 373–399. [CrossRef] [PubMed]
5. Van Breusegem, F.; Dat, F.J. Reactive oxygen species in plant cell death. *Plant Physiol.* **2006**, *141*, 384–390. [CrossRef] [PubMed]
6. Kliebenstein, D.J.; Osbourn, A. Making new molecules—Evolution of pathways for novel metabolites in plants. *Curr. Opin. Plant Biol.* **2012**, *15*, 415–423. [CrossRef] [PubMed]
7. Navarro, J.M.; Flores, P.; Garrido, C.; Martinez, V. Changes in the contents of antioxidant compounds in pepper fruits at different ripening stages, as affected by salinity. *Food Chem.* **2006**, *96*, 66–73. [CrossRef]
8. Ramakrishna, A.; Ravishankar, G.A. Influence of abiotic stress signaling on secondary metabolites in plants. *Plant Signal. Behav.* **2011**, *6*, 1720–1731. [PubMed]
9. Mukherjee, S.K. Medicinal plants of Asteraceae in India and their uses. In *Proceeding of National Seminar*; Gupta, S.K., Mitra, B.R., Eds.; Ramakrishna Mission Ashrama: Kolkata, India, 2006; pp. 43–49.

10. Jayaraman, S.; Manoharam, S.M.; Illanchezian, S. In-vitro antimicrobial and antitumor activities of *Stevia rebaudiana* (Asteraceae) leaf extracts. *Trop. J. Pharm. Res.* **2008**, *7*, 1143–1149. [CrossRef]
11. Kasim, L.S.; Ferro, V.A.; Odukoya, O.A.; Drummond, A.; Ukpo, G.E.; Seidel, V.; Gray, A.I.; Waigh, R. Antimicrobial agents from the leaf of *Struchium sparganophora* (Linn) Ktze, Asteraceae. *J. Microbiol. Antimicrob.* **2011**, *3*, 13–17.
12. Gajić, M. Genus *Cichorium* L. In *Flora of Serbia*; Josifović, M., Ed.; Serbian Academy of Sciences and Arts: Belgrade, Serbia, 1975; Volume 7, pp. 266–268.
13. Lee, K.T.; Kim, J.I.; Park, H.J.; Yoo, K.O.; Han, Y.N.; Miyamoto, K.I. Differentiation-inducing effect of magnolialide, a 1β-hydroxyeudesmanolide isolated from *Cichorium intybus*, on human leukemia cells. *Biol. Pharm. Bull.* **2000**, *23*, 1005–1007. [CrossRef] [PubMed]
14. Mehmood, N.; Zubair, M.; Rizwan, K.; Rasool, N.; Shahid, M.; Ahmad, V.U. Antioxidant, antimicrobial and phytochemical analysis of *Cichorium intybus* seeds extract and various organic fractions. *Iran J. Pharm. Res.* **2012**, *11*, 1145–1151. [PubMed]
15. Pushparaj, P.N.; Low, H.K.; Manikandan, J.; Tan, B.K.; Tan, C.H. Antidiabetc effects of *Cichorium intybus* in streptozotocin-induced diabetic rats. *J. Ethnopharmacol.* **2007**, *111*, 430–434. [CrossRef] [PubMed]
16. Ripoll, C.; Schmidt, B.; Ilic, N.; Poulev, A.; Dey, M.; Kurmukov, A.G. Antinflammatory effects of a sesquiterpene lactone extract from chicory (*Cichorium intybus* L.) roots. *Nat. Prod. Commun.* **2007**, *2*, 717–722.
17. Das, S.; Vasudeva, N.; Sharma, S. *Cichorium intybus*: A concise report on its ethnomedicinal, botanical, and phytopharmacological aspects. *Drug Dev. Ther.* **2016**, *7*, 1–12.
18. Miller, M.C.; Duckett, S.K.; Andrae, J.G. The effect of forage species on performance and gastrointestinal nematode infection in lambs. *Small Rumin. Res.* **2011**, *95*, 188–192. [CrossRef]
19. Wesołowska, A.; Nikiforuk, A.; Michalska, K.; Kisiel, W.; Chojnacka-Wójcik, E. Analgesic and sedative activities of lactucin and some lactucin-like guaianolides in mice. *J. Ethnopharmacol.* **2006**, *107*, 254–258. [CrossRef] [PubMed]
20. Nayeemunnisa, A. Alloxan diabetes-induced oxidative stress and impairment of oxidative defense system in rat brain: Neuroprotective effects of *Cichorium intybus*. *Int. J. Diabetes Metab.* **2009**, *17*, 105–109.
21. Gürbüz, I.; Üstün, O.; Yeşilada, E.; Sezik, E.; Akyürek, N. *In vivo* gastroprotective effects of five Turkish folk remedies against ethanol-induced lesions. *J. Ethnopharmacol.* **2002**, *83*, 241–244. [CrossRef]
22. Gilani, A.H.; Janbaz, K.H. Evaluation of the liver protective potential of *Cichorium intybus* seed extract on acetaminophen and CCl4-induced damage. *Phytomedicine* **1994**, *1*, 193–197. [CrossRef]
23. Kim, J.H.; Mun, Y.J.; Woo, W.H.; Jeon, K.S.; An, N.H.; Park, J.S. Effects of the ethanol extract of *Cichorium intybus* on the immunotoxicity by ethanol in mice. *Int. Immunopharmacol.* **2002**, *2*, 733–744. [CrossRef]
24. Behnam-Rassouli, M.; Aliakbarpour, A.; Hosseinzadeh, H.; Behnam-Rassouli, F.; Chamsaz, M. Investigating the effect of aqueous extract of *Cichorium intybus* L. leaves on offspring sex ratio in rat. *Phytother. Res.* **2010**, *24*, 1417–1421. [CrossRef] [PubMed]
25. Süntar, I.; Küpeli-Akkol, E.; Keles, H.; Yesilada, E.; Sarker, S.D.; Baykal, T. Comparative evaluation of traditional prescriptions from *Cichorium intybus* L. for wound healing: Stepwise isolation of an active component by *in vivo* bioassay and its mode of activity. *J. Ethnopharmacol.* **2012**, *143*, 299–309. [CrossRef] [PubMed]
26. Al-Snafi, A.E. Medicinal importance of *Cichorium intybus*—A review. *IOSR J. Phram.* **2016**, *6*, 41–56.
27. Ksouri, R.; Megdiche, W.; Debez, A.; Falleh, H.; Grignon, C.; Abdelly, C. Salinity effects on polyphenol content and antioxidant activities in leaves of the halophyte *Cakile maritima*. *Plant Physiol. Biochem.* **2007**, *45*, 244–249. [CrossRef] [PubMed]
28. Montefusco, A.; Semitaio, G.; Marrese, P.P.; Iurlaro, A.; de Caroli, M.; Piro, G.; Dalassandro, G.; Lenucci, M.S. Antioxidants in varieties of chicory (*Cichorium intybus* L.) and wild poppy (*Papaver rhoeas* L.) of southern Italy. *J. Chem.* **2015**. [CrossRef]
29. Queslati, S.; Karray-Bouraoui, N.; Attia, H.; Rabhi, M.; Ksouri, R.; Lachaal, M. Physiological and antioxidant responses of *Mentha pulegium* (Pennyroyal) to salt stress. *Acta Physiol. Plant* **2010**, *32*, 289–296. [CrossRef]
30. Bourgou, S.; Kchouk, M.E.; Bellila, A.; Marzouk, B. Effect of salinity on phenolic composition and biological activity of *Nigella sativa*. *Acta Hortic.* **2010**, *853*, 57–60. [CrossRef]

31. Cik, J.K.; Klejdus, B.; Hedbavny, J.; Bačkor, M. Salicylic acid alleviates NaCl-induced changes in the metabolism of *Matricaria chamomilla* plants. *Ecotoxicology* **2009**, *18*, 544–554.
32. Alonso-Amelot, M.E.; Oliveros-Bastidas, A.; Calcagno-Pisarelli, M.P. Phenolics and condensed tannins in relation to altitude in neotropical *Pteridium* spp. A field study in the Venezuelan Andes. *Biochem. Syst. Ecol.* **2004**, *32*, 969–981. [CrossRef]
33. Li, Y.; Gao, J.; Zhang, L.; Su, Z. Responses to UV-B exposure by saplings of the relict species *Davidia involucrata* Bill are modified by soil nitrogen availability. *Pol. J. Ecol.* **2014**, *62*, 101–110. [CrossRef]
34. Stanković, M.; Topuzović, M.; Solujić, S.; Mihajlović, V. Antioxidant activity and concentration of phenols and flavonoids in the whole plant and plant parts of *Teucrium chamaedrys* L. var. *glanduliferum* Haussk. *J. Med. Plants Res.* **2010**, *4*, 2092–2098.
35. Ksouri, R.; Megdiche, W.; Falleh, H.; Abdelly, C. Influence of biological, environmental and technical factors on phenolic content and antioxidant activities of Tunisian halophytes. *C. R. Biol.* **2008**, *331*, 865–873. [CrossRef] [PubMed]
36. Stanković, M.S.; Petrović, M.; Godjevac, D.; Dajić-Stevanović, Z. Screening inland halophytes from the central Balkan for their antioxidant activity in relation to total phenolic compounds and flavonoids: Are there any prospective medicinal plants? *J. Arid. Environ.* **2015**, *120*, 26–32. [CrossRef]
37. Liu, L.; Gitz, C.D.; McClure, W.J. Effects of UV-B on flavonoids, ferulic acid, growth and photosynthesis in barley primary leaves. *Physiol. Plant* **1995**, *93*, 725–733. [CrossRef]
38. Abideen, Z.; Qasim, M.; Rasheed, A.; Adnan, M.Y.; Gul, B.; Khan, M.A. Antioxidant activity and polyphenolic content of *Phragmites karka* under saline conditions. *Pak. J. Bot.* **2015**, *47*, 813–818.
39. Unal, B.T.; Aktas, L.Y.; Guven, A. Effects of salinity on antioxidant enzymes and proline in leaves of barley seedlings in different growth stages. *Bulg. J. Agric. Sci.* **2014**, *20*, 883–887.
40. Alonso-Amelot, M.E.; Oliveros-Bastidas, A.; Calcagno-Pisarelli, M.P. Phenolics and condensed tannins of high altitude *Pteridium arachnoideum* in relation to sunlight exposure, elevation, and rain regime. *Biochem. Syst. Ecol.* **2007**, *35*, 1–10. [CrossRef]
41. Ganzera, M.; Guggenberger, M.; Stuppner, H.; Zidorn, C. Altitudinal variation of secondary metabolite profiles in flowering heads of *Matricaria chamomilla* cv. BONA. *Planta Med.* **2008**, *74*, 453–457. [CrossRef] [PubMed]
42. Stanković, M.S. Total phenolic content, flavonoid concentration and antioxidant activity of *Marrubium peregrinum* L. extracts. *Kragujevac J. Sci.* **2011**, *33*, 63–72.
43. Zlatić, N.M.; Stanković, M.S.; Simić, Z.S. Secondary metabolites and metal content dynamics in *Teucrium montanum* L. and *Teucrium chamaedrys* L. from habitats with serpentine and calcareous substrate. *Environ. Monit. Assess.* **2017**, *189*, 110. [CrossRef] [PubMed]
44. Singleton, V.L.; Orthofer, R.; Lamuela, R.R.M. Analysis of total phenols and other oxidation substrates and antioxidants by means of Folin-Ciocalteu reagent. *Methods Enzymol.* **1999**, *299*, 152–178.
45. Quettier, D.C.; Gressier, B.; Vasseur, J.; Dine, T.; Brunet, C.; Luyckx, M.C.; Cayin, J.C.; Bailleul, F.; Trotin, F. Phenolic compounds and antioxidant activities of buckwheat (*Fagopyrum esculentum* Moench) hulls and flour. *J. Ethnopharmacol.* **2000**, *72*, 35–42. [CrossRef]
46. Takao, T.; Watanabe, N.; Yagi, I.; Sakata, K. A simple screening method for antioxidant and isolation of several antioxidants produced by marine bacteria from fish and shellfish. *Biosci. Biotechnol. Biochem.* **1994**, *58*, 1780–1783. [CrossRef]
47. Kumarasamy, Y.; Byres, M.; Cox, P.J.; Jaspars, M.; Nahar, L.; Sarker, S.D. Screening seeds of some Scottish plants for free radical scavenging activity. *Phytother. Res.* **2007**, *21*, 615–621. [CrossRef] [PubMed]

© 2017 by the authors. Licensee MDPI, Basel, Switzerland. This article is an open access article distributed under the terms and conditions of the Creative Commons Attribution (CC BY) license (http://creativecommons.org/licenses/by/4.0/).

Article

The Effect of Harvesting on the Composition of Essential Oils from Five Varieties of *Ocimum basilicum* L. Cultivated in the Island of Kefalonia, Greece

Gerasimia Tsasi [1,2], Theofilos Mailis [2], Artemis Daskalaki [2], Eleni Sakadani [1], Panagis Razis [1], Yiannis Samaras [1] and Helen Skaltsa [2,*]

[1] Department of Food Technology, Technological Educational Institute (T.E.I.) of Ionian Islands, 28100 Argostoli, Greece; gtsasi@pharm.uoa.gr (G.T.); elenisakadani@hotmail.com (E.S.); p_razis@yahoo.gr (P.R.); ysamaras@teiion.gr (Y.S.)
[2] Department of Pharmacognosy and Chemistry of Natural Products, School of Pharmacy, University of Athens, 15771 Athens, Greece; thema-89@hotmail.com (T.M.); artemisdask@hotmail.com (A.D.)
* Correspondence: skaltsa@pharm.uoa.gr; Tel./Fax: +30-210-727-4593

Received: 28 August 2017; Accepted: 16 September 2017; Published: 18 September 2017

Abstract: Five varieties of *Ocimum basilicum* L. namely lettuce, cinnamon, minimum, latifolia, and violetto were separately cultivated in field and greenhouse in the island Kefalonia (Greece). The effect of successive harvesting to the essential oil content was evaluated. In total 23 samples of essential oils (EOs) were analyzed by GC-FID and GC-MS. Ninety-six constituents, which accounted for almost 99% of the oils, were identified. Cluster analysis was performed for all of the varieties in greenhouse and field conditions, in order to investigate the possible differentiation on the chemical composition of the essential oils, obtained between harvests during growing period. Each basil variety showed a unique chemical profile, but also the essential oil composition within each variety seems to be differentiated, affected by the harvests and the cultivation site.

Keywords: basil varieties; essential oil; GC-MS; harvest; cluster analysis

---

1. Introduction

Basil (*Ocimum basilicum* L.) is an annual herb growing in several regions around the world. The genus *Ocimum* consists of more than 150 species whereof basil is cultivated in many countries as a major essential oil crop [1]. The species most commonly used as spices and medicinal herbs are *O. basilicum* L., *O. americanum* L. (syn. *O. canum* Sims.), *O. gratissimum* L., *O. kilimandscharicum* Guerke, *O. tenuiflorum* L. (syn. *O. sanctum* L.), and *O. africanum* Lour. (syn. *O.* × *citriodorum* Vis.), that is a hybrid of *O. basilicum* and *O. americanum* [2]. The taxonomy of the genus *Ocimum* is very complicated due to the occurrence of interspecific hybridization, polyploidy, aneuploidy, the existence of numerous botanical varieties, and chemotypes, as well as the many synonymous names [3–5]. Plants of the species *O. basilicum* have square stems, fragrant opposite leaves, and whorled flowers on spiked inflorescences [6]. *O. basilicum* L. is widely used in the culinary arts and in the food processing industry [7]. Traditionally, it has been used as medicinal plant for the treatment of headaches, coughs, diarrhea, constipation, warts, worms, and kidney malfunction [8]. A plethora of biological activities have been attributed to basil essential oils like antimicrobial, insecticidal, and recently was found to exhibit in vivo anti-malarial activity [9]. In addition, extracts from leaves and flowers can be used as aroma additives in food, pharmaceutical, and cosmetic industry [10]. Economical data clearly indicate the commercial importance of basil essential oil, since the world production is estimated to 1,200,000 € [8]. Basil essential oils contain a broad array of chemical compounds depending on variations in chemotypes, flower, and leaf colours, aroma and especially the origin of the plant [7]. The essential oil constituents vary among sweet basil cultivars, with linalool, methyl chavicol, eugenol, 1,8-cineole, geranial, neral, methyl cinnamate recognized as main components [11–13]. Several scientists have classified basil to different chemotypes according to the main components of the essential oil. Thereby, Marotti et al. (1996) divided essential oil from Italian basil varieties to three chemotypes "linalool", "linalool-methyl chavicol", and "linalool-eugenol" [14]. According to the plant origin, basils are grouped to European chemotype that has linalool and methyl chavicol as main components and Tropical chemotype having methyl chavicol as a main compound [15]. Concerning the cultivation of basil, the plants can be harvested one to three times during cropping season depending on the climate [16]. Generally, basil is harvested for the leaves that are sold fresh or dried. In the cases that basil is cultivated for dried leaves and extraction of essential oil, the plants are cut just prior the appearance of the flowers [17]. Scientific data concerning how harvest and number of harvests during the cultivation of aromatic plants affect essential oil quality and composition are lacking either in field or greenhouse conditions. To the best of our knowledge, so far Carlo et al. (2013) studied the effect of cut number in quality traits of sweet basil [18]. Also Zheljazkof et al. (2008) studied the effect of harvests in *O. basilicum* L. (cvs. German and Mesten) and *O. sanctum* L. (syn. *O. tenuiflorum* L.) (cv. Local) cultivated in Mississippi [9]. Taking into account that geographic position and the environmental characteristics of the habitat should be taken as factors affecting the chemical composition of the essential oils [19], the current study is the first including total chemical analysis of the essential oils from five basil varieties cultivated and harvested in the field and greenhouse in the island of Kefalonia, (Greece). The aim of our study is the chemical analysis of essential oils from the five varieties of basil considering also the cultivation site as well as how the ability of the plant to rejuvenate after successive cuttings during growing cycle, affects the obtained essential oil yield and composition. In order to simulate the natural growing conditions, successive cuttings were performed in every variety just before flowering stage, as done practically from aromatic plants' growers. Moreover the cluster analysis of all basil varieties for greenhouse and field samples is given.

2. Results and Discussion

2.1. Chemical Analysis

In the greenhouse conditions the essential oil (EO) content of varieties violetto, latifolia, and minimum increased in the second and/or third harvest, whereas in var. lettuce the second harvest produced less essential oil content when comparing to the first and third harvest where the essential oil content was in the same levels. Noteworthy is the fact, that in the field conditions, the essential oil content in the varieties of latifolia, minimum and lettuce decreased gradually after the successive harvests while in var. cinnamon sequential harvests didn't affect overall essential oil yield (Table 1).

Results indicate that different cultivation conditions affect plant's response to successive harvests and consequently the essential oil content, even of the same variety. As such controlled greenhouse conditions favor or have no impact to the total production of essential oil in the varieties latifolia, minimum, and lettuce when comparing to the field conditions.

All obtained essential oils were analyzed with GC/FID and GC/MS. In total, 96 individual constituents were identified representing 98.0–99.9% of the total essential oil. The detailed chemical analysis of the essential oils of the greenhouse samples showed that the major constituents, in var. violetto, were linalool (22.5–26.5%) and *trans*-bergamotene (15.2–20.0%), followed by eugenol (4.5–11.9%). Variety latifolia has as major compounds linalool (17.1–35.6%), eugenol (10.1–23.3%) and *trans*-bergamotene (8.8–16.1%). In var. minimum linalool (27.4–28.3%) and eugenol (14.5–23.4%) were the main constituents while in var. lettuce linalool (21.0–23.6%), methyl chavicol (12.1–17.5%), *trans*-bergamotene (11.4–12.9%) and *epi*-α-cadinol (6.0–9.2%) (Table 2).

Table 1. Samples from *O. basilicum* varieties and essential oil yield.

O. basilicum L. Varieties	Harvest	Cultivation Site	EO (% Fresh Weight)
O. basilicum L. var. *violetto*	OCK1	greenhouse	0.017
	OCK2	greenhouse	0.04
	OCK3	greenhouse	0.08
O. basilicum L. var. *latifolia*	OCP4	greenhouse	0.026
	OCP5	greenhouse	0.14
O. basilicum L. var. *minimum*	OCS6	greenhouse	0.033
	OCS7	greenhouse	0.08
O. basilicum L. var. *lettuce*	OCM8	greenhouse	0.10
	OCM9	greenhouse	0.04
	OCM10	greenhouse	0.15
O. basilicum L. var. *latifolia*	OCP11	field	0.18
	OCP12	field	0.20
	OCP13	field	0.12
	OCP14	field	0.08
O. basilicum L. var. *minimum*	OCS15	field	0.11
	OCS16	field	0.04
	OCS17	field	0.036
O. basilicum L. var *lettuce*	OCM18	field	0.12
	OCM19	field	0.07
	OCM20	field	0.08
O. basilicum L. var. *cinnamon*	OCC21	field	0.21
	OCC22	field	0.21
	OCC23	field	0.20

In the field conditions, the main constituents of var. latifolia were linalool (32.2–49.5%), eugenol (19.8–34.9%), *trans*-bergamontene (5.9–9.5%), and 1,8-cineole (4.2–6.7%). In var. minimun linalool (30.2–52.0%) and eugenol (28.8–36.0%), were the dominating compounds. Var. lettuce has as major components linalool (20.4–26.0%) and methyl-chavicol (45.4–54.1%). The latter constituent was dominating also in the essential oil of var. cinnamon in a range 61.2–75.1% among harvests (Table 3). Between harvests and cultivation sites, the main compounds of the essential oils displayed varying concentrations. The highest concentration for linalool (52.0%) was measured in var. minimum in the field cultivation, in the second harvest in contrast to what has been stated from Zheljazkof et al. (2008) that measures a higher content of linalool in the third cutting of *O. basilicum* varieties [9]. Interesting is the fact that eugenol showed the highest concentration in var. minimum in the first harvest in the field conditions. Noteworthy, from var. minimum methyl-chavicol is lacking or is present in small amounts. Given that methyl-chavicol has a structural resemblance to potential carcinogenic phenylpropanoids such as safrole, chemotypes rich in linalool are preferred for cultivation when used in food and perfume industries [15,20]. Methyl-chavicol showed the highest concentration (75.1%) in var. cinnamon in the field conditions from the first harvest followed by var. lettuce (54.1%) from the first harvest in the same conditions. In addition, it has been observed that the intensity of purple leaf colour is positively correlated in varieties rich in methyl chavicol [21]. However, in compliance with Liber et al. (2011), our results show that the biosynthesis of phenylpropanoids e.g., methyl chavicol, is not characteristic exclusively for the purple morfotypes. Also, green morfotypes (var. lettuce and cinnamon) are rich sources of methyl chavicol as well. [22].

Table 2. Chemical composition of the essential oils from *O. basilicum* L. varieties, violetto, latifolia, minimum and lettuce, cultivated in the greenhouse conditions.

	Compound	RI [a]	OCK1	OCK2	OCK3	OCP4	OCP5	OCS6	OCS7	OCM8	OCM9	OCM10
1	α-thujene	923	-	-	tr	-	-	tr	tr	tr	tr	tr
2	α-pinene	932	0.9	0.1	0.4	0.8	0.2	0.8	0.3	0.5	0.3	0.4
3	camphene	946	-	-	tr	0.1	0.1	0.1	0.1	0.1	tr	tr
4	sabinene	968	0.3	0.1	0.3	0.4	0.2	0.1	0.1	0.2	0.1	0.2
5	β-pinene	973	0.7	-	0.6	0.7	0.4	0.3	0.2	0.5	0.2	0.5
6	myrcene	984	1.0	0.1	0.7	0.8	0.5	0.4	0.3	0.3	0.2	0.4
7	α-phellandrene	1001	-	-	0.2	-	-	-	0.1	-	0.1	0.1
8	δ-3-carene	1004	0.3	-	-	0.4	0.1	0.4	tr	0.2	0.1	-
9	α-terpinene	1011	-	-	0.1	0.1	-	0.1	0.1	0.1	0.1	0.1
10	limonene	1019	-	0.2	-	-	-	-	-	-	-	-
11	p-cymene	1017	-	-	-	-	-	-	0.1	-	-	-
12	1,8-cineole	1027	6.6	2.3	6.3	4.9	7.1	2.7	2.7	5.2	3.5	4.8
13	*trans*-β-ocimene	1043	-	-	tr	1.7	1.7	1.1	1.3	0.9	0.8	1.1
14	γ-terpinene	1051	0.1	-	0.2	0.1	0.1	0.2	0.4	0.3	0.4	0.4
15	α-terpinolene	1078	0.1	0.6	0.1	0.4	0.2	0.3	0.2	0.2	0.2	0.2
16	linalool	1100	26.5	22.5	23.5	17.1	35.6	27.4	28.3	21.0	23.1	23.6
17	allo-ocimene	1133	-	-	-	-	-	-	-	tr	-	-
18	*trans*-epoxy-ocimene	1134	-	-	-	0.2	0.4	0.1	-	-	-	-
19	camphor	1142	-	-	tr	0.2	0.3	0.3	0.3	0.4	0.6	0.5
20	borneol	1166	-	-	0.2	0.5	0.8	0.3	0.3	-	-	-
21	terpinen-4-ol	1172	0.2	0.2	0.2	0.2	0.2	0.5	1.2	-	2.0	1.9
22	α-terpineol	1184	0.7	-	1.4	0.7	1.3	0.4	-	1.5	-	-
23	methyl chavicol	1193	-	0.5	-	-	-	-	1.9	17.5	12.1	16.0
24	myrtenal	1205	-	-	-	-	-	-	-	tr	tr	-
25	octyl acetate	1207	0.2	-	0.2	0.4	0.1	0.2	0.2	-	-	-
26	endo-fenchyl acetate	1213	0.7	-	0.4	-	-	-	-	-	-	-
27	neral	1231	-	-	-	-	-	0.1	-	-	-	-
28	chavicol	1245	-	-	-	-	-	-	-	1.0	0.3	0.8
29	geraniol	1247	-	-	0.7	-	-	-	0.9	-	-	-
30	*trans*-anethole	1280	-	-	-	-	-	-	0.2	-	0.3	-
31	bornyl-acetate	1282	-	-	-	3.2	1.4	1.5	1.1	0.7	0.5	0.6
32	carvacrol	1296	10.0	0.8	2.9	4.4	0.7	4.3	2.7	2.3	1.1	1.5
33	α-cubebene	1334	-	-	0.1	0.1	-	0.1	tr	0.1	-	tr
34	eugenol	1356	8.6	4.5	11.9	10.1	23.3	14.5	23.4	7.2	2.1	9.1
35	α-copaene	1369	0.2	0.2	-	tr	0.1	-	-	0.3	0.2	-

Table 2. Cont.

	Compound	RI [a]	OCK1	OCK2	OCK3	OCP4	OCP5	OCS6	OCS7	OCM8	OCM9	OCM10
36	β-patchoulene	1372	-	-	-	-	-	tr	-	0.1	0.1	0.1
37	β-cubebene	1375	-	-	-	-	-	-	0.1	-	-	-
38	β-elemene	1387	1.4	1.9	1.9	1.3	0.9	3.6	2.3	2.6	3.8	2.4
39	methyl eugenol	1399	0.1	-	0.4	0.2	0.1	0.1	0.1	0.1	0.1	0.1
40	cis-α-bergamotene	1398	-	-	-	-	0.1	-	0.1	-	-	-
41	trans-β-caryophyllene	1409	1.3	1.6	1.9	1.1	0.2	0.8	0.4	0.6	0.6	0.4
42	β-gurjunene	1417	-	0.7	-	-	-	0.5	0.3	-	-	-
43	trans-α-bergamotene	1432	15.6	20.0	15.2	16.1	8.8	3.8	5.1	11.4	12.9	11.7
44	aromadendrene	1437	0.2	0.3	0.3	0.4	0.2	0.6	0.1	0.4	0.5	0.4
45	cis-muurola-3,5-diene	1439	-	-	-	-	0.3	0.6	0.5	-	-	-
46	trans-β-farnesene	1446	0.3	0.4	-	1.3	-	-	-	-	-	0.4
47	α-humulene	1448	1.1	2.6	1.4	0.4	0.9	1.5	0.9	0.4	1.4	0.6
48	epi-bicyclosesquiphellandrene	1453	0.4	0.6	0.5	0.9	0.4	1.0	0.8	0.7	0.9	0.7
49	allo-aromadendrene	1459	-	-	0.2	0.2	-	0.2	0.1	tr	-	-
50	α-amorphene	1460	-	-	0.3	-	-	-	-	tr	-	-
51	germacrene-D	1476	3.7	9.4	5.4	6.7	3.6	7.0	4.5	5.4	8.2	5.4
52	β-selinene	1475	2.9	3.5	-	-	-	-	-	0.2	-	-
53	bicyclogermacrene	1486	-	-	3.0	2.2	1.1	2.5	2.0	1.3	1.2	1.3
54	δ-guaiene	1496	2.7	7.1	3.8	3.1	1.2	3.8	2.1	2.3	3.3	1.7
55	germacrene-A	1499	2.6	5.4	1.5	1.4	1.1	1.1	1.0	0.8	1.0	0.7
56	γ-cadinene	1506	1.9	2.8	1.7	2.7	1.6	3.2	3.4	2.7	3.5	2.8
57	β-sesquiphellandrene	1514	0.7	1.0	0.9	0.9	0.4	0.2	0.2	0.6	0.7	0.6
58	δ-cadinene	1516	0.2	0.3	0.1	0.2	0.1	0.2	0.1	0.1	0.2	0.1
59	α-cadinene	1520	-	-	0.1	0.1	0.2	0.1	0.1	0.1	tr	0.1
60	trans-γ-bisabolene	1530	0.2	-	-	0.1	-	-	-	0.1	0.1	0.1
61	γ-cuprenene	1535	0.1	0.2	0.1	0.1	-	-	-	tr	0.1	tr
62	trans-cadina-1,4-diene	1534	-	-	0.1	0.1	-	0.1	-	0.1	0.2	tr
63	germacrene B	1548	-	-	-	0.1	-	-	-	-	-	-
64	trans-nerodilol	1543	-	-	0.4	-	-	0.2	0.1	0.1	0.1	0.2
65	maaliol	1550	-	1.5	0.5	-	-	-	-	0.1	0.9	0.2
66	caryophyllene oxide	1557	-	-	-	-	-	-	-	tr	-	-
67	spathulenol	1562	-	-	0.1	0.1	-	-	tr	tr	0.1	tr
68	viridiflorol	1595	0.5	0.7	0.9	1.1	-	-	-	0.9	1.1	0.8
69	1,10-di-epi-cubenol	1596	-	-	-	-	0.4	1.0	0.8	-	-	-
70	epi-α-cadinol	1638	3.7	5.9	5.0	6.7	3.2	7.2	6.3	6.2	9.2	6.0

Table 2. *Cont.*

	Compound	RI [a]	OCK1	OCK2	OCK3	OCP4	OCP5	OCS6	OCS7	OCM8	OCM9	OCM10
71	vulgarone B	1641	-	-	-	-	-	-	-	-	-	-
72	α-eudesmol	1651	-	-	-	-	-	-	-	-	-	-
73	α-cadinol	1652	0.2	0.5	0.7	0.8	0.2	0.7	0.5	0.7	0.9	0.6
74	7-epi-α-eudesmol	1659	-	-	-	-	-	-	-	tr	-	-
75	α-bisabolol	1685	0.6	-	0.2	0.2	-	0.1	0.1	0.1	0.1	0.1
76	germacrone	1691	-	-	-	0.3	-	-	-	-	-	-
77	*cis*-farnesol	1692	-	-	0.3	-	-	-	-	0.1	0.2	0.1
78	β-sinensal	1693	-	-	0.4	-	-	0.2	-	0.2	0.2	0.1
79	mint sulfide	1733	-	-	tr	0.1	-	0.1	-	tr	-	-
80	neo phytadiene	1838	-	-	-	0.8	-	-	-	-	-	-
81	nonadecane	1900	-	-	-	tr	-	-	-	-	-	-
82	eicosane	2000	-	-	-	0.4	-	-	-	-	-	-
83	heneicosane	2100	-	-	-	0.2	-	-	-	-	-	-
84	docosane	2200	-	-	-	0.2	-	-	-	-	-	-
85	tricosane	2300	-	-	-	0.1	-	-	-	-	-	-
86	tetracosane	2400	-	-	-	-	-	tr	-	-	-	-
87	pentacosane	2500	-	-	-	-	-	0.2	-	tr	-	-
88	hexacosane	2600	-	-	-	-	-	tr	-	-	-	-
89	heptacosane	2700	0.3	0.5	0.1	-	-	0.2	-	0.1	-	-
90	nonacosane	2900	0.3	0.5	0.1	-	-	-	-	0.1	-	-
91	triacontane	3000	0.2	0.3	0.1	-	-	-	-	0.1	-	-
	Total		98.3	99.8	98.3	98.0	99.8	97.1	98.5	99.2	99.9	99.9

[a] Retention indices were calculated against C9-C23 *n*-alkanes on the HP 5MS capillary column.

Table 3. Chemical composition of the essential oils from *O. basilicum* L. varieties latifolia, minimum, lettuce and cinnamon, cultivated in the field conditions.

	Compound	RI [a]	OCP11	OCP12	OCP13	OCP14	OCS15	OCS16	OCS17	OCM18	OCM19	OCM20	OCC21	OCC22	OCC23
1	α-thujene	922	-	-	-	-	-	-	tr	tr	tr	tr	tr	tr	tr
2	α-pinene	931	0.2	0.2	-	0.5	-	0.3	0.3	0.3	0.3	0.2	0.2	0.5	0.5
3	Camphene	946	0.1	tr	tr	0.1	tr	0.1	0.1	tr	tr	tr	tr	0.2	0.2
4	sabinene	967	0.2	0.2	0.2	0.2	tr	0.2	0.1	0.1	0.2	0.1	0.2	0.4	0.4
5	β-pinene	974	0.3	0.3	0.3	0.5	0.1	0.3	0.3	0.3	0.3	0.3	0.3	0.7	0.8
6	myrcene	984	0.3	0.3	0.5	0.7	0.1	-	0.5	0.2	0.3	0.3	0.4	0.7	0.8
7	3-octanol	988	-	-	-	-	-	-	-	-	-	tr	-	-	-
8	n-decane	1000	-	-	-	-	-	-	-	-	-	0.1	-	-	-
9	α-phellan-drene	1002	-	-	-	-	-	-	tr	-	0.1	tr	tr	0.1	0.1
10	δ-3-carene	1004	tr	-	-	0.1	-	-	tr	tr	-	-	-	-	-

Table 3. Cont.

	Compound	RI [a]	OCP11	OCP12	OCP13	OCP14	OCS15	OCS16	OCS17	OCM18	OCM19	OCM20	OCC21	OCC22	OCC23
11	α-terpinene	1013	-	-	tr	0.1	0.1	-	tr	0.1	0.1	0.1	tr	0.1	0.1
12	p-cymene	1019	-	-	-	-	0.1	-	tr	-	-	0.1	-	-	-
13	limonene	1022	-	-	-	-	0.1	-	0.2	-	-	-	-	-	-
14	1,8-cineole	1026	4.2	4.2	5.1	6.7	1.6	3.8	2.7	5.7	4.8	4.6	4.8	6.8	7.6
15	cis-β-ocimene	1035	-	-	-	-	-	-	-	-	tr	-	-	-	-
16	trans-β-ocimene	1043	1.4	1.1	0.9	0.6	0.8	1.4	0.9	0.7	0.5	0.4	1.9	3.0	2.9
17	γ-terpinene	1053	0.1	0.1	0.1	tr	0.2	0.2	0.2	0.3	0.3	0.4	0.1	0.1	0.1
18	cis-sabinene hydrate	1064	-	-	-	-	-	-	-	-	-	tr	-	tr	tr
19	α-terpinolene	1081	0.2	0.2	0.1	0.2	0.2	0.1	0.3	0.1	0.1	0.1	0.1	0.3	0.3
20	fenchone	1081	-	-	-	-	-	-	-	-	-	tr	-	-	-
21	linalool	1100	32.2	34.0	35.5	49.5	36.8	52.0	30.2	26.0	21.8	20.4	5.7	0.6	1.3
22	allo-ocimene	1132	-	-	-	tr	-	-	-	tr	tr	tr	-	-	-
23	trans-epoxy-ocimene	1134	0.3	0.3	0.1	0.1	0.1	0.2	tr	-	-	-	tr	0.1	0.1
24	camphor	1141	0.3	0.3	0.1	0.1	0.3	0.2	0.3	0.3	0.2	0.3	0.6	2.1	1.9
25	borneol	1165	0.6	0.6	0.5	0.4	0.4	0.2	0.4	-	-	0.2	-	-	tr
26	terpinen-4-ol	1173	0.1	0.1	0.2	0.1	1.2	1.1	0.8	-	tr	2.1	-	-	-
27	α-terpineol	1184	0.7	1.0	1.0	0.8	0.8	0.5	0.9	-	-	-	-	-	-
28	methyl chavicol	1186	-	-	-	-	-	0.3	0.3	45.4	54.1	46.7	75.1	60.2	64.9
29	octyl acetate	1208	0.1	0.1	0.1	0.2	0.2	0.1	0.1	tr	tr	tr	tr	tr	-
30	nerol	1224	-	-	-	-	-	-	0.1	-	-	tr	-	-	-
31	citronellol	1225	-	-	-	-	-	-	0.1	-	-	tr	-	-	-
32	neral	1234	-	-	-	-	-	-	0.1	-	-	-	-	-	-
33	carvone	1239	-	-	-	-	-	-	-	-	tr	-	tr	0.1	0.1
34	chavicol	1243	-	-	-	-	-	0.7	1.5	0.2	0.2	0.5	tr	-	-
35	geraniol	1245	-	-	-	-	0.6	-	-	-	-	-	-	-	-
36	citronellyl formate	1277	-	-	-	-	-	-	-	-	-	-	-	-	tr
37	trans-anethole	1280	-	-	-	-	-	-	-	tr	tr	-	-	tr	tr
38	bornyl acetate	1289	0.8	0.8	0.6	1.2	0.6	0.3	1.3	0.1	0.1	0.2	0.2	0.6	0.5
39	carvacrol	1298	1.0	0.4	0.3	tr	0.7	0.3	0.3	0.7	1.1	0.2	0.2	0.4	0.2
40	δ-elemene	1336	-	-	-	-	-	-	-	-	-	-	tr	tr	tr
41	carvyl acetate	1341	-	-	-	-	-	-	-	-	-	0.1	-	tr	tr
42	α-cubebene	1356	26.9	32.2	34.9	19.8	36.0	28.8	28.9	10.0	2.5	5.6	0.1	0.1	0.1
43	eugenol	1359	-	-	-	-	0.2	tr	-	-	tr	0.1	-	0.1	0.2
44	neryl acetate	1369	0.1	-	-	tr	-	-	-	-	-	-	tr	0.2	-
45	α-copaene	1378	-	-	-	-	-	-	0.8	-	-	0.1	0.6	-	0.2
46	geranyl acetate	1386	-	-	-	-	-	-	-	tr	tr	0.1	-	0.2	-
47	β-bourbonene	1387	-	-	-	-	-	-	-	-	-	tr	0.1	-	0.1
48	β-cubebene	1391	1.0	-	tr	0.7	1.9	1.1	1.1	0.5	0.6	0.8	0.9	1.9	1.3
49	β-elemene	1402	0.1	0.2	0.2	0.1	0.1	-	0.2	tr	tr	0.1	0.1	0.5	0.5
50	methyl-eugenol	1398	-	-	-	-	-	-	-	-	-	-	-	tr	tr
51	α-gurjunene	1401	0.1	-	-	-	-	-	-	tr	tr	0.1	-	tr	tr
52	cis-α-bergamotene	1415	0.2	0.2	0.2	0.2	0.2	0.1	0.2	0.1	0.1	0.1	0.1	0.4	0.3
53	trans-β-caryophyllene	1414	-	-	-	-	-	-	-	-	-	0.1	-	0.1	0.1
54	β-copaene	1414	-	-	-	-	-	-	-	-	-	-	-	-	-

Table 3. Cont.

	Compound	RI [a]	OCP11	OCP12	OCP13	OCP14	OCS15	OCS16	OCS17	OCM18	OCM19	OCM20	OCC21	OCC22	OCC23
55	β-gurjunene	1420	0.4	-	-	-	-	-	0.1	-	-	-	tr	0.2	-
56	trans-α-bergamotene	1431	9.5	7.6	6.5	5.9	2.2	2.1	3.1	4.0	5.1	4.4	0.4	1.0	1.5
57	α-guaiene	1427	-	-	-	-	-	-	-	-	-	-	0.1	0.4	-
58	aromadendrene	1435	0.2	-	-	0.1	-	-	0.1	tr	0.1	0.1	tr	tr	-
59	cis-muurola-3,5-diene	1439	0.3	0.2	0.2	-	0.2	-	0.3	-	-	-	-	-	-
60	trans-β-farnesene	1445	-	-	0.2	0.2	-	-	0.2	-	0.1	-	-	-	-
61	α-humulene	1448	0.8	0.8	0.2	0.2	0.5	0.2	0.3	0.2	0.1	0.4	0.5	1.4	1.0
62	epi-bicyclosesquiphellandrene	1454	0.5	0.4	0.3	0.2	0.4	-	0.6	0.1	0.1	0.3	0.1	0.5	0.4
63	allo-aromadendrene	1456	-	-	2.6	tr	-	0.1	tr	-	-	-	-	-	-
64	α-amorphene	1460	-	-	-	-	-	-	-	-	-	-	-	-	-
65	germacrene-D	1475	4.2	3.5	0.1	1.9	3.0	1.5	2.8	1.2	1.6	2.0	3.2	5.0	3.2
66	β-selinene	1479	-	0.1	0.8	0.1	tr	-	0.1	0.2	tr	0.1	-	-	-
67	bicyclogermacrene	1486	1.0	0.9	0.9	0.4	0.9	-	1.2	0.2	0.3	0.5	0.3	0.3	0.2
68	δ-guaiene	1493	1.4	1.3	0.4	0.5	1.4	-	1.3	0.1	0.3	0.6	0.5	1.4	1.0
69	germacrene-A	1498	1.2	0.7	1.7	0.2	0.5	-	0.4	-	0.1	0.3	0.2	0.5	0.3
70	γ-cadinene	1505	1.9	1.6	0.3	1.5	2.0	0.9	3.2	0.7	0.9	1.2	0.9	1.9	1.6
71	trans-calamenene	1506	-	-	-	-	tr	-	0.2	-	-	-	tr	0.1	-
72	β-sesquiphellandrene	1508	0.5	0.4	tr	0.2	0.1	-	0.1	0.1	tr	0.2	-	0.1	0.1
73	δ-cadinene	1514	0.2	0.1	tr	tr	tr	-	0.1	-	tr	0.1	tr	0.2	0.1
74	α-cadinene	1520	-	tr	-	tr	tr	-	-	-	tr	tr	-	-	-
75	γ-cuprenene	1528	-	0.1	-	-	-	-	tr	-	tr	-	-	tr	tr
76	trans-γ-bisabolene	1530	0.1	-	-	-	-	-	-	-	tr	-	tr	tr	tr
77	germacrene B	1551	0.7	0.5	-	-	-	-	-	-	-	-	-	-	-
78	trans-nerolidol	1557	-	-	-	tr	0.1	-	0.1	tr	tr	0.1	tr	tr	tr
79	maaliol	1565	-	-	-	-	-	-	-	tr	tr	0.1	tr	0.2	0.1
80	spathulenol	1571	-	tr	-	0.1	-	-	tr	-	0.1	tr	-	0.1	tr
81	caryophyllene-oxide	1585	-	-	tr	-	-	-	-	-	tr	tr	-	tr	tr
82	viridiflorol	1592	-	-	-	-	-	-	0.2	0.1	0.2	0.4	0.2	0.6	0.5
83	guaiol	1600	-	-	-	-	-	-	-	-	-	-	-	tr	-
84	1,10-di-epi-cubenol	1616	0.5	0.5	0.5	0.5	0.6	-	1.0	1.8	2.9	3.2	2.3	4.1	3.1
85	epi-α-cadinol	1637	4.5	3.8	3.7	4.4	4.1	1.8	6.6	0.1	0.1	0.3	0.1	0.4	0.3
86	α-cadinol	1653	0.3	0.3	0.3	0.2	0.2	-	0.5	-	-	-	-	tr	tr
87	allo himachalol	1654	-	-	-	-	-	-	-	-	-	-	-	-	-
88	β-bisabolol	1681	-	-	-	-	-	-	0.2	-	-	-	-	-	-
89	α-bisabolol	1683	-	0.1	tr	0.1	0.1	-	0.1	tr	tr	0.1	tr	0.1	0.1
90	cis-farnesol	1680	-	-	-	-	-	-	-	-	tr	-	-	-	-
91	β-sinensal	1686	-	0.1	-	-	-	-	-	-	tr	-	tr	tr	tr
92	pentadecanal	1711	-	-	-	-	-	-	0.1	-	-	-	-	-	tr
93	mintsulfide	1731	-	-	-	-	-	-	-	-	-	-	-	tr	-
94	6,10,14-trimethyl pendadecanone	1836	-	-	-	-	-	-	tr	-	-	-	-	-	-
95	E.E-farnesyl acetone	1904	-	-	-	-	-	-	tr	-	-	-	-	-	-
96	palmitic acid	1961	-	-	-	-	-	-	tr	-	-	-	-	-	-
	Total		99.7	99.8	99.9	99.5	99.7	99.2	95.7	99.9	99.7	98.9	99.9	99.1	99.2

[a] Retention indices were calculated against C9-C23 n-alkanes on the HP 5MS capillary column.

Despite the observation that in the field conditions the total essential oil yield was decreasing after successive harvests the individual components, prevalent in basil, like linalool and eugenol seem to follow an opposite pattern. Linalool in var. latifolia was beyond 32.2% in all harvests and in the last harvest reached 49.5% while in var. minimum the percentage of linalool was in all harvests above 30.0% and in the second harvest reached 52.0% in the total essential oil concentration. Furtermore, although the concentrations of linalool in var. lettuce for both cultivation conditions were similar, compound methyl-chavicol showed a remarkable difference reaching 54.1% of the EO in the second harvest in the field conditions, whereas in the greenhouse conditions the higher concentration measured, was 17.5% in the first harvest (Tables 2 and 3). These data suggest that changes in the environmental conditions can alter the biosynthesis of individual essential oils' components. Ložienė et al. (2004) have reached similar conclusion while studying the influence of environmental and genetic factors in *Thymus pulegioides'* essential oil [23]. Similar conclusions were reached also by Awadh Ali et al. (2017) while studying the EO content of *O. forskolei* and *Teucrium yemense* collected from different regions of Yemen [24]. In parallel, var. cinnamon grown in the field conditions has methyl-chavicol as major constituent reaching 75.1% of the EO content, in the first harvest (Table 3). It seems that the unstable field conditions, in combination with successive harvests affect and promote the metabolic pathway of L-phenylalanine and cinnamic acid [25], in var. latifolia, minimum and lettuce that leads to the production of methyl chavicol and eugenol. Moreover, the combination of the harvest stress with the unstable field conditions seems to promote the biosynthesis of individual components. In addition to this, the observed decrease in the EO yield in the field conditions during harvests can be attributed to the partial evaporation of the essential oil from the plant surface.

2.2. Statistical Analysis

The statistical analysis of our results confirmed the laborious interaction between the essential oil composition, different varieties, and cultivation conditions. After thorough study of the cluster analyses, it is obvious that during the growing period, varieties can alter the composition of individual components of the essential oil by shifting their chemotype [23].

Taking into account the EO content similarities between harvests for the varieties cultivated in the greenhouse, four clusters were formed (Figure 1). Cluster 1 consists of samples OCK1, OCK3, and OCP4, with linalool, *trans*-α-bergamontene, and eugenol as the dominant components. Cluster 2 consists of samples OCP5, OCS6, and OCS7, with linalool and eugenol as the dominant components. Cluster 3 is "simplicifolious" and consists of OCK2 with major constituents linalool and *trans*-α-bergamontene. Finally, cluster 4 consists of OCM8, OCM9, and OCM10 with major components linalool, methyl chavichol, and *trans*-α-bergamontene (Figure 2).

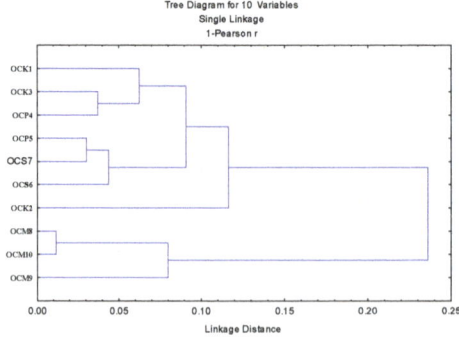

Figure 1. Bidimensional dendrogram representing the similarity in the main components of the essential oils, between 10 harvests of 4 basil varieties growing in the greenhouse conditions.

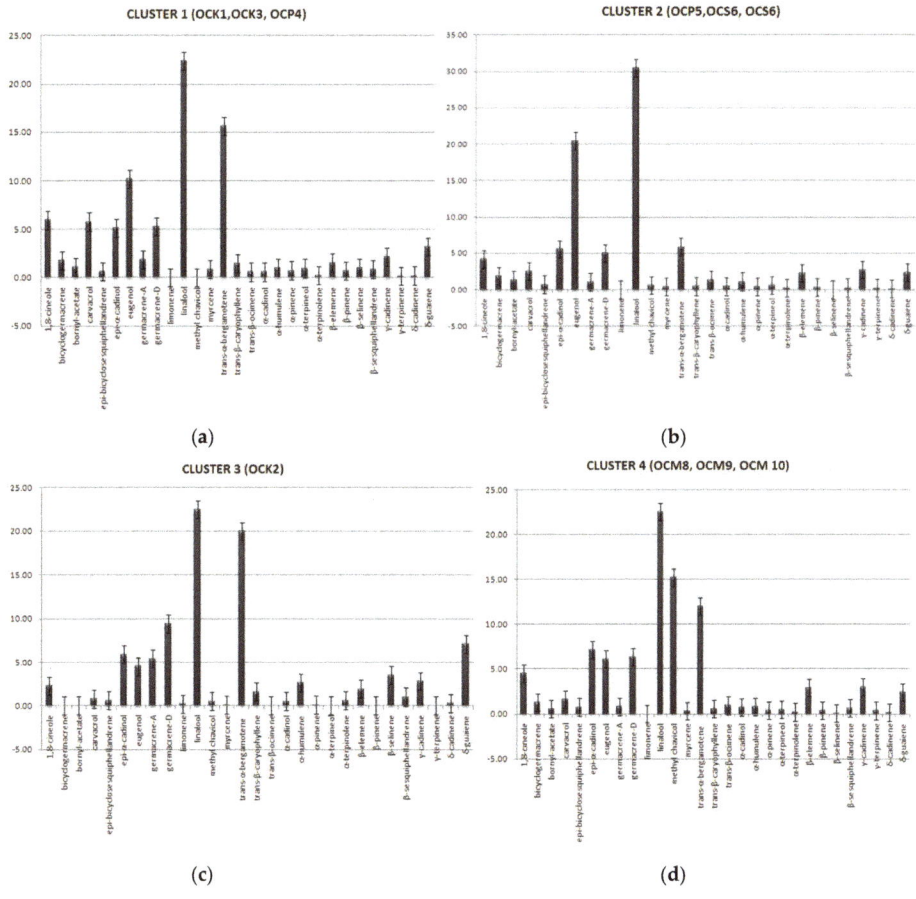

Figure 2. Clusters formatted according to the dominant compounds identified, in varieties violetto, latifolia, minimum and lettuce, cultivated in the greenhouse conditions. (**a**) Dominant compounds in var. violetto (1st and 3rd harvest) and var. latifolia (1st harvest); (**b**) Dominant compounds in var. latifolia (2nd harvest) and var. minimum (1st and 2nd harvest); (**c**) Dominant compounds in var. violetto (2nd harvest); (**d**) Dominant compounds in var. lettuce (1st, 2nd and 3rd harvest).

Cultivation conditions in the field defined the dominant constituents for the formation of four clusters in the dendrogram (Figure 3). The similarities observed between harvests and varieties characterized the clusters as: Cluster 1 consisting of samples OCP11, OCP12, OCP13, OCS15, and OCS17 with linalool, eugenol, and *trans*-α-bergamontene as major components. Cluster 2 consists of OCP14 and OCS16 with linalool and eugenol as major components. Cluster 3 consists of OCM18, OCM19, and OCM20, with methyl chavicol and linalool as major components. Finally, cluster 4 is "simplicifolious" and consists of samples OCC21, OCC22, and OCC23, with methyl-chavicol as the dominant component (Figure 4).

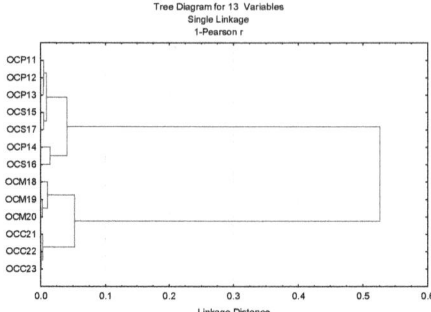

Figure 3. Bidimensional dendrogram representing the similarity in the main components of the essential oils, between 13 harvests of 4 basil varieties growing in the field conditions.

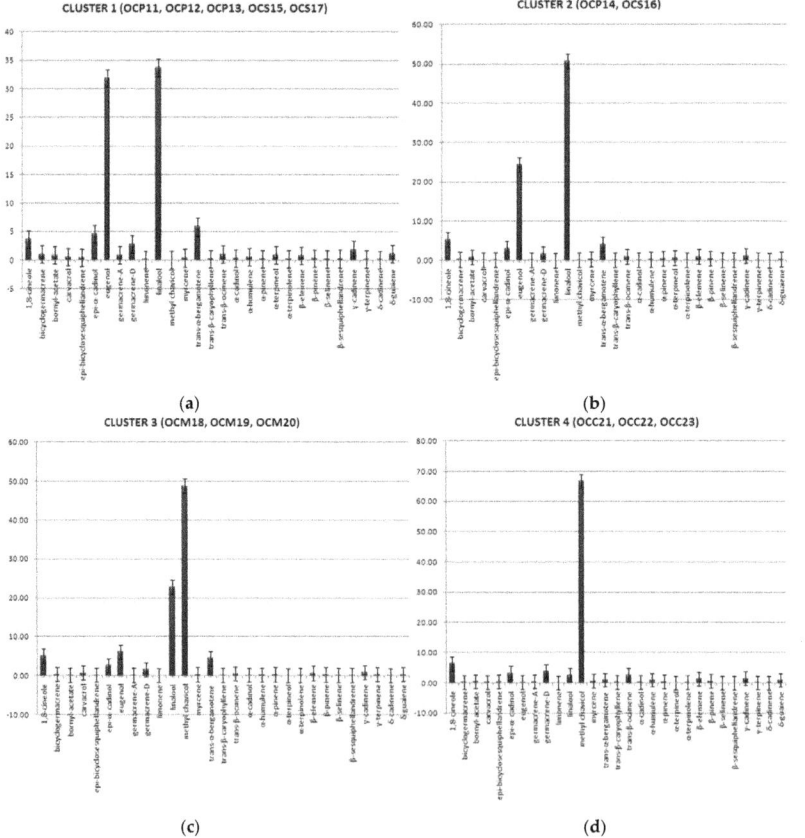

Figure 4. Clusters formatted according to the dominant compounds identified in varieties latifolia, minimum, lettuce and cinnamon, cultivated in the field conditions. (**a**) Dominant compounds in var. latifolia (1st, 2nd and 3rd harvest) and var. minimum (1st and 3rd harvest); (**b**) Dominant compounds in var. latifolia (4th harvest) and var. minimum (2nd harvest); (**c**) Dominant compounds in var. lettuce (1st, 2nd, and 3rd harvest); (**d**) Dominant compounds in var. cinnamon (1st, 2nd and 3rd harvest).

Considering that in both field and greenhouse the conditions were same for all of the varieties, the observed variations can be attributed to the ability of each variety to react to the stressing factor of harvesting. The grouping obtained for the examined basil varieties in greenhouse and field conditions, provides significant data about the effect of harvesting and cultivation site in the chemical composition of basil essential oils. A comparison of Cluster 1 between field and greenhouse conditions reveals that in var. latifolia, between harvests, shifts are observed in the chemotype by alterations in the biosynthesis of essential oils' components. These results are in compliance with the study of Božović et al. (2017), in EOs obtained from *Calamintha nepeta* (L.) Savi subsp. *glandulosa*, indicating that the main compounds were stable, but the ratio between them varied greatly according to the growth stage [26]. Especially in var. lettuce it is obvious that apart from the harvest factor, the cultivation site also affects the chemical profile of the obtained EO. Observation of cluster 3 from field cultivation and cluster 4 from greenhouse cultivation reveals a different chemical profile in the EOs obtained for the same variety. As such, these results imply that the cultivation site should also be considered as factor affecting the composition of EOs, apart from the number of harvests. The variations of the chemical profile of the five basil varieties are in accordance with the results of Verma et al. (2013) where different *Ocimum* populations belonging to the "basilicum" group are distributed to different clusters [27]. The alteration in the chemical profile of the essential oils obtained from *Foeniculum vulgare* Miller between harvests was also indicated by Garzoli et al. (2017). Researchers showed that while *o*-cymene and *a*-phellandrene were dominating compounds in the first two harvests, EO from the last harvest was characterized by a high content of methyl-chavichol (estragole) [28]. Cluster analysis has been proved as a useful tool in the attempt of an evaluation of chemotypes among essential oil bearing plants. However, in *O. basilicum's* case, the numerous synonymous names, varieties, and inerspecific hybridization along with the several factors that affecting essential oils' composition make this process extremely difficult [29].

The obtained data can serve as a guide for cultivation conditions and the harvesting of basil varieties in this geographical region in order to obtain maximum essential oil yield along with certain metabolites or mixtures of them in the composition of the essential oil. Comparing to the literature data about basil chemical composition, current results are in accordance about the yield of essential oil produced. Additionally, they are very promising concerning the composition of them to important constituents like linalool, eugenol, and methyl chavicol. Environmental conditions, like temperature, humidity, and soil conditions, in the island of Kefalonia affect the chemical profile of the cultivated basil varieties and favor the production of bioactive compounds like linalool [30]. Practically, this implies that essential oils from aromatic plants cultivated in the climatic conditions of Kefalonia have the quality characteristics needed for commercial exploitation.

3. Materials and Methods

3.1. Cultivation Conditions and Plant Material

Certified *Ocimum basilicum* L. (basil) seeds, belonging to varieties lettuce, cinnamon, minimum, latifolia, and violetto, were purchased from the local market. Basil seeds of the five varieties were sown into sowing boxes filled with universal soil for sowing seeds in an environmentally controlled growth chamber. Environmental conditions were: photoperiod 12/12, air temperature 18/24 °C, and air relative humidity 50/70% (day/night). When the plants formed true leaves were transplanted into small plastic pots. When they reached 12–15 cm height, they were transplanted to their final position either into plastic 20 L pots in the greenhouse or directly to the land in the experimental field. For each variety 20 plants were planted in each cultivation site. For the field conditions, basil transplants were planted into rows with 0.50 m in-row and 1.00 m between row spacing, covering a total area of 15 m^2. Plants were irrigated every second day with an irrigation tape providing 5–7 L of water, in both field and greenhouse conditions. During the growing period, no pests or diseases were observed apart from the inability of var. cinnamon to grow in the greenhouse conditions and that var. violetto failed to grow in the field conditions, thus only the results from one cultivation site are given. Harvesting of the plants was performed almost two months after transplantation, prior to flowering stage [17]. The field cultivation conducted in the experimental field of the Department of Food Technology (Technological Educational Institute of Ionian Islands) in the island of Kefalonia (Figure 5). Plant material was hand harvested by removing the apical portion of the herbage 20 cm above ground and leaving 1–2 internodes in order to promote rejuvenation from the sleeping buds. For each variety and cultivation site, 2–3 harvests were performed depending on the variety and its ability to regenerate stems and flowers. Practically, this includes a period of 30–40 days between each cutting. In every collection, plant material was collected the same time of the day in order to eliminate day variations to the yield of essential oil, transferred to the laboratory and fresh weight was measured, and the essential oil yield to fresh weight was calculated. In total, 23 samples were collected extracted and analysed, namely 10 essential oils obtained from 4 four different varieties cultivated in the greenhouse, and 13 essential oils obtained from four varieties cultivated in the field (Table 1). The plant materials have been deposited in the herbarium of the Department of Food Technology (Technological Educational Institute of Ionian Islands) under the number 737.

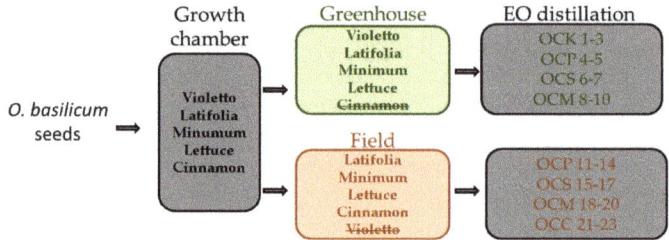

Figure 5. Diagrammatic presentation of the cultivation of basil seeds belonging to five varieties in field and greenhouse, along with the relevant obtained essential oils (EOs).

3.2. Essential Oil Distillation

The collected plant material from the five varieties was separately subjected to hydro distillation in a modified Clevenger-type apparatus according to the Hellenic Pharmacopoeia [31]. Each plant material sample (about 2.0 kg) was extracted for 3 h. The essential oil content (% v/w) was estimated on a fresh weight basis. The oil samples were dehydrated over anhydrous Na_2SO_4. The obtained essential oils were of yellowish colour and a pleasant odor, and were deposited in vials at $-20\ °C$ prior to further chemical analyses.

3.3. GC Analysis

Quantification was performed using gas-chromatography coupled with flame ionization detection (GC-FID). Analyses were carried out on a Perkin Elmer Clarus 500 gas chromatograph with FID, fitted with a fused silica Rtx-5 MS capillary column (30 m × 0.25 mm (i.d.), film thickness: 0.25 μm). The column temperature was programmed from 60 °C to 250 °C at a rate of 3 °C/min. The injector and detector temperatures were programmed at 230 °C and 280 °C, respectively. 0.5 μL of each sample were diluted in 500 μL GC grade n-pentane and 2 μL of the obtained solution was further injected in the GC apparatus. Identification of constituents was achieved by calculating the arithmetic indices relative to linear alkanes from C_9–C_{23} and comparing with data from GC-MS identifications.

3.4. GC-MS Analysis

GC-MS analyses were performed on a Hewlett-Packard 5973-6890 system operating in EI mode (70 eV) equipped with a split/splitless injector (220 °C), a split ratio 1/10, using a fused silica HP-5 MS capillary column (30 m × 0.25 mm (i.d.), film thickness: 0.25 μm). The temperature program for HP-5

MS column was from 60 °C (5 min) to 280 °C at a rate of 4 °C/min. Helium was used as a carrier gas at a flow rate of 1.0 mL/min. Injection volume for all samples, diluted as previously described, was 2 μL.

3.5. Identification of Components

Retention indices for all compounds were determined according to the Van der Dool approach [32] with reference to a homologous series of normal *n*-alkanes from C_9 to C_{23}. The identification of the components was based on a comparison of their mass spectra with those of Wiley and NBS Libraries and those described by Adams (2007), as well as by comparison of their retention indices with literature data [33]. Component relative percentages were calculated based on GC-FID peak areas without using correction factors.

3.6. Statistical Analysis

Data obtained from the chemical analysis of the essential oils from all varieties were analyzed by Multivariate analysis method using Statistica software version 7.0 Statsoft Company. Cluster analyses were performed based on the similarity between harvests from all varieties and their constituent distribution. These analyses were performed on complete data sets. The unweighted pair-group average linkage clustering method based on Pearson distances was used to measure the similarities between each measured unit.

Author Contributions: Y.S. and H.S. conceived and design the experiments of the selection of the varieties, cultivation and evaluation experimental results, respectively. In addition, H.S. supervised the chemical analyses. E.S. performed the cultivation of the basil varieties in the greenhouse and field, the harvesting and the extractions of all essentials oils. T.M. and A.D performed the chemical analyses of the obtained EOs. P.R. performed the statistical analysis of the results. G.T analyzed the data obtained from the chemical analysis, collection of results andwriting of the article.

Conflicts of Interest: The authors declare no conflict of interest.

References

1. Sajjadi, S.E. Analysis of the essential oils of two cultivated basil (*Ocimum basilicum* L.) from Iran. *Daru* **2006**, *14*, 128–130.
2. Paton, W.; Harley, R.M.; Harley, M.M. *Ocimum*—An overview of relationshipsand classification. In *Basil: The Genus Ocimum*; Holm, Y., Hiltuen, R., Eds.; Hawood Academic: Amsterdam, The Netherlands, 1999; pp. 1–33.
3. Paton, A.; Putievsky, E. Taxonomic problems and cytotaxonomic relationships between and within varieties of *Ocimum basilicum* and related species (Labiatae). *Kew Bull.* **1996**, *51*, 509–524. [CrossRef]
4. Putievsky, E.; Paton, A.; Lewinsohn, E.; Ravid, U.; Haimovich, D.; Katzir, I.; Saadi, D.; Dudai, N. Crossability and relationship between morphological and chemical varieties of *Ocimum basilicum* L. *J. Herbs Spices Med. Plants* **1999**, *6*, 11–24. [CrossRef]
5. Labra, M.; Miele, M.; Ledda, B.; Grassi, F.; Mazzei, M.; Sala, F. Morphological characterization: Essential oil composition and DNA genotyping of *Ocimum basilicum* L. cultivars. *Plant Sci.* **2004**, *167*, 725–731. [CrossRef]
6. Darrah, H.H. *The Cultivated Basils*; Buckeye Printing Company: Winter Haven, FL, USA, 1980; p. 82.
7. Beatovic, D.; Krstic-Miloševic, D.; Trifunovic, S.; Šiljegovic, J.; Glamoclija, J.; Ristic, M.; Jelacic, S. Chemical composition, antioxidant and antimicrobial activities of the essential oils of twelve *Ocimum basilicum* L. cultivars grown in Serbia. *Rec. Nat. Prod.* **2015**, *9*, 62–75.
8. Khalid, K.A. Influence of water stress on growth, essential oil, and chemical composition of herbs (*Ocimum* sp.). *Int. Agrophys.* **2006**, *20*, 289–296.
9. Zheljazkov, V.D.; Cantrell, C.L.; Tekwani, B.; Khan, S.I. Content, composition, and bioactivity of the essential oils of three basil genotypes as a function of harvesting. *J. Agric. Food Chem.* **2008**, *56*, 380–385. [CrossRef] [PubMed]
10. Simon, J.E.; Quinn, J.; Murray, R.G. Basil: A source of essential oils. In *Advanced in New Crops*; Janick, J., Simon, J.E., Eds.; Timber Press: Portland, OR, USA, 1999; pp. 484–489.

11. Nurzyńska-Wierdak, R. Morphological and chemical variability of *Ocimum basilicum* L. (Lamiaceae). *Mod. Phytomorphol.* **2013**, *3*, 115–118.
12. Koba, K.; Poutouli, P.W.; Raynaud, C.; Chaumont, J.P.; Sanda, K. Chemical composition and antimicrobial properties of different basil essential oils chemotypes from Togo. *Bangladesh J. Pharmacol.* **2009**, *4*, 1–8. [CrossRef]
13. Singh, S.; Singh, M.; Singh, A.K.; Kalra, A.; Yadav, A.; Patra, D.D. Enhancing productivity of Indian basil (*Ocimum basilicum* L.) through harvest management under rainfed conditions of subtropical north India plains. *Ind. Crops Prod.* **2010**, *32*, 601–606. [CrossRef]
14. Marotti, M.; Piccaglia, R.; Giovanelli, E. Differences in essential oil composition of basil (*Ocimum basilicum* L.) Italian cultivars related to morphological characteristics. *J. Agric. Food Chem.* **1996**, *44*, 3926–3929. [CrossRef]
15. Telci, I.; Bayram, E.; Yılmaz, G.; Avcı, B. Variability in essential oil composition of Turkish basils (*Ocimum basilicum* L.). *Biochem. Syst. Ecol.* **2006**, *34*, 489–497. [CrossRef]
16. Topalov, V.D. Basil. In *Essential Oil and Medicinal Plant*; Topalov, V.D., Ed.; Hr. G. Danov Press: Plovdiv, Bulgaria, 1962; pp. 200–209.
17. Makri, O.; Kintzios, S. *Ocimum* sp. (Basil): botany, cultivation, pharmaceutical properties, and biotechnology. *J. Herbs. Spices Med. Plants* **2007**, *13*, 123–150.
18. Carlo, N.; Silvia, S.; Stefano, B.; Paolo, S. Influence of cut number on qualitative traits in different cultivars of sweet basil. *Ind. Crops Prod.* **2013**, *44*, 465–472. [CrossRef]
19. Lakušić, D.; Ristić, M.; Slavkovska, V.; Lakušić, B. seasonal variations in the composition of the essential oils of rosemary (*Rosmarinus officinalis*, Lamiaceae). *Nat. Prod. Commun.* **2013**, *8*, 131–134. [PubMed]
20. Public Statement on the Use of Herbal Medicinal Products Containing Estragole. Available online: http://www.ema.europa.eu/docs/en_GB/document_library/Scientific_guideline/2010/04/WC500089960.pdf (accessed on 17 September 2017).
21. Saran, P.L.; Tripathy, V.; Meena, R.P.; Kumar, J.; Vasara, R.P. Chemotypic characterization and development of morphological markers in *Ocimum basilicum* L. germplasm. *Sci. Hortic.* **2017**, *215*, 164–171. [CrossRef]
22. Liber, Z.; Carović-Stanko, K.; Politeo, O.; Strikić, F.; Kolak, I.; Milos, M.; Satovic, Z. Chemical characterization and genetic relationships among *Ocimum basilicum* L. cultivars. *Chem. Biodivers.* **2011**, *8*, 1978–1989. [CrossRef] [PubMed]
23. Ložienė, K.; Venskutonis, P.R. Influence of environmental and genetic factors on the stability of essential oil composition of *Thymus pulegioides*. *Biochem. Syst. Ecol.* **2005**, *33*, 517–525. [CrossRef]
24. Awadh Ali, N.A.; Chhetri, B.K.; Dosoky, N.S.; Shari, K.; Al-Fahad, A.J.A.; Wessjohann, L.; Setzer, W.N. Antimicrobial, Antioxidant, and Cytotoxic Activities of *Ocimum forskolei* and *Teucrium yemense* (Lamiaceae) Essential Oils. *Medicines* **2017**, *4*, 17.
25. Nykanen, I. the effect of cultivation conditions on the composition of basil oil. *Flavour Fragr. J.* **1989**, *4*, 125–128. [CrossRef]
26. Božović, M.; Garzoli, S.; Sabatino, M.; Pepi, F.; Baldisserotto, A.; Andreotti, E.; Ramagnoli, C.; Mai, A.; Manfredini, S.; Ragno, R. Essential oil extraction, chemical analysis and anti-Candida activity of *Calamintha nepeta* (L.) Savi subsp. *glandulosa* (Req.) Ball—New approaches. *Molecules* **2017**, *22*, 203.
27. Verma, R.S.; Padalia, R.C.; Chauhan, A.; Thul, S.T. Exploring compositional diversity in the essential oils of 34 *Ocimum* taxa from Indian flora. *Ind. Crops Prod.* **2013**, *45*, 7–19. [CrossRef]
28. Garzoli, S.; Božović, M.; Baldisserotto, A.; Sabatino, M.; Cesa, S.; Pepi, F.; Vincentini, C.B.; Manfredini, S.; Ragno, R. Essential oil extraction, chemical analysis and anti-Candida activity of *Foeniculum vulgare* Miller—New approaches. *Nat. Prod. Res.* **2017**. [CrossRef] [PubMed]
29. Sharopova, F.S.; Satyal, P.; Awadh, A.N.A.; Pokharel, S.; Zhang, H.; Wink, M.; Kukanieve, M.A.; Setzer, W.N. The Essential Oil Compositions of *Ocimum basilicum* from three different Regions: Nepal, Tajikistan, and Yemen. *Chem. Biodivers.* **2016**, *13*, 241–248. [CrossRef] [PubMed]
30. Aprotosoaie, A.C.; Hăncianu, M.; Costache, I.I.; Miron, A. Linalool: A review on a key odorant molecule with valuable biological properties. *Flavour Fragr. J.* **2014**, *29*, 193–219. [CrossRef]
31. Scientific Committee of the Hellenic Pharmacopoeia. *Hellenic Pharmacopoeia*, 5th ed.; National Organization for Medicines of Greece: Athens, Greece, 2002; Chapter 28.12.

32. Van den Dool, H.; Kratz, P.D. A generalization of the retention index system including linear temperature programmed gas-liquid partition chromatography. *J. Chromatogr. A* **1963**, *11*, 463–471. [CrossRef]
33. Adams, R.P. *Identification of Essential Oils Components by Gas Chromatography/Quadrupole Mass Spectroscopy*; Allured Publishing Corp: Carol Stream, IL, USA, 2007.

© 2017 by the authors. Licensee MDPI, Basel, Switzerland. This article is an open access article distributed under the terms and conditions of the Creative Commons Attribution (CC BY) license (http://creativecommons.org/licenses/by/4.0/).

Communication

Caffeoylquinic Acids from the Aerial Parts of *Chrysanthemum coronarium* L.

Chunpeng Wan [1,2,†], Shanshan Li [1,†], Lin Liu [1], Chuying Chen [1,2] and Shuying Fan [1,*]

1   College of Agronomy, Jiangxi Agricultural University, Nanchang 330045, China; lemonwan@126.com (C.W.); liss0824@126.com (S.L.); linliu0960@126.com (L.L.); ccy0728@126.com (C.C.)
2   Collaborative Innovation Center of Post-Harvest Key Technology and Quality Safety of Fruits and Vegetables in Jiangxi Province, Jiangxi Agricultural University, Nanchang 330045, China
* Correspondence: chunpengwan@jxau.edu.cn or fansy12@126.com; Tel.: +86-791-8381-3185
†   These authors contributed equally to this work.

Academic Editor: Milan S. Stankovic
Received: 6 January 2017; Accepted: 9 February 2017; Published: 17 February 2017

Abstract: To elucidate the chemical compositions of the aerial parts of *Chrysanthemum coronarium* L., the ethanol extracts of *Ch. coronarium* L. were firstly isolated by the MCI-gel resin column. The caffeoylquinic acid-rich fractions were further purified by various chromatographic columns including silica gel, Sephadex LH-20, and semi-preparative HPLC to yield the compounds. The purified compounds were characterized by ^{1}H-Nuclear Magnetic Resonance (^{1}H-NMR), ^{13}C-NMR, and high resolution electrospray ionisation mass spectral (HR-ESI-MS) spectroscopy. Seven caffeoylquinic acid (CQA) compounds were isolated from this plant. Their structures were clarified by spectrometric methods and identified as 3-*O*-caffeoylquinic acid (**1**), 5-*O*-caffeoylquinic acid (**2**), 4-*O*-caffeoylquinic acid (**3**), 3,4-di-*O*-caffeoylquinic acid (**4**), 1,5-di-*O*-caffeoylquinic acid (**5**), 3,5-di-*O*-caffeoylquinic acid (**6**), and 4,5-di-*O*-caffeoylquinic acid (**7**). Caffeoylquinic acids were the major constituents present in the aerial parts of *Ch. coronarium* L. All of the isolates except for compounds **2** and **6** were reported for the first time from this species. Moreover, compounds **3**–**5**, and **7** were identified from the *Chrysanthemum* genus for the first time.

Keywords: *Chrysanthemum coronarium* L.; aerial parts; caffeoylquinic acids

1. Introduction

Chrysanthemum coronarium L., commonly known as "Tonghao", is used as an edible vegetable and medicinal plant, and belongs to the genus of *Chrysanthemum* (Compositae) [1]. Many previous studies have reported the isolation, identification, and the biological activities of the plant. The results have shown that flavonoids [2], phenolic acids [3], sesquiterpene lactone [4], monoterpene [5], diterpene [6], glycosyldiglycerides [7], alkaloid [8], phytosterol [9], heterocyclic compounds [8], polyacetylenes [9], and essential oils [10,11] were the major chemical constituents present in the plant. *Ch. coronarium* L. has shown a variety of biological activities including the elimination of phlegm, plant allelopathy, nematicidal activity, cytotoxic activity, antioxidant and free radical scavenging, insect antifeedant, hepatic protection, and antimicrobial properties [1]. Flavonoids and phenolic acids are responsible for the plant allelopathy and antioxidant and free radical scavenging activities [3]; polyacetylenes and essential oil are responsible for the insect antifeedant [12] and antimicrobial activities [10]; and terpene, in particular sesquiterpene lactone, is responsible for the cytotoxic [13] and antimicrobial activities [6].

Flavonoids and polyphenols are the characteristic constituents of *Ch. coronarium* L., which are also the main bioactive constituents of the Compositae family [2,3]. Caffeoylquinic acids (CQAs) are cinnamate conjugates, which are biosynthesized through the phenylpropanoid pathway. These phenolic compounds are generally involved in plant disease-resistance responses to biotic or abiotic stress [14]. Preliminary studies on the chemical constituents of the aerial parts of *Ch. coronarium* L. indicated that the caffeoylquinic acids were the major components in the plant, however, only three CQA compounds were elucidated until now; 3,5-di-*O*-caffeoyl-4-succinylquinic acid, chlorogenic acid, and 3,5-di-*O*-caffeoylquinic acid [3].

Herein, the isolation and structure elucidation of caffeoylquinic acids (CQAs) from the aerial parts of *Ch. coronarium* was achieved.

2. Results

Caffeoylquinic acid derivatives showed typical UV spectra peaks at 327, 298 (sh), and 246 nm [15]. The HPLC profile of the ethanolic extract (TH) and of its five purified fractions (THA to THE) indicated that THA−THC were caffeoylquinic acid derivative-rich fractions (Figure 1). Many components in the THB fraction overlapped with the THA and THC fractions based on the HPLC profile. Thus, the THA and THC fractions were further isolated to yield the pure compounds.

Figure 1. HPLC chromatogram of ethanol extract (TH) and MCI fractions (THA–THE) of *Ch. coronarium* L.

The structures of the isolated caffeoylquinic acid derivatives were elucidated by HR-ESI-MS analysis, 1D-NMR data (Tables 1 and 2), and comparison of these data with the literature. Compounds **1**–**7** (Figure 2) were identified as 3-*O*-caffeoylquinic acid (**1**) [16], 5-*O*-caffeoylquinic acid (**2**) [17], 4-*O*-caffeoylquinic acid (**3**) [16], 3,4-di-*O*-caffeoylquinic acid (**4**) [18,19], 1,5-di-*O*-caffeoylquinic acid (**5**) [20], 3,5-di-*O*-caffeoylquinic acid (**6**) [21], and 4,5-di-*O*-caffeoylquinic acid (**7**) [18,19]. Among them, compounds **1**, **3**–**5**, and **7** were isolated from this species for the first time. Moreover, compounds **3**–**5**, and **7** were reported from the genus *Chrysanthemum* for the first time.

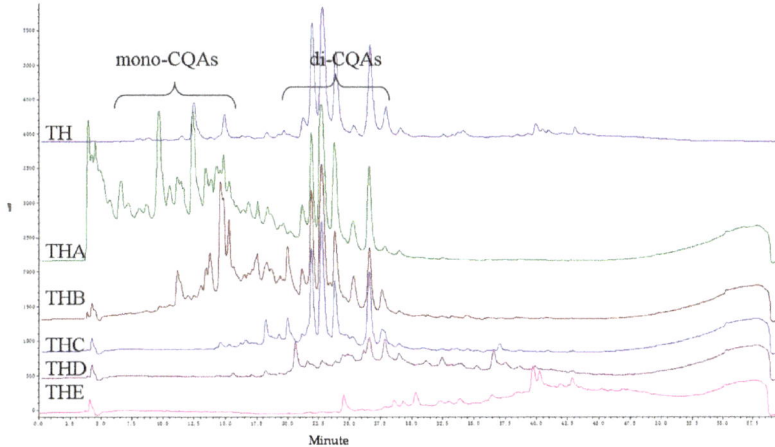

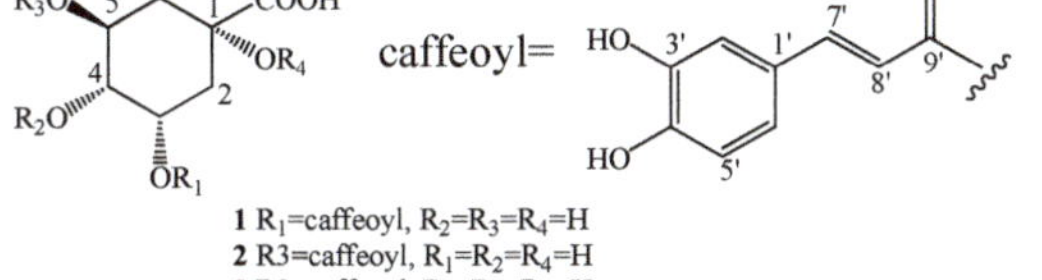

1 R_1=caffeoyl, R_2=R_3=R_4=H
2 R_3=caffeoyl, R_1=R_2=R_4=H
3 R_2=caffeoyl, R_1=R_3=R_4=H
4 R_1=R_2=caffeoyl, R_3=R_4=H
5 R_1=R_2=H, R_3=R_4=caffeoyl
6 R_1=R_3=caffeoyl, R_2=R_4=H
7 R_2=R_3=caffeoyl, R_1=R_4=H

Figure 2. The chemical structures of the compounds **1**–**7** isolated from *Ch. coronarium* L.

Table 1. ^{1}H-NMR (^{1}H-Nuclear Magnetic Resonance, 400 MHz, MeOH-d_4) characteristics of the caffeoylquinic acid derivatives **1–7** isolated from aerial parts of *Ch. coronarium* L.

No.	1 δH (J Hz)	2 δH (J Hz)	3 δH (J Hz)	4 δH (J Hz)	5 δH (J Hz)	6 δH (J Hz)	7 δH (J Hz)
2	1.93–2.22 (2H, m)	2.04–2.25 (2H, m)	1.98–2.23 (2H, m)	2.07–2.35 (2H, m)	2.54 (1H, dd, 3.5, 10.2) 2.43 (1H, m)	2.12–2.32 (2H, m)	2.14–2.30 (2H, m)
3	5.36 (1H, brd, 2.9)	4.18 (1H, brd, 2.9)	4.29 (1H, brs)	5.61 (1H, brd, 3.4)	4.28 (1H, brd, 3.5)	5.36 (1H, brd, 5.6)	4.40 (1H, brs)
4	3.65 (1H, dd, 2.8, 8.5)	3.75 (1H, dd, 2.3, 8.0)	4.80 (1H, dd, 2.3, 9.0)	4.96 (1H, dd, 3.1, 9.4)	3.76 (1H, dd, 3.0, 8.1)	3.96 (1H, dd, 3.2, 7.3)	5.10 (1H, d, 8.5)
5	4.14 (1H, ddd, 3.6, 8.5, 8.5)	5.35 (1H, ddd, 3.5, 8.0, 8.0)	4.27 (1H, ddd, 9.0, 9.0, 4.5)	4.35 (1H, ddd, 4.2, 9.4, 9.4)	5.37 (1H, ddd, 3.7, 8.1, 8.1)	5.42 (1H, ddd, 3.2, 7.3, 7.3)	5.63 (1H, ddd, 4.2, 8.5, 8.5)
6	1.93–2.22 (2H, m)	2.04–2.25 (2H, m)	1.98–2.23 (2H, m)	2.07–2.35 (2H, m)	2.05 (1H, dd, 11.1, 13.8) 2.43 (1H, m)	2.12–2.32 (2H, m)	2.14–2.30 (2H, m)
2'/2''	7.04 (1H, d, 1.4)	7.05 (1H, brs)	7.06 (1H, d, 1.4)	7.02/7.00 (each 1H, d, 1.8)	7.04 (1H, brs)	7.05 (each 1H, s)	7.00/6.97 (each 1H, s)
5'/5''	6.78 (1H, d, 8.0)	6.78 (1H, d, 8.0)	6.78 (1H, d, 8.0)	6.76/6.72 (each 1H, d, 7.9)	6.78/6.76 (each 1H, d, 8.1)	6.77/6.75 (each 1H, d, 8.1)	6.73/6.71 (each 1H, d, 8.1)
6'/6''	6.94 (1H, dd, 1.4, 8.0)	6.95 (1H, d, 8.0)	6.97 (1H, dd, 1.4, 8.0)	6.90/6.85 (each 1H, dd, 1.8, 7.9)	6.96/6.94 (each 1H, d, 8.1)	6.96/6.94 (each 1H, m)	6.89/6.87 (each 1H, d, 8.1)
7'/7''	7.58 (1H, d, 15.9)	7.56 (1H, d, 15.9)	7.63 (1H, d, 15.9)	7.57/7.52 (each 1H, d, 15.9)	7.58/7.55 (each 1H, d, 15.9)	7.60/7.56 (each 1H, d, 15.9)	7.57/7.49 (each 1H, d, 15.9)
8'/8''	6.30 (1H, d, 15.9)	6.27 (1H, d, 15.9)	6.37 (1H, d, 15.9)	6.27/6.23 (each 1H, d, 15.9)	6.27/6.24 (each 1H, d, 15.9)	6.33/6.24 (each 1H, d, 15.9)	6.26/6.17 (each 1H, d, 15.9)

Table 2. ^{13}C-NMR Data for Compounds 2, 5–6 (100 MHz, MeOH-d_4).

No.	Compounds		
	2	5	6
1	74.7	80.9	73.3
2	36.8	35.7	34.6
3	69.9	69.4	71.1
4	72.1	72.8	69.2
5	70.5	71.6	70.6
6	37.4	36.9	36.2
7	175.6	174.8	175.9
1'	126.4	127.8/127.8	126.5/126.4
2'	113.8	115.3/115.3	114.2/113.9
3'	145.3	147.6/147.4	145.3/145.3
4'	148.1	149.7/149.7	148.1/148.0
5'	115.1	116.5/116.5	115.1/115.1
6'	121.6	123.1/123.1	121.6/121.6
7'	145.7	147.4/146.9	145.9/145.6
8'	113.8	115.2/115.1	113.7/113.7
9'	167.3	168.7/168.0	167.4/167.0

3. Discussion

Compounds **1–3** were obtained as white power. The HR-ESI-MS yielded a quasi-molecular ion peak [M-H]$^-$ at m/z 353.08. The UV spectrum showed λ_{max} at 328, 298 (shoulder), and 246 nm, which suggested that compounds **1–3** were single caffeoyl substituted quinic acid derivatives. The ^{1}H-NMR spectrum (400 MHz, MeOH-d_4) of compounds **1–3** showed caffeoyl signals at δ 7.56, 7.58, 7.63 (1H, d, J = 15.9 Hz, H-7'), 6.27, 6.30, 6.37 (1H, d, J = 15.9, H-8'), 7.04, 7.05, 7.06 (1H, H-2'), 6.94, 6.95, 6.97 (1H, H-6'), 6.78, 6.78, 6.78 (1H, H-5'), and quinic acid signals at [**1**: δ 5.36 (1H, brd, J = 2.9, H-3), 3.65 (1H, dd, J = 8.5, 2.8, H-4), 4.14 (1H, ddd, J = 8.5, 8.5, 3.6, H-5), 1.93–2.22 (4H, m, H-2, H-6); **2**: δ 4.18 (1H, brd, J = 2.9, H-3), 3.75 (1H, dd, J = 8.0, 2.3, H-4), 5.35 (1H, ddd, J = 8.0, 8.0, 3.5, H-5), 2.04–2.25 (4H, m, H-2, H-6); **3**: δ 4.29 (1H, brs, H-3), 4.80 (1H, dd, J = 9.0, 2.3, H-4), 4.27 (1H, ddd, J = 9.0, 9.0, 4.5, H-5), 1.98–2.23 (4H, m, H-2, H-6)]. The substituted position of caffeoyl can be determined by the analysis of the chemical shift and coupling constants of the oxygenated methine protons of the quinic acid core. Once the oxygenated methine of quinic acid was acylated by caffeoyl, the proton signal will shift downfield. The coupling constant of the downfield shifted proton was then applied to the acylation position. Generally, the H-3 signal has a small coupling constant and shows a brd or brs type peak, the H-4 signal showed a dd type peak with coupling constants at 8.0–9.0 Hz and 2.0–3.0 Hz, while the H-5 signal showed a ddd type peak with coupling constants at 8.0–9.0 Hz, 8.0–9.0 Hz, and 3.0–5.0 Hz. Based on these rules, the structures of compounds **1–3** were determined as depicted.

Compounds **4–7** were obtained as white powder. The ESI-MS yielded a quasi-molecular ion peak [M-H]$^-$ at m/z 515.11, and the UV spectrum showed λ_{max} at 327, 298 (shoulder), and 245 nm, suggesting that compounds **4–7** were double caffeoyl substituted quinic acid derivatives. The ^{1}H-NMR spectrum showed similar signal patterns to compounds **1–3**, but one more caffeoyl signal was observed. The ^{1}H-NMR spectrum of compounds **4–7** showed two sets of caffeoyl signals [7.49–7.60 (each 1H, d, J = 15.9 Hz, H-7' and H-7''), 6.97–7.05 (each 1H, H-2' and H-2''), 6.85–6.96 (each 1H, H-6' and H-6''), 6.71–6.78 (each 1H, H-5' and H-5''), 6.17–6.33 (each 1H, d, J = 15.9 Hz, H-8' and H-8'')], and quinic acid signals (see Table 1). Similar to compounds **1–3**, the acylation positions were determined by the chemical shift and coupling constants of the oxygenated methine protons of quinic acid. As only the H-5 signal of compound **5** was observed with a downfield shift, another substituted position was tentatively assigned to C-1 of quinic acid.

The chemical compositions of the plant are characterized by flavonoids and phenolic acids [22], which showed typical UV spectra based on the HPLC-DAD. In the current study, seven CQAs including three mono-CQAs and four di-CQAs were isolated and identified. Previously, two phenolic acids as plant growth inhibitors were isolated from *Ch. coronarium* L. and identified as isoferulic acid and methyl parahydroxybenzoats [12]. Ferulic acid methyl ester was also detected in *Ch. coronarium* L., which showed low-density lipoprotein (LDL) oxidation inhibited activity [8]. Only three quinic acid derivatives, namely, chlorogenic acid (5-*O*-caffeoylquinic acid, **2**), 3,5-di-*O*-caffeoylquinic acid (**6**), and 3,5-di-*O*-caffeoyl-4-succinylquinic acid were detected by the HPLC method [3]. This is the first report of the isolation of CQAs except for compounds **2** and **6**, while compounds **3–5**, and **7** were identified from the Chrysanthemum genus for the first time. Additionally, the di-CQAs are the major phenolic acid constituents present in this plant.

4. Materials and Methods

4.1. Plant Material

The aerial parts of *Ch. coronarium* L. were purchased from a local market, Nanchang City, Jiangxi Province, China, and identified by Prof. Shuying Fan (College of Agronomy, Jiangxi Agricultural University, Nanchang, China). A voucher specimen (TH-2015041) has been deposited in the Department of Horticulture, College of Agronomy, Jiangxi Agricultural University (Nanchang, Jiangxi, China).

4.2. Equipment and Reagents

^{1}H and ^{13}C-NMR were detected on a Varian 400 MHz spectrometer in CD_3OD with Tetramethylsilane (TMS) as the internal standard. HR-ESI-MS data were obtained on a 6538 Ultra High Definition (UHD) Accurate-Mass Q-TOF LC/MS system (Agilent, Santa Clara, CA, USA). High performance liquid chromatography (HPLC) was performed on a Hitachi Elite Chromaster system including a 5110 pump, 5210 autosampler, 5310 column oven, a 5430 diode array detector, and operated by EZChrom Elite software. Luna C18 (2) column (5 μm, 4.6 × 250 mm) for analysis and Luna C18 (2) column (10 μm, 10 × 250 mm) for HPLC preparation were purchased from Phenomenex Inc (Torrance, CA, USA). The HPLC grade solvents were purchased from Sigma (Sigma, St. Louis, MO, USA). All analytical solvents were purchased from Tansoole (Shanghai, China). Silica gel (250 mesh; Qingdao Haiyang Chemical Co., LTD, Qingdao, China) was used as normal phase, whereas YMC Pack ODS-A (50 μm; YMC) was used as reversed phase column material. MCI gel CHP20P (75–150 μm; Mitsubishi Chemical Corp, Tsukuba, Japan) and Sephadex LH-20 (GE Healthcare, Uppsala, Sweden) were also used for column chromatography.

4.3. Extraction and Chromatography

The fresh aerial parts of *Ch. coronarium* L. (20 kg) were dried in air, yielding a crude dry material which amounted to about 2.2 kg. The dried material (2.0 kg) was ground and extracted using an ultrasonic-assisted method with 95% ethanol (3 × 50 L) at 45 °C for 2 h. The dried ethanol extract (TH, 118.5 g) was subjected to MCI gel column chromatography (4.0 × 25 cm), eluted with water, 10% methanol, 30% methanol, 50% methanol, 70% methanol, and 90% methanol, respectively (each, 2.0 L). Lastly, the MCI gel column was washed with acetone. Six fractions were yielded (THA–THF).

4.4. Purification of the Caffeoylquinic Acid Derivatives

The THA fraction (13.2 g) was subjected to ODS C18 column chromatography (3.0 × 25 cm) eluting with 5% methanol, 15% methanol, 25% methanol, 35% methanol, and 50% methanol (each 1.0 L), respectively. Five fractions (THA-1–THA-5) were obtained after being pooled according to their HPLC profiles.

Fraction THA-3 (3.3 g) was further subjected to Sephadex LH-20 (2.0 × 100 cm) elution with MeOH to furnish fractions THA-3A–3D. Fraction THA-3C was purified by semi-preparative HPLC (10 μm, 10 × 250 mm), eluting with MeOH-H_2O (0–23 min: 18:82 to 45:55; v/v, 3 mL/min) and yielding compounds **1** (12.5 mg), **2** (26.8 mg), and **3** (14.5 mg).

The THC fraction (6.57 g) was subjected to silica gel chromatography (4.0 × 26 cm) using CHCl$_3$-MeOH (100:1 to 2:1, v/v) for elution to yield six fractions THC-**1–6** according to their TLC profiles. THC-**6** (1.2 g) was further subjected to Sephadex LH-20 (2.0 × 100 cm) elution with MeOH to furnish fractions THC-6A–6G. THC-6D was purified by semi-preparative HPLC (10 µm, 10 × 250 mm), eluting with an isocratic elution of MeOH-H$_2$O (34:66; v/v, 3 mL/min) yielding compounds **4** (8.5 mg) and **6** (9.8 mg). THC-6F was purified by semi-preparative HPLC (10 µm, 10 × 250 mm), eluting with an isocratic elution of MeOH-H$_2$O (35:65; v/v, 3 mL/min) yielding compounds **5** (13.5 mg) and **7** (14.6 mg) (Figure 3).

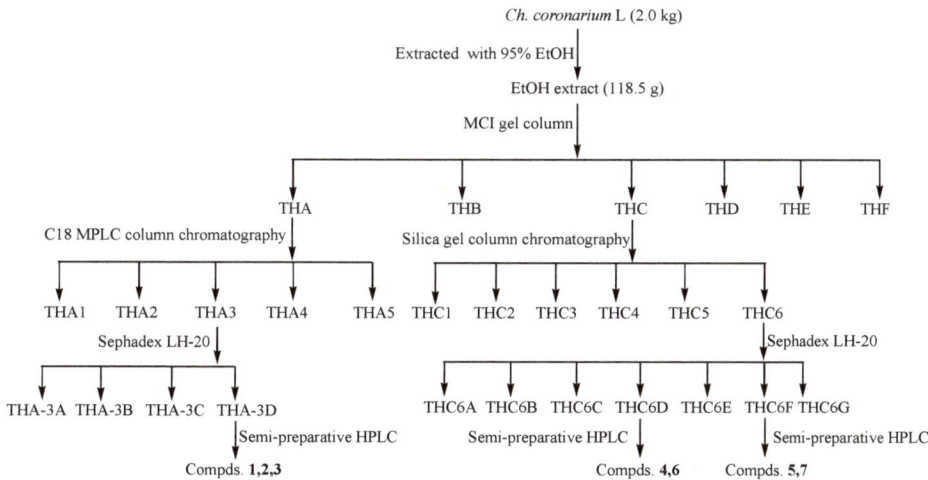

Figure 3. The procedure for the extraction and isolation of compounds from *Ch. coronarium* L.

5. Conclusions

Caffeoylquinic acids (CQAs) were the major phenolic constituents present in the aerial parts of *Ch. coronarium* L. Seven CQAs including three mono-CQAs and four di-CQAs were isolated from this plant. They were 3-*O*-caffeoylquinic acid (**1**), 5-*O*-caffeoylquinic acid (**2**), 4-*O*-caffeoylquinic acid (**3**), 3,4-di-*O*-caffeoylquinic acid (**4**), 1,5-di-*O*-caffeoylquinic acid (**5**), 3,5-di-*O*-caffeoylquinic acid (**6**), and 4,5-di-*O*-caffeoylquinic acid (**7**), respectively. All of the isolates except for **2** and **6** were isolated from this species for the first time. Moreover, compounds **3–5**, and **7** were identified from the Chrysanthemum genus for the first time.

Acknowledgments: This project was supported by the National Natural Science Foundation of China (31360487) and the Natural Science Foundation of Jiangxi Province (20132BAB204016).

Author Contributions: Chunpeng Wan and Shuying Fan conceived and designed the experiments; Chunpeng Wan, Shanshan Li, Lin Liu, and Chuying Chen performed the experiments; Chunpeng Wan and Shuying Fan analyzed the data; Chunpeng Wan and Shuying Fan wrote the paper.

Conflicts of Interest: The authors declare no conflict of interest.

References

1. Dokuparthi, S.K.; Manikanta, P. Phytochemical and pharmacological studies on *Chrysanthemum coronarium* L.: A review. *J. Drug Discov. Ther.* **2015**, *3*, 11–16.
2. Ibrahim, L.F.; El-Senousy, W.M.; Hawas, U.W. NMR spectral analysis of flavonoids from *Chrysanthemum coronarium*. *Chem. Nat. Compd.* **2007**, *43*, 659–662. [CrossRef]
3. Chuda, Y.; Suzuki, M.; Nagata, T.; Tsushida, T. Contents and cooking loss of three quinic acid derivatives from garland (*Chrysanthemum coronarium* L.). *J. Agric. Food Chem.* **1998**, *46*, 1437–1439. [CrossRef]

4. El-Masry, S.; Abou-Donia, A.H.; Darwish, F.A.; Abou-Karam, M.A.; Grenz, M.; Bohlmann, F. Sesquiterpene lactones from *Chrysanthemum coronarium*. *Phytochemistry* **1984**, *23*, 2953–2954. [CrossRef]
5. Song, M.C.; Kim, D.H.; Hong, Y.H.; Kim, D.K.; Chung, I.S.; Kim, S.H.; Baek, N.I. Terpenes from the aerial parts of *Chrysanthemum coronarium* L. *Agric. Chem. Biotechnol.* **2003**, *46*, 118–121.
6. Ragasa, C.Y.; Natividad, G.M. An Antimicrobial Diterpene from *Chrysanthemum coronarium*. *Kimica* **1998**, *14*, 17–20.
7. Song, M.C.; Yang, H.J.; Lee, D.G.; Kim, D.K.; Ahn, E.M.; Woo, Y.M.; Baek, N.I. Glycosyldiglycerides from the Aerial Parts of Garland (*Chrysanthemum coronarium*). *J. Korean Soc. Appl. Biol. Chem.* **2009**, *52*, 88–91. [CrossRef]
8. Song, M.C.; Yang, H.J.; Jeong, T.S.; Kim, K.T.; Baek, N.I. Heterocyclic compounds from *Chrysanthemum coronarium* L. and their inhibitory activity on hACAT-1, hACAT-2, and LDL-oxidation. *Arch. Pharm. Res.* **2008**, *31*, 573–578. [CrossRef] [PubMed]
9. Song, M.C.; Kim, D.H.; Hong, Y.H.; Yang, H.J.; Chung, I.S.; Kim, S.H.; Baek, N.I. Polyacetylenes and Sterols from the Aerial Parts of *Chrysanthemum coronarium* L. (Garland). *Front. Nat. Prod. Chem.* **2005**, *1*, 163–168. [CrossRef]
10. Alvarez-Castellanos, P.P.; Bishop, C.D.; Pascual-Villalobos, M.J. Antifungal activity of the essential oil of flowerheads of garland chrysanthemum (*Chrysanthemum coronarium*) against agricultural pathogens. *Phytochemistry* **2001**, *57*, 99–102. [CrossRef]
11. Basta, A.; Pavlović, M.; Couladis, M.; Tzakou, O. Essential oil composition of the flowerheads of *Chrysanthemum coronarium* L. from Greece. *Flavour Fragr. J.* **2007**, *22*, 197–200. [CrossRef]
12. Mahahiro, T.; Kazuhiro, C. Novel plant growth inhibitors and an insect antifeedant from *Chrysanthemum coronarium* L. *Agric. Biol. Chem.* **1984**, *48*, 1367–1369.
13. Lee, K.D.; Park, K.H.; Kim, H.; Kim, J.H.; Rim, Y.S.; Yang, M.S. Cytotoxic Activity and Structural Analogues of Guaianolide Derivatives from the Flower of *Chrysanthemum coronarium* L. *Agric. Chem. Biotechnol.* **2003**, *46*, 29–32.
14. Mondolot, L.; La, Fisca, P.; Buatois, B.; Talansier, E.; De Kochko, A.; Campa, C. Evolution in caffeoylquinic acid content and histolocalization during *Coffea canephora* leaf development. *Ann. Bot.* **2006**, *98*, 33–40. [CrossRef] [PubMed]
15. Wan, C.; Yu, Y.; Zhou, S.; Tian, S.; Cao, S. Isolation and identification of phenolic compounds from *Gynura divaricata* leaves. *Pharmacogn. Mag.* **2011**, *7*, 101–108. [PubMed]
16. Nakatani, N.; Kayano, S.; Kikuzaki, H.; Sumino, K.; Katagiri, K.; Mitani, T. Identification, Quantitative Determination, and Antioxidative Activities of Chlorogenic Acid Isomers in Prune (*Prunus domestica* L.). *J. Agri. Food Chem.* **2000**, *48*, 5512–5516. [CrossRef]
17. Wan, C.; Yuan, T.; Cirello, A.L.; Seeram, N.P. Antioxidant and α-glucosidase inhibitory phenolics isolated from highbush blueberry flowers. *Food Chem.* **2012**, *135*, 1929–1937. [CrossRef] [PubMed]
18. Shi, S.; Huang, K.; Zhang, Y.; Zhao, Y.; Du, Q. Purification and identification of antiviral components from *Laggera pterodonta* by high-speed counter-current chromatography. *J. Chromatogr. B* **2007**, *859*, 119–124. [CrossRef] [PubMed]
19. Chen, J.; Mangelinckx, S.; Ma, L.; Wang, Z.; Li, W.; De Kimpe, N. Caffeoylquinic acid derivatives isolated from the aerial parts of *Gynura divaricata* and their yeast α-glucosidase and PTP1B inhibitory activity. *Fitoterapia* **2014**, *99*, 1–6. [CrossRef] [PubMed]
20. Carnat, A.; Heitz, A.; Fraisse, D.; Carnat, A.P.; Lamaison, J.L. Major dicaffeoylquinic acids from *Artemisia vulgaris*. *Fitoterapia* **2000**, *71*, 587–589. [CrossRef]
21. Kodoma, M.; Wada, H.; Otani, H.; Kohmoto, K.; Kimura, Y. 3,5-Di-O-caffeoylquinic acid, an infection-inhibiting factor from *Pyrus pyrifolia* induced by infection with *Alternaria alternata*. *Phytochemistry* **1998**, *47*, 371–373. [CrossRef]
22. Wan, C.; Liu, Q.; Zhang, X.; Fan, S. A Review of the Chemical Composition and Biological Activities of the Edible and Medicinal Plant *Chrysanthemum coronarium* L. *Mod. Food Sci. Technol.* **2014**, *30*, 282–288.

© 2017 by the authors. Licensee MDPI, Basel, Switzerland. This article is an open access article distributed under the terms and conditions of the Creative Commons Attribution (CC BY) license (http://creativecommons.org/licenses/by/4.0/).

Article

Genus-Specific Real-Time PCR and HRM Assays to Distinguish *Liriope* from *Ophiopogon* Samples

Eva Masiero [1], Dipanwita Banik [1,2], John Abson [3], Paul Greene [3], Adrian Slater [1] and Tiziana Sgamma [1,*]

[1] Biomolecular Technology Group, Allied Health Sciences, Faculty of Health and Life Sciences, De Montfort University, Leicester LE1 9BH, UK; eva.masiero@dmu.ac.uk (E.M.); banikdipanwita@yahoo.com (D.B.); ads@dmu.ac.uk (A.S.)
[2] CSIR-North East Institute of Science & Technology, Jorhat 785006, Assam, India
[3] Environmental Industries and Business, Brooksby Melton College, Brooksby Campus, Brooksby, Leicestershire LE14 2LJ, UK; johnabson42@gmail.com (J.A.); pgreene@BrooksbyMelton.ac.uk (P.G.)
* Correspondence: tiziana.sgamma@dmu.ac.uk; Tel.: +44-116-201-3925; Fax: +44-116-257-7892

Received: 20 September 2017; Accepted: 19 October 2017; Published: 26 October 2017

Abstract: *Liriope* and *Ophiopogon* species have a long history of use as traditional medicines across East Asia. They have also become widely used around the world for ornamental and landscaping purposes. The morphological similarities between *Liriope* and *Ophiopogon* taxa have made the taxonomy of the two genera problematic and caused confusion about the identification of individual specimens. Molecular approaches could be a useful tool for the discrimination of these two genera in combination with traditional methods. Seventy-five *Liriope* and *Ophiopogon* samples from the UK National Plant Collections of *Ophiopogon* and *Liriope* were analyzed. The 5′ end of the DNA barcode region of the gene for the large subunit of ribulose-1,5-bisphosphate carboxylase/oxygenase (*rbcLa*) was used for the discrimination of the two genera. A single nucleotide polymorphism (SNP) between the two genera allowed the development of discriminatory tests for genus-level identification based on specific PCR and high-resolution melt curve (HRM) assays. The study highlights the advantage of incorporating DNA barcoding methods into plant identification protocols and provides simple assays that could be used for the quality assurance of commercially traded plants and herbal drugs.

Keywords: *Ophiopogon*; *Liriope*; rbcL; DNA barcoding; high-resolution melt curve (HRM) analysis

1. Introduction

Plants belonging to the genera *Liriope* Lour. and *Ophiopogon* Ker Gawl. are collectively known by the English common name liriopogon [1–3]. The collective name itself indicates the close relationship and morphological similarities between the two genera and the potential for misidentification. Liriopogon are widely cultivated as ornamentals and ground cover plants for garden landscaping due to their hardiness, and pest and disease resistance. However, mishandling, mislabelling, and mismanagement of nursery practices can lead to sexual propagation of cultivars, hybridisation, plant substitution, and finally degradation of the morphological/phenotypic identity of the cultivars [4].

Tubers of a few species of both *Liriope* and *Ophiopogon* are used in traditional medicines across East and South Asia for the treatment of respiratory ailments, diabetes, thirst, and as an aphrodisiac [5]. In the Chinese and Korean traditions, substitution of *Liriope* for *Ophiopogon* is permissible, although the Chinese Pharmacopoeia considers them to be separate remedies [6]. In contrast, the Japanese Pharmacopoeia stipulates that the traditional medicine "Bakumondo" must be derived from *O. japonicus* tubers, i.e., material derived from *Liriope* is not a legal substitute. The close similarity in the morphological characteristics of their leaves and tubers makes it difficult to distinguish between members of the two genera in both the horticultural and phytopharmaceutical industries [4]. Methods for discrimination of samples from the two genera are therefore important for quality assurance in these commercial sectors.

Authentication of plant material used for herbal medicines typically relies on chemical analysis. Liriopogons are characterised by their content of steroidal saponins and homoisoflavonoids [6–8]. TLC methods are straightforward and suitable for multiple samples. A TLC assay to distinguish the two genera has been developed, but is limited by low sensitivity and resolution [9]. More precise analysis of the saponin and flavonoid components has been achieved by HPLC-UV [10] and HPLC-UV-ELSD [8], but these require a long run time for each individual sample. Recent comparison of the two genera by LC-MS/MS also showed that differences in the profiles of steroidal saponins and homoisoflavonoids could be used to discriminate between *Ophiopogon* and *Liriope* [6].

DNA-based tests have emerged as a powerful system for authentication of medical plants and commercial herbal products [11–13]. Many of these target "DNA barcode" regions of the plant genome. DNA barcoding is a technique for identifying biological specimens using short DNA sequences from either the nuclear or organelle genome, termed DNA barcodes. In plants, the major DNA barcode candidates are the plastid *matK*, *rbcL*, and *trnH-psbA* loci and the nuclear ribosomal ITS region (nrITS) [14–17]. DNA tests have been developed to distinguish *Liriope* from *Ophiopogon*, including the use of SCAR [18] and EST-SSR [19] markers. A barcoding approach targeting a SNP in the 3′ region of the rbcL region was developed by Ito et al. (2015) [20]. Digestion with the restriction enzyme *Hinc*III cut amplicons from *Liriope* into two fragments, but left *Ophiopogon* amplicons intact. This is an effective assay, but the digestion and gel electrophoresis steps are time-consuming and limit the throughput of the assay.

The current study proposes a new strategy for the identification of *Ophiopogon* and *Liriope* species by designing specific real-time PCR and high-resolution melt curve (HRM) assays targeting genus-specific single nucleotide polymorphisms (SNPs) in the *rbcL* barcode region. These techniques are ideally suited for the design of quick, reliable, robust, and affordable identification assays that could be incorporated into industrial quality control procedures for herbal medicines [13].

This project arose from a collaboration to verify the identity of accessions in the UK National Collections of *Ophiopogon* and *Liriope* by DNA barcoding [21]. However, to date, there are few examples of DNA techniques being applied to the classification of cultivated ornamental plants. An identification test based on DNA barcoding would be much faster than the traditional botanical methods of identification, which require growing the target plants to the flowering stage, in parallel with control plants. This new affordable method could also be useful for nurseries and plant collections and the wider horticultural community of professional and amateur gardeners.

2. Results

2.1. DNA Barcoding of the rbcLa Region of Liriope and Ophiopogon Accessions

The plastid *rbcLa* barcode region of 75 National Collection *Ophiopogon* and *Liriope* specimens was analysed (Table S1). Genomic DNA was extracted from all the samples and amplified by conventional PCR using *rbcLa* generic primers. The expected fragment of about 700 bp was clearly visualised in all of the *Ophiopogon* and *Liriope* samples (Figure 1).

In order to confirm the genus and the species, the *rbcLa* amplicons were sequenced from the *rbcLa* primer. A multiple sequence alignment was generated, combining sequences from the GenBank database with the newly generated sequences (Figure S1). The alignment showed very little sequence variation between species of the same genus, with just three single nucleotide polymorphic (SNP) positions observed. One was found to vary between the two genera, with the guanine predominantly present in the *Ophiopogon* samples substituted by an adenine in the *Liriope* samples (Figure 2).

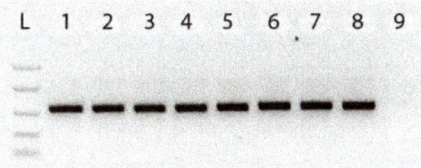

Figure 1. Agarose gel electrophoresis of PCRs using *rbcL* generic primers. Gel lanes: L. Easy Ladder I (Bioline) 1. *Ophiopogon* Sample 678; 2. *Ophiopogon* Sample 679; 3. *Ophiopogon* Sample 680; 4. *Ophiopogon* Sample 682; 5. *Liriope* Sample 626; 6. *Liriope* Sample 627; 7. *Liriope* Sample 631; 8. *Liriope* Sample 632; 9. Negative (no template) control.

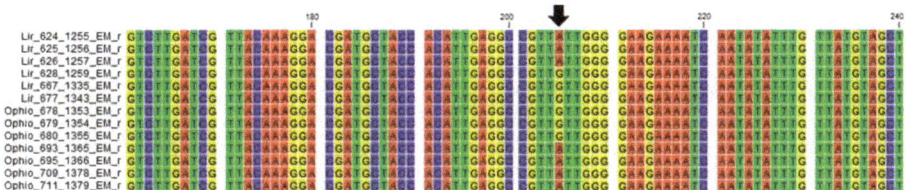

Figure 2. Fragment of a multiple alignment of the rbcL region from a selection of *Liriope* and *Ophiopogon* samples, highlighting a number of atypical samples. *Liriope* Samples 624, 625, and 626 match the consensus *Liriope* sequence. *Ophiopogon* Samples 678, 679, and 680 match the *Ophiopogon* consensus. *Liriope* Samples 628, 667, and 677 and *Ophiopogon* Samples 693, 695, 709, and 711 are atypical in the SNP position highlighted by the black arrow.

Only four *Ophiopogon* samples (693, 695, 709, and 711) and three *Liriope* samples (628, 667, and 677) did not fit this SNP pattern, having instead an adenosine and a guanine, respectively (Figure 2). One explanation is that these accessions had been misidentified. Three of the atypical *Ophiopogon* accessions had been classified as the same species, *O. japonicus*, but the three atypical *Liriope* accessions were originally classified as different species: *L. graminifolia*, *L muscari*, and *L spicata*. Morphological analysis of these specimens was not able to resolve this, since the vegetative aerial parts share common morphological characters (Figure S2).

In order to resolve the anomaly, the identity of these specimens was determined by DNA barcoding of the nrITS region (data not shown). The nrITS sequences of all seven specimens confirmed the genus assignment indicated by the *rbcLa* SNP, i.e., the plants did appear to have been mislabelled or misidentified. The consistent genus-specificity of the SNP confirmed by these results presented an opportunity to discriminate the *Ophiopogon* and *Liriope* genera by designing specific PCR primers to target this SNP.

2.2. The rbcL Feature Provides a Target for Simple Genus Discriminatory Tests

In a study conducted by Ito et al. (2015) [21] it was reported that the two genera showed a single nucleotide variation in the *rbcLb* region and they designed a restriction-enzyme-based assay to target their SNP in order to discriminate the two genera. Their SNP is about 120 bp downstream from the one reported here. In order to develop a new and more robust assay for discriminating the two genera, two sets of genus-specific primers were designed to incorporate both SNPs. Thus, the *Ophiopogon*-specific forward primer was designed to end with the guanine base, while the reverse primer was designed to end with a cytosine corresponding to the guanine in the plus strand. Conversely, the *Liriope*-specific forward primer was designed to end with the adenosine base, while the reverse primer was designed to end with an adenosine corresponding to the thymidine in the plus strand (Figure 3).

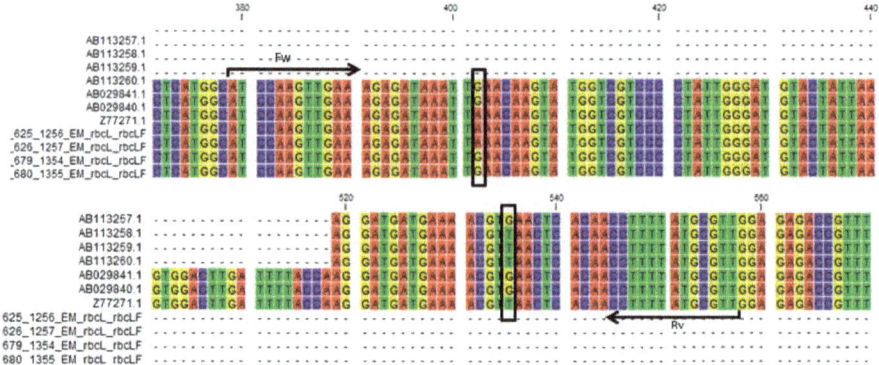

Figure 3. Graphical representation of the location of genus-specific primers. The figure shows a multiple alignment with 4 *Liriope* and 4 *Ophiopogon* *rbcLa* consensus sequences joined to the *rbcLb* sequences published by Ito et al. (2012) using three *rbcL* sequences from the database that bridge the *rbcLa* and *rbcLb* regions. The black arrow indicates where the forward and reverse primers were designed. The black boxes indicate the two SNPs incorporated into the 3′ position of the primer sequences.

Conventional PCR was performed with samples from the target and non-target genera in order to confirm the specificity of the primers (Figure 4). The annealing temperature of the PCR was optimised for each specific primer pair such that a prominent amplicon was produced with the correct template, but no product was visible with a template from the opposite genus. For example, in Figure 4a, DNA from the confirmed *Ophiopogon* Samples 678 and 679 (Lane 1 and 2) produce clear bands with the Ophiopogon-specific primers, but no bands are seen in these lanes in Figure 4b.

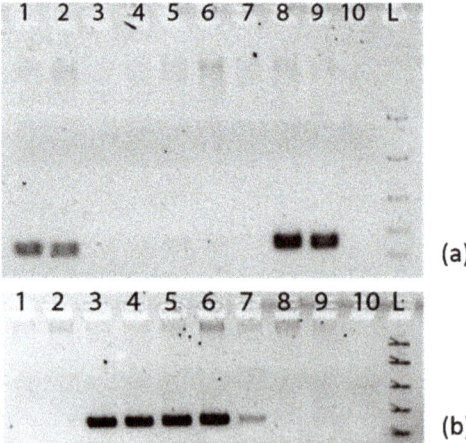

Figure 4. Agarose gel electrophoresis of PCR using rbcL_Ophiopogon and rbcL_Liriope specific primers. (**a**) rbcL_Ophiopogon specific primers. (**b**) rbcL_Liriope specific primers. Gel lines: 1. *Ophiopogon* Sample 678; 2. *Ophiopogon* Sample 679; 3. *Ophiopogon* Sample 693; 4. *Ophiopogon* Sample 695; 5. *Ophiopogon* Sample 709; 6. *Ophiopogon* Sample 711; 7. *Liriope* Sample 624; 8. *Liriope* Sample 628; 9. *Liriope* Sample 677; 10. Negative (no template) control; L. Easy Ladder I (Bioline).

The assay was then used to test those samples that showed a different nucleotide base compared to their original classification. The *Ophiopogon*-specific primers clearly showed that Samples 693,

695, 709, and 711 did not belong to the genera to which they had been first allocated (Figure 4a), and the *Liriope*-specific primers confirmed that these samples belonged to the *Liriope* genus. In the same way, the specific assays confirmed that Samples 628 and 677 belong in the *Ophiopogon* genus and not in the *Liriope* genus to which they were originally assigned (Figure 4b).

To confirm the accuracy of the initial sampling of these specimens, a repeat collection of samples was carried out and the repeat DNA extractions were tested with the specific PCR assay. The results again confirmed that Samples 693, 695, and 709 belong to the *Liriope* genus and 628, 667, and 677 to *Ophiopogon* (Figure S3a,b). Our new discriminatory test, using genus-specific primers, permits us to identify *Ophiogon* and *Liriope* samples with an easy and economic system by conventional PCR.

The specific primers targeted two genus-specific SNPs within a short region of the *rbcL* barcode sequence and were designed so that they were also suitable for use in real-time PCR analysis. The speed, simplicity, and sensitivity of real-time PCR assay are ideally suited to industrial quality control tests [13]. Real-time PCR was performed using the genus-specific primers after optimising the thermocycling programs and primer concentrations. The amplification plots for the *Liriope*-specific primers showed a marked difference in C_t value (around 12 cycles) between *Liriope* and *Ophiopogon* samples (Figure 5a). The *Ophiopogon*-specific primers were less efficient, but careful optimisation of the annealing temperature allowed a difference in C_t values between the two genera of around 10 cycles to be achieved. (Figure 5b).

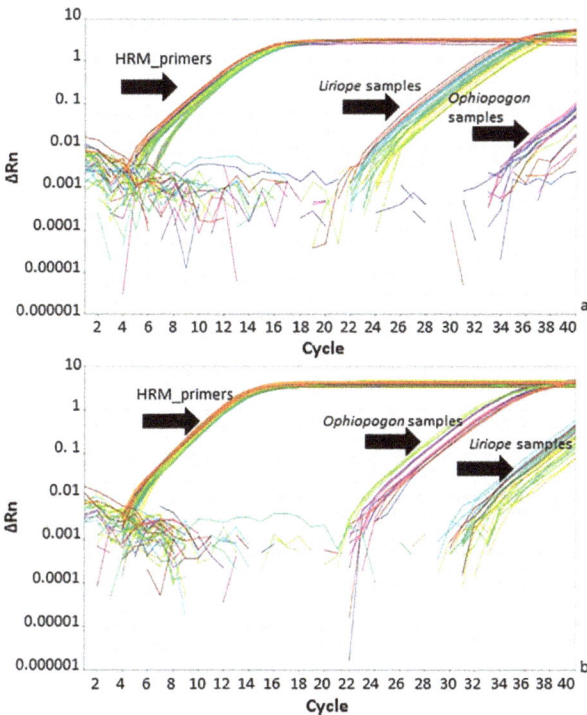

Figure 5. Real-time PCR amplification using *Ophiopogon*-specific, *Liriope*-specific and generic high-resolution melt curve (HRM) primers. (**a**) Amplification plot of *Liriope*-specific and HRM primers. (**b**) Amplification plot with *Ophiopogon*-specific and HRM primers. The black arrows indicate the primer/template combinations: HRM_primers: *Ophiopogon* and *Liriope* templates with the generic primers; *Liriope* samples: *Liriope* templates with specific primers; *Ophiopogon* samples: *Ophiopogon* templates with specific primers.

In order to normalise the C_t values to allow for differences in the amount of DNA template, the HRM primers described in the next section were tested for their suitability as generic/universal primers. Real-time PCR with the HRM_primers did not show any significant variation in the C_t values obtained from samples from the two genera. The C_t value obtained with the generic primers was subtracted from the specific primer C_t to obtain a ΔC_t value for comparison between different DNA samples. In order to identify an unknown sample as *Liriope* or *Ophiopogon*, the ΔC_t value for a reference sample was subtracted from the unknown to produce a $\Delta\Delta C_t$ value. It is expected that the $\Delta\Delta C_t$ value from the correct specific primers will be ≤ 2.0, whilst that for an incorrect genus would be >7.0. A $\Delta\Delta C_t > 7.0$ is arbitrarily chosen as the threshold because it represents the equivalent of detection of the correct template at a dilution of one molecule in one thousand. The results in Table 1 clearly show that the $\Delta\Delta C_t$ values correspond to the genus identity of each sample.

Table 1. Results of real-time PCR assay using *Ophiopogon*- and *Liriope*-specific primers. The C_t value obtained for each specific primer pair was normalised by subtracting the C_t value obtained for the same sample with the generic (HRM) primers, giving the ΔC_t (genus-specific − generic) value. The ΔC_t (genus-specific − generic) value for a reference standard was then subtracted from the ΔC_t genus-specific value for each sample, giving the $\Delta\Delta C_t$ [(genus-specific − generic)$_{sample}$ − (genus-specific − generic)$_{standard}$] value. The reference standard for *Liriope* was chosen as Sample 624 and for *Ophiopogon* was Sample 678.

Sample Number	ΔC_t *Liriope*-Specific	ΔC_t *Ophiopogon*-Specific	$\Delta\Delta C_t$ *Liriope*-Specific	$\Delta\Delta C_t$ *Ophiopogon*-Specific
624	12.41	23.53	0	8.61
626	13.28	24.21	0.87	9.29
628	25.51	13.00	13.10	−1.92
633	12.28	23.64	−0.13	8.72
634	11.67	22.91	−0.74	7.99
667	24.74	15.17	12.33	0.25
677	24.46	13.87	12.05	−1.05
678	24.34	14.92	11.92	0
679	25.89	15.45	13.48	0.53
693	13.76	25.68	1.35	10.76
695	14.43	25.62	2.02	10.70
709	13.03	24.33	0.62	9.41
711	12.56	22.55	0.15	7.63

Thus, for example, each known or suspected *Liriope* sample produces a $\Delta\Delta C_t$ value close to zero with the *Liriope* primers, whilst the *Ophipogon* primers produces a corresponding $\Delta\Delta C_t$ value above seven.

2.3. Identification of Liriope and Ophiopogon Samples by Using the HRM System

The *rbcLa* SNP also provides a useful target for developing a single tube assay to discriminate between the two genera using HRM analysis. This technology can discriminate between sequences containing a difference of a single base if it has a significant effect on the melting temperature. HRM primers were designed to the conserved regions of the *rbcL* sequence on either side of the SNP (Figure 6).

After the optimisation of the PCR conditions by conventional PCR (Figure S4), the HRM curves produced from *Liriope* and *Ophiopogon* samples were compared. The results revealed that there was a difference in Tm of 2 °C, allowing the two genera to be easily distinguished from each other. A difference plot of the melting curves showed two distinct variants that corresponded to samples from the two genera (Figure S5).

The reliability of the assay was assessed by a blind experiment in which the identities of ten samples were unknown to the operator. Alignment of the melting curves allowed the two variants to be easily discriminated. In Variant 1, four unknown samples were matched with the *Ophiopogon* controls, whilst 6 unknown samples in Variant 2 matched the *Liriope* controls (Figure 7). These ten samples included the seven misidentified samples.

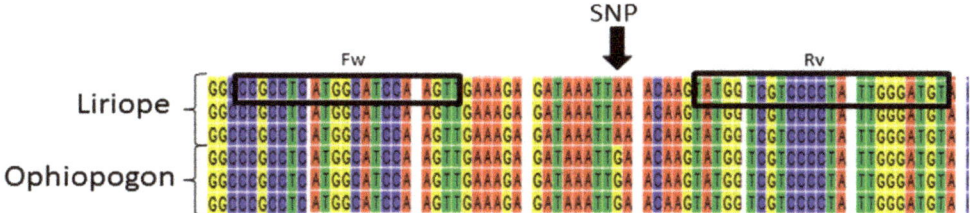

Figure 6. Schematic representation of the location of the HRM primers. The black arrow indicate the position of the SNP.

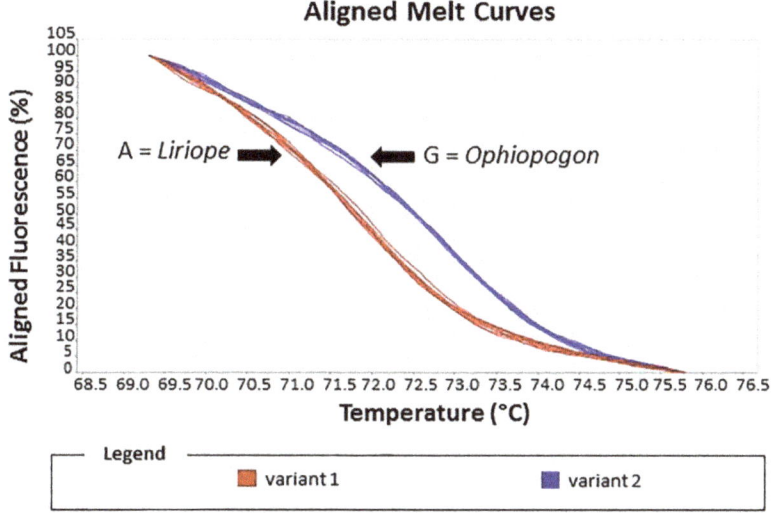

Figure 7. HRM assay of reference standards and test samples. The melting curve plot showed two distinct variants: Variant 1, *Liriope* standards; Variant 2, *Ophiopogon* standards. Test samples were assigned to the *Liriope* or *Ophiopogon* genus according to which variant curve they matched.

This assay confirmed the correct genus of these samples, supporting the results of the genus-specific real-time PCR assay.

3. Discussion

Sequencing of the *rbcLa* barcode region of 75 samples of *Ophiopogon* and *Liriope* from the UK National Plant Collections of these two genera provided a large dataset for analysis. Multiple alignment of the sequences revealed that the region is very highly conserved, with only three SNPs observed, one of which distinguishes the genus *Liriope* from the genus *Ophipogon*. This reflects the findings of [22] who reported the high degree of conservation of the *rbcLb*, though they observed five genus-specific SNPS in this downstream section of the gene. The *rbcLb* region has been noted as being more variable than the *rbcLa* region in a number of plant groups [23]. Ito et al. (2015) [20] targeted one of these genus-specific *rbcLb* SNPs using a restriction-enzyme-digestion-based approach for the identification of the two genera. This SNP was around 100 bases downstream of ours in the *rbcLb* region, so proved ideal for the design of pairs of specific real-time PCR primers.

Our analysis of the sequence data indicated that our *rbcLa* SNP might not be entirely genus-specific because three *Liriope* and four *Ophiopogon* samples had the "wrong" base at this position. There are a number of possible biological explanations for this including homoplasy and hybridisation, but human errors of identification or labelling appeared more likely. Sequencing of the nrITS barcode region of these specimens showed that the genus identification agreed with the *rbcL* SNP, indicating that they had been misidentified (data not shown). This confirmed that the SNP was entirely genus-specific across the collection.

In order to develop rapid, reliable *Ophiopogon* and *Liriope* identification tests, the SNP identified in our study and that targeted by Ito et al. (2015) [20] allowed the design of genus-specific primers for a simple PCR-based test. The primers were designed to incorporate the variable base at the 3' end of the primer for conventional and real-time PCR. This strategy has been used in the design of a number of PCR tests for the authentication of herbal medicines [13,15]. The specificity of the primers was confirmed by conventional PCR; bands were only seen with template DNA from the corresponding genus.

The genus-specific primers were also designed for use in real-time PCR. The *Liriope*-specific primers could distinguish DNA from the two genera by a difference in C_t value of 12 cycles, when normalised with the HRM generic primers. The *Ophiopogon* primers showed a similarly large difference in C_t values after the annealing temperature of the PCR was optimised for specificity. It is noticeable that the C_t values obtained with the specific primer pairs were considerably higher than the C_t obtained with the generic HRM primers. This is the result of a number of factors. The design of the specific primers is constrained by the position of the polymorphic base at the 3' end. The only flexibility in design is variation of the total length of the primer. In addition, the optimisation of specificity often requires that the annealing temperature of the PCR is higher than optimal for amplification efficiency. Nevertheless, the results indicate that the design constraints and sub-optimal conditions do not affect the ability of the assay to identify *Ophiopogon* and *Liriope* samples in a quick and consistent way. The value of the assay was highlighted when it was used to rapidly confirm the genus of the misidentified accessions after resampling.

Under optimal conditions, HRM assays can discriminate between sequences containing a difference of a single base and can rapidly and accurately identify species from a diverse range and quality of materials [13]. In our study, the HRM assay also proved to be a simple and reliable method for the identification of the two genera. The results grouped the samples into two distinct variants due to base-pair mismatching between the two species causing a Tm shift of 2 °C. For testing the accuracy of this assay, a blind experiment was performed using a range of samples including some of the misidentified sample. The results showed a clear discrimination of the two variants. The results confirmed the re-classification of those samples into the correct genus, supporting the results with the specific primers.

All together these results proved the specificity and reliability of both techniques in the identification of *Ophiopogon* and *Liriope* samples. The PCR assays are limited by the requirement to design-specific primers for each known target plant and likely adulterant. The HRM assay has the ability to detect unknown contaminants provided they share the same genetic sequences and could be used to analyse admixtures in a single tube. However, genus-specific PCR primers and HRM are both powerful assays for a rapid genus-level screen without having to go through the entire DNA barcoding process for the identification of *Ophiopogon* and *Liriope* species. These two assays could be good tools for the discrimination of genus, species, or cultivars based on individual SNPs.

4. Materials and Methods

4.1. Plant Material and Total DNA Extraction

Fresh leaves were collected from 75 different species of *Ophiopogon* and *Liriope* at Brooksby Melton College (Melton Mowbray, Leicestershire, UK) from the UK National Plant Collections for *Ophiopogon* and *Liriope*. Details of the genus, species, and accession number are in Table S1. Samples were stored at −80 °C. DNA was extracted from 100 mg of frozen material, previously ground to a fine powder in liquid nitrogen with mortar and pestle, using DNeasy Plant Mini Kit (Qiagen Inc., Germantown, MD, USA) following the manufacturers' guidelines.

4.2. PCR Protocols

PCRs were carried out using different primers as detailed in Table 2.

Table 2. List of primers with relative Ta and predict band size.

Primers	Sequences	Annealing Temperature (Conventional PCR)	Annealing Temperature (Real-Time PCR)	Amplicon Size (bp)
rbcLFw	ATGTCACCACAAACAGAGACTAAAGC	52 °C	N/A	700
rbcLRv	GTAAAATCAAGTCCACCRCG			
rbcL_Liriope_Fw	ATCCAAGTTGAAAGAGATAAATTA	61 °C	61 °C	180
rbcL_Liriope_Rv	AACGCATAAAAGGTTGTGAGTTA			
rbcL_Ophiopogon_Fw	ATCCAAGTTGAAAGAGATAAATTG	62.5 °C	64.5 °C	180
rbcL_Ophiopogon_Rv	AACGCATAAAAGGTTGTGAGTTC			
HRM_rbcL_Fw	CGCCTCATGGCATCCAAGT	61 °C	61 °C	80
HRM_rbcL_Rv	AATAGGGGACGACCATACTTG			

PCR reaction mixes contained 1X MyTaq Red Mix (Bioline), 0.2 µM of each forward and reverse primer, and 1 µL of gDNA as template. A G-Storm GS1 Thermal Cycler (G-Storm Ltd., Somerton, UK) was used with the following program:

- *rbcLa* PCR: initial denaturation step of 5 min at 95 °C followed by 35 cycles consisting of 30 s at 95 °C, 20 s at 52 °C, and 50 s at 72 °C, with a final extension period of 5 min at 72 °C.
- *Ophiopogon*-specific *rbcL* PCR: initial denaturation step of 5 min at 95 °C followed by 40 cycles consisting of 30 s at 95 °C, 30 s at 62.5 °C, and 45 s at 72 °C, with a final extension period of 5 min at 72 °C.
- *Liriope*-specific *rbcL* PCR: initial denaturation step of 5 min at 95 °C followed by 35 cycles consisting of 30 s at 95 °C, 30 s at 61 °C, and 30 s at 72 °C, with a final extension period of 5 min at 72 °C.
- HRM (generic) *rbcL* PCR: initial denaturation step of 5 min at 95 °C followed by 35 cycles consisting of 30 s at 95 °C, 30 s at 60 °C, and 30 s at 72 °C, with a final extension period of 5 min at 72 °C.

PCR products were run on 2% (*w*/*v*) agarose, 1X TBE gels with 1 µL SYBR® Safe DNA Gel Stain (Invitrogen, Paisley, UK) at 100 V for 30 min and analysed in a Gel Doc™ EZ Gel Documentation System (BioRad, Oxford, UK).

4.3. DNA Sequence Analysis

Published *Liriope* and *Ophiopogon* rbcL DNA sequences were obtained from the National Center for Biotechnology Information (NCBI) GenBank database (http://www.ncbi.nlm.nih.gov/). A multi alignment was generated using CLC Main Workbench 7.5.1 software (Qiagen, Germantown, MD, USA).

4.4. Real-Time PCR Analysis

Each real-time PCR reaction contained 1 µL of gDNA, 1X Sensifast SYBR green Hi-Rox mix (Bioline), 0.1 µM of each forward and reverse primer in a total volume of 10 mL made up with sterile distilled water. A StepOnePlus™ Real-Time PCR thermocycler machine (Applied Biosystem) was used.

Amplification conditions were as follows: 95 °C for 2 min followed by 40 cycles of 5 s at 95 °C and 30 s at the primer specific Ta (Table 1). The melting curve was obtained by melting the amplified template from 65 to 95 °C increasing the temperature by 0.5 °C per cycle. No-template controls were included. Three technical replicates were used for each sample. Internal StepOne software (Applied Biosystems) was used for the analysis of the results [24].

4.5. High-Resolution Melt Curve Analysis (HRM) Methods

HRM primers were designed to match the conserved sequences of the *rbcL* gene on either side of the genus-specific SNP (Figure 7). Each HRM real-time PCR reaction contained 1 µL of gDNA, 1X MeltDoctor™ HRM Master Mix (Applied Biosystem), 0.1 µM of each HRM_rbcL_forward and reverse primer (Table 1) in a total volume of 10 µL made up with sterile distilled water. A StepOnePlus™ Real-Time PCR thermocycler machine (Applied Biosystem) was used. Amplification conditions were as follows: 95 °C for 10 min followed by 40 cycles of 15 s at 95 °C and 30 s at 60 °C. The fluorescent data for PCR amplification was recorded during the extension step. The final melting curve was obtained by melting the amplified template 65 to 95 °C increasing the temperature by 0.3 °C per cycle with a 15 s hold time for each acquisition step. No-template controls were included. Three technical replicates were used for each sample.

HRM software (Applied Biosystem) has been used to analyse the results. For each sample, a melting curve plot, a melting peak plot and difference plot was generated [25].

Supplementary Materials: The following are available online at www.mdpi.com/2223-7747/6/4/53/s1. Figure S1: Multiple alignment of the *rbcLa* region of all of the *Liriope* and *Ophiopogon* samples tested, Figure S2: Agarose gel electrophoresis of PCR products of a re-sampled specimens using *Ophiopogon*- and *Liriope*-specific primers, Figure S3: Agarose gel electrophoresis of PCR products using HRM primers.

Acknowledgments: The authors are grateful to Joanne Wilkes for role in the cultivation of the *Liriope* and *Ophiopogon* national collections. They also acknowledge the role of Plant Heritage in coordinating and supporting the UK National Plant Collection scheme. This project was funded in part by De Montfort University HEIF funds.

Author Contributions: J.A. and P.G. curated the national collections of *Liriope* and *Ophiopogon*, J.A. and A.S. conceived the project; T.S. and A.S. designed and supervised the experimental strategy, and edited the paper; E.M. performed the experiments, analysed the data, and wrote most of the paper; D.B. photographed and sampled the collections, and wrote part of the paper.

Conflicts of Interest: The authors declare no conflict of interest. The funding sponsors had no role in the design of the study; in the collection, analyses, or interpretation of data; in the writing of the manuscript; or in the decision to publish the results.

Abbreviations

nrITS	nuclear ribosomal internal transcribed spacer
rbcL	large subunit of ribulose-1,5-bisphosphate carboxylase/oxygenase gene
HRM	high-resolution melt curve assay

References

1. Fantz, P.R. Species of *Liriope* Cultivated in the Southeastern United States. *HortTechnology* **2008**, *18*, 343–348.
2. Fantz, P.R. Names and Species of *Ophiopogon* Cultivated in the Southeastern United States. *HortTechnology* **2009**, *19*, 385–394.
3. Nesom, G.L. Overview of *Liriope* and *Ophiopogon* (Ruscaceae) naturalized and commonly cultivated in the USA. *Phytoneuron* **2010**, *56*, 1–31.
4. Lattier, J.D.; Ranney, T.G. Identification, Nomenclature, Genome Sizes, and Ploidy Levels of *Liriope* and *Ophiopogon* Taxa. *HortScience* **2014**, *49*, 145–151.
5. Mabberley, D.J. *A Portable Dictionary of Plants, Their Classification and Uses*; Cambridge University Press: Cambridge, UK, 2008.

6. Wu, Y.; Dong, Z.; Wu, H.; Ding, W.; Zhao, M.; Shi, Q.; Wang, Q. Comparative studies on Ophiopogonis and Liriopes based on the determination of 11 bioactive components using LC–MS/MS and hierarchical clustering analysis. *Food Res. Int.* **2014**, *57*, 15–25. [CrossRef]
7. Wang, K.; Zhang, H.; Shen, L.; Wang, W. Novel steroidal saponins from *Liriope graminifolia* (Linn.) Baker with anti-tumor activities. *Carbohydr. Res.* **2011**, *346*, 253–258. [CrossRef] [PubMed]
8. Li, N.; Che, Y.; Zhang, L.; Zhang, J.; Zhou, Y.; Jiang, Y.; Tu, P. Fingerprint analysis of Ophiopogonis Radix by HPLC-UV-ELSD coupled with chemometrics methods. *J. Chin. Pharm. Sci.* **2013**, *22*, 55–63. [CrossRef]
9. Tang, X.; Cheng, Z.; Yu, B. Qualitative and quantitative studies on Chinese drug Maidong. *China J. Chin. Mater. Med.* **1999**, *24*, 390–445.
10. Zeng, P.; Zhou, H.; Zheng, Y.; Xu, X.; Fu, S. Simultaneous determination of three homoisoflavonoids in Ophiopogon japonicus by HPLC. *China J. Chin. Mater. Med.* **2012**, *37*, 71–74.
11. Mishra, P.; Kumar, A.; Nagireddy, A.; Mani, D.N.; Shukla, A.K.; Tiwari, R.; Sundaresan, V. DNA barcoding: An efficient tool to overcome authentication challenges in the herbal market. *Plant Biotechnol. J.* **2016**, *14*, 8–21. [CrossRef] [PubMed]
12. Coutinho Moraes, D.F.; Still, D.W.; Lum, M.R.; Hirsch, A.M. DNA-Based Authentication of Botanicals and Plant-Derived Dietary Supplements: Where Have We Been and Where Are We Going? *Planta Med.* **2015**, *81*, 687–695. [CrossRef] [PubMed]
13. Sgamma, T.; Lockie-Williams, C.; Kreuzer, M.; Williams, S.; Scheyhing, U.; Koch, E.; Slater, A.; Howard, C. DNA Barcoding for Industrial Quality Assurance. *Planta Med.* **2017**, *83*, 1117–1129. [CrossRef] [PubMed]
14. Kazi, T.; Hussain, N.; Bremner, P.; Slater, A.; Howard, C. The application of a DNA-based identification technique to over-the-counter herbal medicines. *Fitoterapia* **2013**, *87*, 27–30. [CrossRef] [PubMed]
15. Howard, C.; Bremner, P.D.; Fowler, M.R.; Isodo, B.; Scott, N.W.; Slater, A. Molecular identification of Hypericum perforatum by PCR amplification of the ITS and 5.8S rDNA region. *Planta Med.* **2009**, *75*, 864–869. [CrossRef] [PubMed]
16. Singh, H.K.; Parveen, I.; Raghuvanshi, S.; Babbar, S.B. The loci recommended as universal barcodes for plants on the basis of floristic studies may not work with congeneric species as exemplified by DNA barcoding of Dendrobium species. *BMC Res. Notes* **2012**, *5*, 42. [CrossRef] [PubMed]
17. Cameron, K.M.; Chase, M.W.; Whitten, W.M.; Kores, P.J.; Jarrell, D.C.; Albert, V.A.; Yukawa, T.; Hills, H.G.; Goldman, D.H. A phylogenetic analysis of the Orchidaceae: Evidence from *rbcL* nucleotide. *Am. J. Bot.* **1999**, *86*, 208–224. [CrossRef] [PubMed]
18. Li, G.; Park, J.Y. SCAR markers for discriminating species of two genera of medicinal plants, *Liriope* and *Ophiopogon*. *Genet. Mol. Res.* **2012**, *29*, 2987–2996. [CrossRef] [PubMed]
19. Li, G.; Ra, W.; Park, J.-W.; Kwon, S.-W.; Lee, J.; Park, C.; Park, Y. Developing EST-SSR markers to study molecular diversity in *Liriope* and *Ophiopogon*. *Biochem. Syst. Ecol.* **2011**, *39*, 241–252. [CrossRef]
20. Ito, M.; Sato-Masumoto, N.; Kobayashi, F.; Matsumura, K. Distinguishing *Ophiopogon* and *Liriope* tubers based on DNA sequences. *J. Nat. Med.* **2015**, *69*, 555–564. [CrossRef] [PubMed]
21. Abson, J. Botany, medicine and mistaken identity. *Plant Heritage J.* **2017**, *25*, 41–43.
22. Shiba, M.; Yamaji, H.; Kondo, K.; Ichiki, H.; Sakakibara, I.; terabayashi, S.; Amagaya, S.; Aburada, M.; Miyamoto, K. Discrimination of Maidong Derived from *Ophiopogon* and *Liriope* Species by *rbcL* Sequences, and Their Chemical Components and Tuber Anatomy. *Nat. Med.* **2004**, *58*, 15–21.
23. Dong, W.; Cheng, T.; Li, C.; Xu, C.; Long, P.; Chen, C.; Zhou, S. Discriminating plants using the DNA barcode *rbcLb*: An appraisal based on a large data set. *Mol. Ecol. Res.* **2013**, *14*, 336–343. [CrossRef] [PubMed]
24. Applied Biosystems StepOne™ and StepOnePlus™ Real-Time PCR Systems Reagent Guide. Available online: https://assets.thermofisher.com/TFS-Assets/LSG/manuals/cms_046739.pdf (accessed on 25 October 2017).
25. Applied Biosystems High Resolution Melting Getting Started Guide. Available online: https://tools.thermofisher.com/content/sfs/manuals/cms_050347.pdf (accessed on 25 October 2017).

© 2017 by the authors. Licensee MDPI, Basel, Switzerland. This article is an open access article distributed under the terms and conditions of the Creative Commons Attribution (CC BY) license (http://creativecommons.org/licenses/by/4.0/).

MDPI
St. Alban-Anlage 66
4052 Basel
Switzerland
Tel. +41 61 683 77 34
Fax +41 61 302 89 18
www.mdpi.com

Plants Editorial Office
E-mail: plants@mdpi.com
www.mdpi.com/journal/plants

www.ingramcontent.com/pod-product-compliance
Lightning Source LLC
LaVergne TN
LVHW071943080526
838202LV00064B/6667